Computational Modeling in Biomedical Engineering and Medical Physics

Computational Modeling in Biomedical Engineering and Medical Physics

ALEXANDRU MOREGA
Faculty of Electrical Engineering and Faculty of Medical Engineering, University Politehnica of Bucharest, Bucharest, Romania
"Gh. Mihoc — C. Iacob" Institute of Statistical Mathematics and Applied Mathematics, the Romanian Academy, Bucharest, Romania

MIHAELA MOREGA
Faculty of Electrical Engineering and Faculty of Medical Engineering, University Politehnica of Bucharest, Bucharest, Romania

ALIN DOBRE
Faculty of Electrical Engineering and Faculty of Medical Engineering, University Politehnica of Bucharest, Bucharest, Romania

ELSEVIER

ACADEMIC PRESS
An imprint of Elsevier

Academic Press is an imprint of Elsevier
125 London Wall, London EC2Y 5AS, United Kingdom
525 B Street, Suite 1650, San Diego, CA 92101, United States
50 Hampshire Street, 5th Floor, Cambridge, MA 02139, United States
The Boulevard, Langford Lane, Kidlington, Oxford OX5 1GB, United Kingdom

British Library Cataloguing-in-Publication Data
A catalogue record for this book is available from the British Library

Library of Congress Cataloging-in-Publication Data
A catalog record for this book is available from the Library of Congress

ISBN: 978-0-12-817897-3

For Information on all Academic Press publications
visit our website at https://www.elsevier.com/books-and-journals

Publisher: Mara Conner
Acquisitions Editor: Chris Katsaropoulos
Editorial Project Manager: Gabriela D. Capille
Production Project Manager: Sojan P. Pazhayattil
Cover Designer: Christian J. Bilbow

Typeset by MPS Limited, Chennai, India

Contents

Preface

This book is about physical, mathematical, and numerical modeling in engineering problems of biomedical investigation, procedures, and therapies. The elect viewpoint is the continuous media, subject to processes and interactions that happen contiguously, in space, and continuously, in time. The emphasis is set on the quantitative representation of the underlying phenomenological couplings (electrical, mechanical, hemodynamic, and thermal), unifying vehicle being the thermodynamic analysis, and the principle of the constructal organization and evolution of animated and inanimate systems—structuring, morphology, and evolution. The accent is set on the physical understanding and on the liberty to choose between approximate and exact calculations—from scale analysis, as a preliminary step that produces bird's eye view estimations, to numerical methods that provide accuracy to the predicted results.

Computational modeling for biomedical engineering

Biomedical engineering (used here), medical bioengineering, engineering in medicine and biology, or else are mere as many names for a reach and complex field of research and development that belongs in the same time to biology, medicine, medical physics, and engineering of various orientations, given by its pioneers in the field first, the historical stage and the participating disciplines from which they came. Regardless of the name given, however, the "engineering" part leads us to think about its significance: creation, engineering, practical realization, under the sign of the rigor of the measure, and space-time dimensional predetermination of the object, the produced system.

Measurement and sizing are based on and develop quantitative cause–effect relationships that give confidence in the feasibility, realism, and robustness (object and system) of the engineering creation. And quantitative relations are, in fact, mathematical formulations. The computational methods—theories, algorithms, software, and hardware systems, for short numerical modeling—which are developed and used for the analysis and solution of the mathematical representations of biomedical systems and in the design of devices, have reached the level of development at which physical–mathematical–numerical modeling becomes a tool and a way of thinking common to engineering, biology, and medicine. Numerical modeling thus becomes a powerful and valuable means in understanding and predicting medical applications—underlying physical phenomena, procedures, therapies, and scanning technologies.

Solvable physical–mathematical models

The solvable formulations of the "thought" experiments are the trait of a union of the examples offered in this book and ensure consistency with the physical experiment counterpart. Even if it would be possible to verify experimentally the numerical results of the thought experiments, there is no need to perform it except for confirming their consequences. They may complete the physical experiments that, even when indulged, cannot be performed because, for instance, of ethical concerns. Not in the least, numerical modeling is also a design tool that may contribute to optimize the equipment's development, reduce the design cycle and costs, and augment the safety margins. Numerical experiments *in simulacra* to *real-life* conditions provide a wealth of knowledge and qualify as an exceptionally useful tool. In the paradigm of hypothesis-driven research, the thought experiments expressed through physical, mathematical, numerical modeling are the means used to study the models presented throughout this book. This perspective strengthens and confers predictability to medical procedures and techniques, complementary to diagnosis.

Shape, structure, and rhythm

The fundamental question that arises when attempting to describe the animate and inanimate systems may be seen as an evolutionary design, which is governed by the laws of physics, free to morph and to evolve toward greater access. A compact and robust phrasing of what is observed in nature is propositioned by the Constructal law of physics, which perceives the living systems in motion, driven by power, subjected finite-size constraints, endowed with the freedom to change, and abiding the *time arrow for evolution*. The Constructal law does not rely on the *time arrow of empiricism*, which starts with nature and the unsolved observation, for example, the allometric laws, nor does it rely on abstract constructs emptied of physics, for example, the fractal geometry. Life has to be maintained, and the constructal theory explains that this is a predictable result of evolutionary design: the growth and the evolution refer to architectures that change and flow. They apply to organs and vascularized tissues that have the same objectives and perform under the same constraints.

From the drawing to the numerical modeling—computational domains

The starting point in numerical modeling is a good sketch of the thought experiment—the drawing. It has to bear the main features of the physical experiment counterpart. To this end, the physical systems are conceptualized into "solid models" that are constructed using medical images or CADs. These are then *Finite Element Method* (FEM)-discretized to become computational domains.

Modeling multidisciplinary processes and interactions

The cardiovascular system is a prominent subject. It balances complex linked electromechanical processes and energy interactions that concur in ensuring the hemodynamic flow. The electrical activity of the heart—action potentials traveling through atria and ventricles—elicits the myocardium contraction and relaxation, which provide the pressure field that drives the hemodynamic flow. From the nonlinear dynamics of the excitable cardiac strand to the electric field diffusion through the thorax—that charts the electrical activity of the heart measured noninvasively on the thorax surface—numerical modeling is the evocative, insightful support of choice that predictively complements the usual diagnosis of the cardiovascular system's state and condition.

More likely to be used as a first stage, expressive, general measurement principle, bioimpedance (BI), may be used *per se* or complemented with several other techniques, from surface plasmons to echography. BI measurements recast the complexity of multiphysics processes into time series and signal spectra. It provides concise and meaningful information that, sometimes, is difficult (if possible) to retrieve otherwise. There is considerable interest in modeling the BI applied to monitor the cardiovascular activity at the scale of the system, body, or parts of it, and "Gedanken" experiments are key to understanding the physics insights perceived through a global, dynamic quantity—an impedance.

Much attention is devoted to magnetic nanoparticles, useful in various processes—separation and purification of cell and macromolecule, immunoassays, controlled drug delivery, electromagnetic hyperthermia, magnetic resonance imaging, gene therapy, etc. Magnetic drug targeting is among them, and selected models are discussed in the context, to identify their potential and possible adverse effects.

Magnetic stimulation is used for noninvasive nerve stimulation, transcranial magnetic stimulation, motor evoked potentials, and neuropsychiatric applications. Although its physics are simple, magnetic stimulation is still a sensitive procedure, as it requires the precise location of the envisaged targeted region inside the body, good focus strength of the magnetic stimulus, while minimizing its side effects. The knowledge of the EMF stimuli distribution inside the body and the optimization of the magnetic applicators helps the progress of this medical procedure. Several analytic and numerical models for the stimulation of peripheral nerves and the transcranial and lumbar magnetic stimulation are presented. These add to some considerations on the magnetotherapy known in physical therapy for its multiple curative effects and reduced costs.

Thermal medical procedures are proven to have benefits in many severe diseases and the quest for their development and optimization is of major concern. Heat transfer is the underlying mechanism, and the work to produce it is generally done by different physics—for instance, electromagnetic fields and ultrasounds. Diffusive-convective and radiation processes occur, and their knowledge is key to sizing the applicators, electrodes, and adjusting the medical protocols. Mathematical and numerical simulation of several thermotherapy methods—radiofrequency, microwave, magnetic, ultrasound hyperthermia and ablation, and their numerical solution—are included. Moreover, the numerical modeling gains in its predictive value when it is related to patient-specific computational domains.

Alexandru M. Morega, Mihaela Morega, Alin A. Dobre

Bucharest, Romania

August 2020

Acknowledgments

The work was conducted in the Laboratory for Engineering in Medicine and Biology, Faculty of Electrical Engineering, University Politehnica of Bucharest, Romania. Several of our students contributed with enthusiasm and expertise during the various stages of this project. We treasure the support of Daniel Mocanu, Cristina Săvăstru, Alina M. Săndoiu, Roxana M. Baerov, Ştefania Preda, Loredana C. Manea, Mihaela C. Ipate, Dumitru M. Marius, and Diana Stanciu. We also appeal to the readers to communicate to us any errors that might persist in this edition.

Our gratitude extends to Chris Katsaropoulos, Gabriela Capille, Swapna Praveen, and Sojan P. Pazhayattil from Academic Press for the excellent expertise, advice, and kind support.

CHAPTER 1

Physical, mathematical, and numerical modeling

1.1 Experiments and numerical simulation

A "thought experiment"[1] is a conceptual experiment that relies on hypotheses, theories, or principles aiming at thinking through its predictions. Even if it would be possible there is no need to perform it except for validating its consequences. In this paradigm[2] (concepts or thought patterns, including theories, research methods, postulates) of hypothesis-driven research, thought experiments embodied through physical, mathematical, numerical modeling is the vehicle used to study a number of models that are presented throughout this book.

From a converging perspective, there is an underlying concern in complying with ethical norms and regulations of physical experiments. For example, Art. 7 Line 2 of the EU Directive 86/609/EEC of November 24, 1986, Bruxelles, and its updates regarding the protection of the animals subjected to experiments or of other scientific interests states that "An experiment shall not be conducted if there exists another reasonable and practical method to satisfactorily obtain the pursued result without implying the usage of animals" (Ruhdel, 2007; Hartung, 2014).

Much has been done to comply with this directive and its subsequent revisions. Recent progresses in mathematical algorithms, numerical analysis, and the unprecedented development of hardware and software tools capable of implementing them make possible the numerical simulation of complex problems. Thus numerical modeling has become a powerful and valuable mean in understanding and predicting medical applications—underlying physical phenomena, procedures, therapies, and scanning technologies. The "numerical experiment" based on physical and mathematical modeling and on numerical simulation may complement the physical experiment that even when permissible cannot be performed always because, for instance, of ethical concerns. It may provide also information otherwise inaccessible through physical experiments; for instance, it may depict the distribution of the electric field inside the body during the MRI scanning. It is also a design tool aimed, for instance, to optimize

[1] German: Gedankenexperiment (Perkowitz, 2010) or Gedankenerfahrung (Brown, 2019)

[2] *Paradigm* comes from Greek παράδειγμα (paradeigma), "pattern, example, sample" from the verb παραδείκνυμι (paradeiknumi), "exhibit, represent, expose" from παρά (para), "beside, beyond" from δείκνυμι (deiknumi), "to show, to point out" (Paradigm, 2018).

Computational Modeling in Biomedical Engineering and Medical Physics
DOI: https://doi.org/10.1016/B978-0-12-817897-3.00001-4

the scanner development, reduce the design costs, and enhance its safety margins. Numerical experiments *in simulacra* to *real-life* circumstances provide a wealth of knowledge, and numerical simulation qualifies as an extremely useful tool.

1.2 The system and its boundary

The starting point in any experimental or theoretical analysis has to be the precise definition of the *system*, which is—for a simple, comprehensive definition—a collection of matter in a region of interest (ROI) in space, to be observed, investigated, and measured. In particular in numerical modeling, the system substantiates and hosts the thought experiment. With the system comes its *environment*, or *surroundings*, with which the system interacts through work, heat, and mass transfer. The entity that separates the system and its environment is thus the *boundary*, which belongs in the same time to the system and to its surroundings. In fact the *state* of the system (i.e., the collection of values that the state quantities have at a specific time moment) depends on the interactions with its environment, which are perceived or stated at the boundary level. The boundary is traditionally assimilated to a surface and not to another system. Mathematically it is a two-dimensional geometric variety of zero thickness, that is, a surface, if the system is a three-dimensional volume, a contour if the system is two-dimensional, and two points if the system is one-dimensional. Whereas in solid-body mechanics the system and its boundary may be evident, in electromagnetism, fluid mechanics, heat transfer, the system emerges once its boundary is drawn and its interactions with the surroundings (the boundary conditions) are set.

The history or the film of the states that an evolving system may "visit" while undergoing internal and or external interactions is called the evolution *path*. The state of a system, at any time, is defined through the ensemble of local quantities called *thermodynamic properties*. For instance, physical quantities such as temperature, pressure, electric potential, energy, and densities are thermodynamic properties. Their values do not depend on the history of the system that evolves in time and depend strictly on the instantaneous conditions in which they are measured or computed. Physical quantities such as work, heat, and mass transfer interactions are not thermodynamic properties. The thermodynamic state properties variation depends on the initial and final states of the system that are associated with the path. The quantities that are not thermodynamic properties depend on the initial and final states and on the path between them—they are *quantities of interaction*.

Of concern in defining a system is the continuity, in the mathematical sense, of the underlying properties across its boundary. Their discontinuity may trigger analytical difficulties. For instance, the gradient of the temperature, (a scalar property, which is proportional to the heat flux density, may not be well defined on the boundary should this scalar be a discontinuous function there.

As the interactions with the environment are part of the state of the system and its evolution, it is important to recognize whether the boundary is crossed by mass flow or *permeable*. An *impermeable* boundary defines a *closed* system. Open systems, or flow systems, are those systems whose boundaries are permeable, that is, crossed by mass fluxes.

1.3 First law analysis: energy, heat, and work interactions

The first law of thermodynamics introduces the energy of the system, E [J], an extensive quantity, dependent on the amount of substance, as thermodynamic property, the heat, Q [J], and the work, W [J], as interactions. It states that when the system evolves from an initial state, $(\cdot)_1$, to the final state, $(\cdot)_2$, the change in the energy of the system is a measure of the interactions that it undergoes

$$Q_{1\to 2} - W_{1\to 2} = E_2 - E_1. \tag{1.1}$$

The system here is a control volume and the signs suggest that the system receives heat and executes work with respect to its environment. For a process that characterizes the transition between two close states with small interactions, the first law Eq. (1.1) is written as follows:

$$\delta Q - \delta W = dE, \tag{1.2}$$

where $d(\cdot)$ is the total, exact differential operator and $\delta(\cdot)$ denotes a small variation. Using Eq. (1.2) , the per-unit-time basis form of Eq. (1.3) becomes

$$\dot{Q} - \dot{W} = \frac{dE}{dt}, \tag{1.3}$$

where $\dot{W}$ [W] is power (work transfer rate), and $\dot{Q}$ [W] is heat transfer rate (thermal power rate).

The change in energy, $E_2 - E_1$, distinguishes between the macroscopically discernible forms of energy storage and a form of energy storage, that is, unidentifiable microscopically, denoted by U [J], which is called for this reason as *internal energy*, and

$$\underbrace{E_2 - E_1}_{\text{Energy}} = \underbrace{U_2 - U_1}_{\text{Internal}} + \underbrace{\frac{1}{2}m\left(v_2^2 - v_1^2\right)}_{\text{energy}} + \underbrace{mg(z_2 - z_1)}_{\text{Potential}} + \underbrace{(E_2 - E_1)_i}_{\text{Other forms of energy storage}}. \tag{1.4}$$

Eq. (1.4) singles out several forms of energy: internal, kinetic, potential, and other macroscopic forms of energy storage, $(E_2 - E_1)_i$, which may include electric energy, magnetic energy, chemical energy, and so on. Divided through the volume of the system, these energies become energy densities, which are intensive, local quantities, independent on the amount of substance in the system.

In the limits of reversible transformations, without heat transfer, the infinitesimal work transfer interaction in Eq. (1.2), δW_{rev}, is the scalar product force—displacement (deformation)

$$\delta W_{rev} \overset{\delta Q=0}{=} -\sum_i Y_i dX_i, \tag{1.5}$$

where Y_i are the components (projections) of a generalized force with respect to the generalized coordinate, of X_i components. The minus sign denotes that the interaction is performed by the environment upon the system, and not conversely. The subscript $(\cdot)_{rev}$ denotes that the process is reversible, that is, it is without heat transfer. Such transformations may be either *adiabatic*, $\dot{Q} = 0$, or *diaterman*, $\frac{\partial T}{\partial n} = 0$ (n is the coordinate associated with the outward pointing normal to the surface).

Eqs. (1.3) and (1.5) then yield

$$Y_i = -\frac{\partial E}{\partial X_i}, \tag{1.6}$$

which shows off that the generalized force is actually the gradient, (a vector quantity) of the energy (a scalar quantity). This is the *theorem of generalized forces* and it provides a method to compute forces that produce work. If the control volume is a continuum medium then Eq. (1.6) yields the specific, per volume basis (or body) force as the gradient of an energy density.

For isothermal internal processes, where the internal work interactions are of electrical and or magnetic nature, the theorem of generalized forces states that the electrical and, or magnetic body forces are the derivatives (the gradients) of the electrical and, or magnetic energy density, respectively. For instance, for magnetic linear media, the magnetic energy density, e_{mag} [J/m^3], is as follows (Mocanu, 1981):

$$e_{mag} = \frac{\mathbf{BH}}{2}, \tag{1.7}$$

where **B** [T] is the magnetic flux density and **H** [A/m] is magnetic field strength. Eqs. (1.6) and (1.7) then yield the magnetic body force

$$\mathbf{f}_{mag} = \nabla e_{mag}. \tag{1.8}$$

This approach is used for instance in the magnetic drug targeting analysis (Chapter 6: Magnetic Drug Targeting) to evaluate the magnetization body forces.

The first principle, Eq. (1.3), written for an *open control volume* yields the following (Bejan, 1988):

$$\frac{dE}{dt} = \dot{Q} - \dot{W} + \sum_{inlets} \dot{m}(e + Pv) - \sum_{outlets} \dot{m}(e + Pv), \tag{1.9}$$

where $\dot{m}$ [kg/s] is the mass flow rate, e [J/kg] is the energy density, P [Pa] is the pressure, v [m^3/kg] is the specific volume, and the symbolic sums account for the inlet and outlet mass transfer through all permeable ports on the boundary.

For a system of volume V_Σ, that is, bounded by the closed surface Σ, Eq. (1.9) becomes

$$\int_{V_\Sigma} \frac{\partial(\rho e)}{\partial t} dv = \oint_\Sigma \mathbf{q} \cdot \mathbf{n} dA - \dot{W} + \oint_\Sigma \rho h \mathbf{v} \cdot \mathbf{n} dA, \tag{1.10}$$

where ρ [kg/m^3] is the mass density, $\mathbf{q}$ [W/m^2] is the heat flux rate, $\mathbf{v}$ [m/s] is the velocity, and $h = u + Pv$ [J/kg] is the specific enthalpy. It is assumed that the only form of energy storage is the specific internal energy, u [J/kg].

The closed surface integrals that replace the sums in Eq. (1.10) suggest that the mass transport, the heat transfer, and the work transfer with the environment may occur (may be distributed) everywhere all over the boundary. These surfaces or flux integrals may be replaced with volume integrals by using Gauss integral (divergence) theorem (Annex 1) to yield the following relation between the integrands

$$\frac{\partial(\rho e)}{\partial t} = -\nabla \cdot \mathbf{q} - w''' - P\nabla \cdot \mathbf{v}, \tag{1.11}$$

where mass conservation principle, $\frac{\partial \rho}{\partial t} = -(\rho\nabla)\cdot\mathbf{v}$, is used.

In Eq. (1.11) w''' [W/m^3] stands for a specific heat generation rate—the "work" needed for the electrical current density to flow through the electroconductive medium is converted into heat, through Joule–Lentz effect. For instance, in an electrokinetic problem $w''' = \mathbf{EJ}$, where $\mathbf{E}$ [V/m] is the electric field strength, and $\mathbf{J}$ [A/m^2] is the electrical conduction current density.

It is worth noting that the first principle, Eq. (1.11), or the *energy equation*, in a particular form or another, is the mathematical model skeleton of the heat transfer used throughout this book.

Electromagnetic power transferred through the boundary (at the electrical terminals)

Depending on the nature of the work interactions, $\dot{W}$ in Eq. (1.3) may have different expressions. For instance, for a closed system situated in an external electromagnetic field (EMF) whose only interaction with its environment is of electromagnetic nature the work interaction at the boundary is as follows (Mocanu, 1981):

$$\underbrace{\oint_{\Sigma} (\mathbf{E} \times \mathbf{H})\mathbf{n}_i dA}_{\dot{W}_{\Sigma,em}} = \underbrace{\oint_{\Sigma} \mathbf{S}\mathbf{n}_i dA}_{\dot{W}_{\Sigma,em}} = \underbrace{\int_{V_\Sigma} \mathbf{E}\mathbf{J} dv}_{\dot{Q}} + \underbrace{\int_{V_\Sigma} \left(\mathbf{E}\frac{\partial \mathbf{D}}{\partial t} + \mathbf{H}\frac{\partial \mathbf{B}}{\partial t} \right) dv}_{\frac{dE_{em}}{dt}} . \tag{1.12}$$

This integral relation states the instantaneous balance between the total electromagnetic work rate, $\dot{W}_{\Sigma,em}$, the total internal thermal power (heat rate) by Joule effect, $\dot{Q}$, and the total change in internal electromagnetic energy, $\frac{dE_{em}}{dt}$. In Eq. (1.12), $\mathbf{D}$ [C/m^2] is the electric flux density, $\mathbf{S}$ [VA/m^2] is Poynting's vector, and $\mathbf{n}_i$ is the interior normal to the boundary, Σ. It is worth noting that the electromagnetic energy is actually the sum of two distinct contributions, presented usually as the *electrical energy density* (only electric field quantities), $e_{el} = \frac{\mathbf{DE}}{2}$, and magnetic energy density (only magnetic field quantities), $e_{mag} = \frac{\mathbf{BH}}{2}$. These forms are of importance in EMF problems as their gradients provide for the electric and magnetic body forces (Chapter 6: Magnetic Drug Targeting).

Eq. (1.12) indicates two EMF-produced heat sources for inclusion in the first law energy balance: an *Ohmic heat source* related to the electrical conduction (by Joule effect) and a *Hertzian heat source*, related to the displacement electrical current (by dielectric heating)

$$p_{Joule} = \mathbf{EJ},\ p_{dielectric} = \mathbf{E}\frac{\partial \mathbf{D}}{\partial t} = \mathbf{E}\frac{\partial}{\partial t}(\varepsilon_0 \mathbf{E} + \mathbf{P}) = \frac{1}{2}\varepsilon_0 \frac{\partial \mathbf{E}^2}{\partial t} + \mathbf{E}\frac{\partial \mathbf{P}}{\partial t}, \tag{1.13}$$

where $\mathbf{P}$ [C/m^2] is the electrical polarization.

In nonlinear dielectric media with hysteresis, the per-cycle integral of $p_{dielectric}$ for a cyclic excitation $\oint_{Cycle} \mathbf{E} d\mathbf{P} = Q_{dielectric,cycle}$ is the heat released per cycle through polarization–depolarization (Warburg theorem; Warburg, 1881). A similar discussion shows off a similar heat source for magnetic nonlinear media with hysteresis, $\oint_{Cycle} \mathbf{H} d\mathbf{B} = Q_{magnetic,cycle}$, which occurs in microwave (MW) hyperthermia [Chapter 8: Hyperthermia and Ablation (Thermotherapy Methods)].

For a harmonic excitation, with $\omega = 2\pi f$ (f is frequency), the simplified complex representation of the local form of the power balance Eq. (1.12) is as follows (Mocanu, 1981):

$$-\,div(\underline{\mathbf{S}}) = \underbrace{-\,div(\underline{\mathbf{E}} \times \underline{\mathbf{H}}^*)}_{S_{elmag}} = \underbrace{j\omega\mu \underline{\mathbf{H}\mathbf{H}}^*}_{P_{mag}} + \underbrace{\sigma \underline{\mathbf{E}\mathbf{E}}^*}_{P_J} \underbrace{-\,j\omega\varepsilon \underline{\mathbf{E}\mathbf{E}}^*}_{P_{Hertz}} \tag{1.14}$$

$$= j\omega\mu \underline{\mathbf{H}\mathbf{H}}^* + \underbrace{(\sigma - j\omega\varepsilon)}_{\underline{\varepsilon}'} \underline{\mathbf{E}\mathbf{E}}^*,$$

where P_{mag} [VA] is the magnetic power, P_{Hertz} [VA] is the Hertzian power, P_J is the resistive (Joule) power [W], and $\underline{\varepsilon}'$ is the complex permittivity. The bar below indicates complex quantities, the upper symbol $(\cdot)^*$ indicates complex conjugated quantities, and $j=\sqrt{-1}$. The *specific absorption rate* (SAR) analysis in Chapter 8: Hyperthermia and Ablation (Thermotherapy Methods), will single out the contributions and combined effect of the two power sources, resistive and dielectric, respectively, in the EMF-induced local hyperthermia pending the EMF–substance interaction.

1.4 Multidisciplinary (multiphysics) problems

The problems presented throughout this book are multidisciplinary (multiphysics), with multiple concurring and interacting "physics" with variable degrees of couplings: hemodynamic activity monitoring (arterial hemodynamic flow, structural mechanics of the arm, piezoelectric, and capacitive processes); magnetic drug targeting (hemodynamic flow with magnetic field interactions, structural response of the blood vessels); thermography (hemodynamic and heat flow); hyperthermia (RF and MW EMFs, hemodynamic, heat flow); thermal ablation (hemodynamic flow, EMF, heat transfer); EMF dosimetry (EMF, hemodynamic flow, heat transfer); and bioimpedance methods (hemodynamic flow, EMF).

In general the steps to take in modeling multiphysics problems and medical engineering problems make no exception to the rule, start with the definition of the system and its boundary, its "fabric" (structure and composition), and the recognition and description of the internal and external constraints and interactions to which this is subjected as describable by the laws of Physics. Multiple phenomena of different nature may occur, may interact, and may influence each other in a complex cause–effect web of conditionalities and dependencies.

More than often, such coupled problems of electromagnetism, heat and mass transfer, structural mechanics, and others have different time and space scales, which may raise concerns in formulating consistent, solvable mathematical models using finite computational resources.

Mathematical modeling is used to find the underlying states of the evolution path that the system pursues assuming the macroscopic, continuum media hypothesis for the system and the physical phenomena to which this it subjected. The evolution and interactions occur then contiguously in space and continuously time. The accompanying internal transformations and the internal and external interactions are presented through a set of physical laws that provide for a quantitative, deterministic cause–effect set of mathematical equations, differential, integral, algebraic, that make the mathematical model. Which particular physics occur and what specific forms (differential, integral, algebraic) may have the mathematical equations that describe them is key, if the model is to provide for a realistic image of the underlying physics. In particular these laws of Physics provide the divergence and curl sources of the vector fields.

As already stated, this first step in building the physical model goes hand in hand with a correct definition of the system and its boundary. An initial sketch with the pencil on the paper is successively morphed into the physical system, the physics and the couplings that describe the internal and external interactions are stated and possibly rescaled in space and time, such that the mathematical model, which then builds upon, may produce solutions that comply with and represent within acceptable accuracy limits the reality. At some point, eventually, the resulting mathematical model may require reconsidering the physical model for reasons of affordable, convenient complexity, efficiency, accuracy, and predictive validity. Each of the models presented in this work went through such process, but usually the final, resulting physical–mathematical–numerical simulation results are presented.

1.5 Mathematical models

Complete and independent, coherent, and noncontradictory system of laws

Completeness, independency, and coherency of the noncontradictory system of laws that present a physical problem should be the concern when the physical model, the physical laws, is stated and the mathematical model [the partial differential equation (PDEs)] is abstracted. *Completeness* means that all relevant laws are included. *Independence* is needed to ascertain that the equations are independent, in the sense that they may not be obtained from each other. *Coherency* requires the usage of a system of units containing a set of fundamental or base units from which all other units in the system are derived (Coherence, 2018). *Noncontradictory* means the equations are not adversarial.

A qualitative analysis of the system of equations that represents the physical model and perhaps its dynamic properties (Michel and Wang, 1995) is then advisable. This may be performed starting with the *fundamental theorem of the vector fields* (Annex 1) and the material properties that relate the quantities through constitutive laws. An example may be the dynamic EMF problem in immobile, nonlinear, homogeneous, and isotropic media. Maxwell laws, used to formulate the physical model, include the magnetic flux law (a scalar equation) and the electromagnetic induction (Faraday) law (a vector equation). It may be easily verified that the magnetic flux law can be deduced from Faraday's law (the divergence of a curl is identically zero) up to a time-independent additive constant (Purcell, 1984). Therefore the magnetic flux law could be discarded form the system, without any difficulty—mathematically there are as many equations left as the number of unknown are. However, if the working conditions are such that the electric and magnetic fields are separable (e.g., in stationary cases) than the magnetic flux law has to be included in the physical model of the magnetic field.

Boundary conditions (external interactions) and initial conditions (initial state)

The boundary conditions define the interactions of the system with the environment during the entire process. The initial conditions define the initial state of the system that executes a process. Initial and boundary conditions play an essential role in providing for the uniqueness of the solution: prior to actually solving the problem it is necessary to verify, prove the existence and uniqueness of the solution of the mathematical mode.

This stage may be difficult, not always possible, and for only few cases theorems of existence and uniqueness are available, for instance, Laplace–Poisson problems with Dirichlet, Neumann, and Robin boundary conditions (Tikhonov and Samarskii, 1963). For scalar fields, *Dirichlet* condition defines the primitive, the unknown to be solved for first, on the boundary—the same for the system and its environment. For vector fields, this condition states that the boundary contains the field, or the field vanishes there, for example, magnetic insulation for the magnetic vector potential. *Neumann* condition defines the normal component of a flux, for example, either zero to state that the boundary is insulated with respect to that flux, or nonzero, to define a boundary source. A *Robin* condition provides for a linear combination between the scalar and the normal component of its flux. It may define a convection boundary condition (Newton law, in convection heat transfer) or a contact resistance for electrokinetic problems, and others.

As it will be seen next, a necessary and sufficient number of boundary conditions have to be given for consistency with the PDE mathematical model. When no theorem is available to ascertain boundary conditions that ensure the existence and uniqueness of the solution, consistent boundary conditions may be observed out of experiments, symmetries, and conservation laws. To this end, conservation laws of momentum, mass, energy, fluxes, charges, are utilized or even statistical models and methods are used to recast experimental data. This holds true, for instance, for Navier–Stokes momentum equation.

In such situations, the *number* of boundary conditions required to obtain a solution is decided first. Then an approximate but efficient method consists in considering the principal part of the PDE. For instance, for Navier–Stokes equation the principal part is the Laplacean, a second-order partial differential operator. This criterion is not always *sufficient*. For instance, Neumann problem $\mathbf{n}\cdot\nabla u = f(x, y)$ (flux boundary conditions only) for Laplace equation, $\nabla^2 u = 0$, has a solution if and only if, supplementary, a necessary closure condition for the fluxes is satisfied, that is, $\oint_\Sigma f(x, y)dA = 0$, which means that the total flux across the boundary is zero—actually it equals the total internal scalar source). However, even if this condition is verified, the solution is still not unique: it can be proven that it is unique but to the limit of an additive arbitrary constant (Mocanu, 1981).

The initial conditions define the *initial state* of the system, and its initial interactions with the environment. As with the boundary conditions, their knowledge is required to predict the state of system at any time later on, during the process. The initial conditions have to be set inside the system and on its boundary, for the unknown, primitive function, and its derivatives, up to the highest but one order of the time derivative in the PDE. In fact, with respect to time, the PDE poses a Cauchy problem.

Whereas time and space are treated as variables in Mathematics, in Physics their significance is quite different. Therefore the mathematical modeling techniques used to solve boundary values and initial physical problems ought to consider this aspect.

Initial values problems

A class of only initial conditions problems exits and these are described by ordinary differential equations (ODEs). The electrical circuit problems are such examples (Mocanu, 1979). Conveniently the electrical circuits are treated as complex systems constructed by interconnecting dipolar, elementary, ideal, passive circuit elements, each representing and underlying electromagnetic process, on one hand, and active circuit elements, that is, the power sources, on the other hand. Thus the resistance presents the electrical conduction (heat source by Joule effect); the inductance synthesizes the electromagnetic induction (magnetic energy); the capacitance electrical models the electrical charge conservation (electrical energy). The active elements are voltage and current sources. In this approach, the terminals of the dipolar elements are their boundaries. The physics inside these idealized systems are present through specific working conditions (equations) between electrical currents and voltages at the terminals. These equations may be algebraic, for resistances, and differential time derivatives or integral time integrals for the inductances and capacitances (reactive elements). Inductances and capacitances are then susceptible of initial state conditions, voltages and currents, respectively, which actually represent internal electric or magnetic fluxes, respectively. Their connectivity, or the topological assembly of these circuit elements, closes the definition of the external interactions between the circuit elements. Topology provides the big picture of the macro system, the electrical network, which connects ideal elements that may be in internal and external disequilibrium. This disequilibrium manifests itself through conjugated fluxes (terminal electrical currents) and gradients (terminal voltages). Kirchhoff's laws are used to present them through integral–differential, algebraic systems of ODEs. To this adds the initial conditions—as many as reactive elements are. The ODEs system may be reduced to a single, higher order ODE, of the order equal to the number of existing reactive elements plus one. Summing up a consistent Cauchy problem may thus be formulated. Cauchy problem solvers are available for both linear and nonlinear ODEs (Vetterling et al., 1997).

Boundary and initial values problems

The physicals model for a system that executes a process under internal and external constraints lead to the mathematical models represented through PDEs, built out of the mathematical equations that present the physical laws, the boundary and initial conditions. Two general classes of problems may be distinguished: *well-posed* and *ill-posed* problems.

In Hadamard's definition (Hadamard, 1902, 1923), a problem is well-posed if (1) it has a solution (here, a physical solution); (2) the solution is unique; and (3) the solution depends continuously on the problem data. If any of these conditions is not fulfilled, then the problem is ill posed.

In practice, even though the existence of a solution is an important requirement for exact data, the condition (1) is sometimes satisfied provided the concept of solution is relaxed.

Condition (2) is more important. Should a problem admit multiple solutions then the question that arises is which of them is relevant in a particular situation. One may decide to use additional information in order to restrict the set of admissible solutions. It is very probable that in a practical application the existing measurement data even if, in the limit, available in an infinite number of points does not completely determine the solution. Although condition (2) is fulfilled in the continuous formulation sense of the problem when data are known everywhere, the nonunicity of the solution may occur due to the discretization of the computational domain, when numerical methods are used to solve the problem.

Condition (3) is motivated by the fact that in applications the problem data are obtained thorough measurements prone to errors. It is wishfully expectable that small errors do not amplify the errors in the solution. If not observed, condition (3) may produce significant numerical concerns because the numerical methods may become unstable. This difficulty may be partially alleviated by the usage of the *regularization techniques* (Vetterling et al., 1997). However, no mathematical artifact may fully "cure" the intrinsically unstable nature of ill-posed problems that are not complying with condition (3). In general regularization methods may recover partial information on the solutions and their application is actually an accepted compromise between the accuracy and the stability of the solution.

From the perspective of the input data, the mathematical models distinguish between *direct* and *inverse* problems. In direct problems the structure of the system, the materials and their properties, the internal sources and the boundary interactions are known. In inverse problems, either the structure of the system, or the materials and their properties, or the sources are unknown. In fact they make the object of the analysis. Examples of inverse problems are the electrical potential cardiac mapping, the ultrasound tomography, the thermography, the electrical impedance tomography, and so on.

Inverse problems are usually ill-posed and their solutions may be constructed on the "skeletons" produced by direct problems. For instance, in cardiac mapping the transfer operator (e.g., a rectangular transfer matrix, in numerical simulation) that projects the electrical potential from the thorax surface onto the epicardium surface may be obtained by solving a companion direct problem, which assumes the type of electrical source (monopole, dipole) and its localization (Mocanu, 2002). The solution to this problem requires the usage of inversion methods, singular value decomposition based algorithms, regularization methods, and so on (Vetterling et al., 1997).

The direct problems of heat and mass transfer, electromagnetism, structural mechanics and transport are frequently problems of *equilibrium*, *eigenvalues*, or *transmission* type.

Equilibrium problems are boundary value problems only. The physical quantities are constant in time and the associated PDE is stationary, that is, there are no time derivatives (Fig. 1.1). Examples of equilibrium problems are for instance the stationary EMFs (electrostatic, magnetostatic, electrokinetic, stationary magnetic field), stationary heat and mass (diffusion and/or convection) transfer, potential and stationary flows.

In Fig. 1.1 ∂D is the boundary, $L[\cdot]$ is the stationary PDE operator, f is the unknown primitive quantity, g is the inhomogeneity (the field "source," a known quantity), $B_i[\cdot]$ is the boundary condition, a known boundary operator, and g_i is the boundary inhomogeneity, a known quantity, e.g., a flux. The subscript $(\cdot)_i$ refers to part i of the boundary.

Eigenvalues problems are boundary value problems formulated in the first place for certain linear operators, but more than often they have a physical significance. For a linear operator, $L[\cdot]$, λ_i is the eigenvalue of index i, $M_i[\cdot]$ is its associated eigenfunction or eigenvector, and $E_i[\cdot]$ is its trace on the boundary (Fig. 1.2).

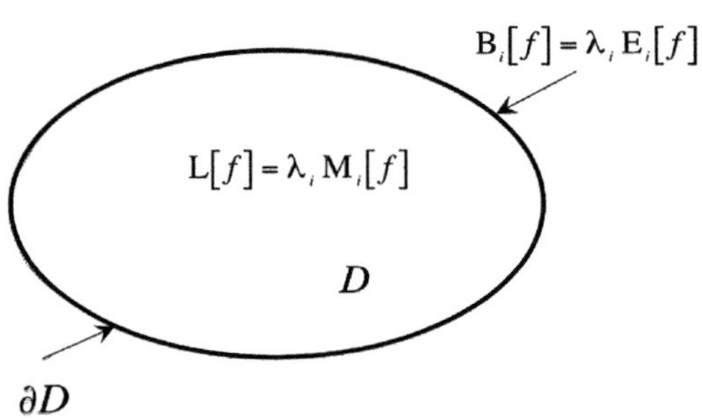

Figure 1.1 Equilibrium problems. The mathematical model.

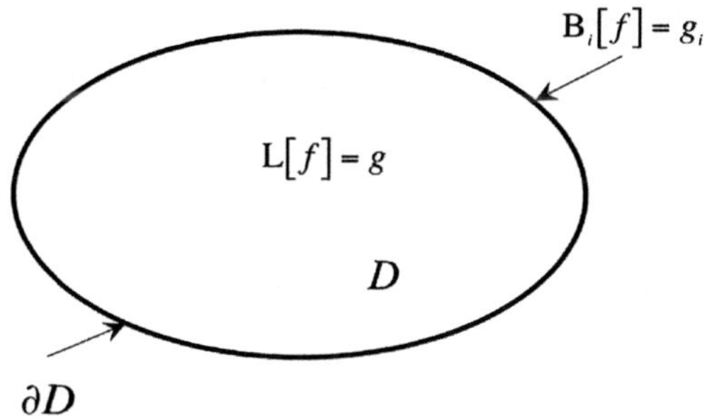

Figure 1.2 Eigenvalues problems. The mathematical model.

Spectral analysis is concerned with the eigensystem of the problem (Vetterling et al., 1997; Canuto and Quateroni, 1984; Ehrenstein and Peyret, 1989; Fox and Parker (1968); Morega and Nishimura, 1996; Peyret, 2002). In general the boundary conditions are imposed and the problem to be solved is stationary. Many mathematical problems that are reducible to stationary problems belong to Sturm–Liouville problem class (Gottlieb and Orszag, 1977): Laplace, Poisson, and Helmholtz.

Special functions, polynomials (Jacobi, Lagrange, Legendre, Cebâşev, Laguerre, Hermite, Gegenbauer, etc.), and functions (Bessel, Fourier, Mathieu, etc.) are used to form bases for projective analytical and numerical methods. In particular the numerical method use to solve many of the problems presented in this work is the finite element method (FEM) in Galerkin formulation Fletcher (1984), which uses Lagrange polynomials to represent and approximate the unknown function and the geometry of the computational domain (Peyret and Taylor, 1983; Bathe and Wilson, 1976).

Transmission problems are boundary and initial value problems presented through PDEs with space and time derivatives: diffusion (parabolic) and propagation (hyperbolic).

The general form of a two-dimensional second-order PDE, with constant coefficients is as follows:

$$\underbrace{a\phi_{xx} + 2b\phi_{xy} + c\phi_{yy}}_{\text{principal part}} + d\phi_x + e\phi_y + f\phi = g(x, y), \tag{1.15}$$

and is qualitatively similar to the general equation of a conical surface

$$\underbrace{ax^2 + 2bxy + cy^2}_{\text{principal part}} + dx + ey + f = 0, \tag{1.16}$$

with the following identification: $x \leftrightarrow \frac{\partial}{\partial x}, y \leftrightarrow \frac{\partial}{\partial y}$.

Using the terminology for the conics and corresponding to the principal part of the complete second-order differential operator Eq. (1.15), three main types of PDEs are obtainable through changes of variables, from (x,y) to (ξ,η), Table 1.1.

Table 1.1 Partial differential operators and related mathematical models.

Principal part	PDE type	Canonical form obtained after change of variables	Analytic problem
$b^2 - ac < 0$	Elliptic	$\phi_{\xi\xi} + \phi_{\eta\eta} = h(\xi, \eta) + D\phi$	Laplace/Poisson
$b^2 - ac = 0$	Parabolic	$\phi_{\xi\xi} - \phi_{\eta\eta} = h(\xi, \eta)$	Helmholtz
$b^2 - ac > 0$	Hyperbolic	$\phi_{\xi\xi} - \phi_{\eta\eta} = h(\xi, \eta) + D\phi$	Helmholtz

Several common physical processes may lead to static and stationary diffusion models (elliptic PDEs, or Laplace and Poisson problem, in Chapters 4: Electrical activity of the heart, and Chapter 6: Magnetic drug targeting), quasisteady and unsteady diffusion models (parabolic PDEs or Helmholtz problems, in Chapters 5: Bioimpedance methods, and Chapter 8: Hyperthermia and ablation (Thermotherapy methods), propagation models (hyperbolic PDEs or Helmholtz problems, in Chapter 8: Hyperthermia and ablation (Thermotherapy methods), steady and unsteady diffusion—transport problems (diffusion—convection PDEs, Navier—Stokes problem, in Chapters 5,8). Navier—Stokes equations have a parabolic character because of the nonzero diffusion term. However, depending on the specific situation, which from the physical point of view means the rheological model of the blood, the flow rate, and the vessel size that together define the Reynolds group, these equations are hyperbolic when they are convection dominated, and parabolic when they are diffusion dominated. From the mathematical point of view, the convection operator is hyperbolic, the diffusion operator is elliptic, and the time operator is parabolic.

Furthermore it is crucial to note that if the coefficients are variable then the PDE type may vary locally, that is, elliptic, parabolic, or hyperbolic. If the PDE is nonlinear then the type of problem to solve numerically may depend on the linearization technique that is utilized. These consequences may raise concerns in the selection of the solver, as solvers are optimized for classes of problems described through algebraic systems of equations.

1.6 Numerical solutions to the mathematical models

Many numerical methods were proposed and successfully used to solve the PDEs produced by the mathematical models, and it is for the researcher to decide which of them to use. Among them, FEM (Chapter 3: Computational Domains) has reached the level of versatility and numerical accuracy where complex mathematical models representing coupled physical processes and complex computational domains, such as those constructed using CAD or image-based construction techniques, may be solved successfully using available hardware and software resources. The mathematical models are solved here (Chapters 4—8) using FEM in Galerkin formulation, which may be introduced as a spectral method too (Peyret and Taylor, 1983). It may also be considered as belonging to the weighted residual class of methods. As with all numerical methods, the PDE numerical solution is calculated in a finite number of points that are obtained by dividing the computational domain into elements that form the *discretization mesh*. The unknowns are related to the grid nodes (vertices), which are a subset of the mesh nodes, called grid nodes. It should be noted that when automatic (Delaunay, 1934) mesh generation algorithms are used only the geometry of the computational domain (curvature, edges, voids, etc.) counts in the discretization of the computational domain, and the underlying physics is of less concern. Either structured or unstructured meshes may be used. Each of them has pros and cons.

The solution to the mathematical problem, the PDE is then represented locally by using a set of trial analytic functions that, endowed with certain smoothness and regularity properties, form a basis for the representation of the exact solution to the problem. For instance, they form a closed set defined on the Hilbert space of the functions of integrable square, and it is desirable that they are orthonormal too (Peyret and Taylor, 1983). Sturm–Liouville singular problems are resources for the set of functions that may form a basis, and the boundary conditions of the mathematical problem, in general the principal part of the PDE may be used to single out one of the available solutions. In general lower order polynomials are preferred—here, Lagrange polynomials (Orszag, and Gotlieb, 1980).

FEM methods may use node elements for electromagnetic (vector) fields too. In this situation the vector field quantities are described with their components at the vertices, that is, their scalar projections are assigned to each vertex. However, using the coordinates of the vectors, give rise to difficulties in implementing the boundary conditions and satisfying the continuity of the numerical solution (Webb, 1993). To overcome these difficulties, *edge elements* were proposed instead by Takahashi et al. (1992). Their usage does not necessarily eliminate the unphysical, spurious modes though Schroeder and Wolf (1994). These may be avoided only by a proper finite element formulation (Mur, 1998).

1.7 Coupled (multiphysics) problems

In many circumstances, the systems are sieges of multiple concurring irreversible-flow transport phenomena—mass, heat, electrical charges, etc. From a thermodynamic perspective, unlike the Gibbsian formulation for systems in internal and external equilibrium, these interactions are characterized through interactions rates. This finding neither introduces new physical insight nor explains, *per se*, the couplings between the interaction rates and the local thermodynamic properties of the system subject of such interactions (Bejan, 1988). The analytic forms of these couplings are just postulated, proposed by the *irreversible thermodynamics* through relations between gradients of local thermodynamic properties and fluxes that express the interaction rates. The particular forms of these "assumed" relations are found empirically or based on a different theory. For example, the diffusion (conduction) laws such as Fourier, Ohm, and Fick empirical laws.

In this respect, Onsager relations (Bejan, 1988) provide for a succinct mathematical representation of these relations of reciprocity, which presents a unified analytically consistent expression of irreversible-flow processes. Most notably, and expected after all, material properties such as the electrical conductivity, mass diffusivity, thermal conductivity, viscosity, and thermoelectric power have their "trace" as "coefficients" in these relations, relating the fluxes with the gradients.

Couplings between the PDEs that present the physics may occur through the material properties (e.g., the properties are functions of temperature, Chapters 7: Magnetic stimulation, and Chapter 8: Hyperthermia and ablation (Thermotherapy methods), heat source terms (e.g., Joule, dielectric heating, or body forces (e.g., magnetic forces in MDT, Chapter 6: Magnetic Drug Targeting).

The physical and mathematical models then sum up all laws of Physics needed to describe these phenomena in a consistent body of equations that has then to be solved. It should be recognized that, again, a qualitative analysis has to be performed in the first place to infer the strengths of such couplings with the aim to reduce the complexity of the problem to be solved without losing the physical reality of the solution and the meaningfulness of the results. The evaluation of the *time* and *space scales*, *penetration depths* is part of this analysis.

1.8 Time and space scales

The physical phenomena that define a specific problem may evolve at different paces. Their time scales may be estimated *a priori,* analytically, in the order of magnitude sense, considering the time-dependent PDEs of their associated mathematical models. The underlying mechanisms of evolution range from unsteady diffusion (conduction), to propagation or transport (advection, convection) processes, and an earlier stage qualitative analysis aims at evaluating their orders of magnitude, which may differ considerably. Constitutive, matter properties, system structure, and its relative motion contribute in defining their size.

Diffusion time scales are governing in unsteady conduction heat transfer, mass transfer, and low-frequency EMFs. For instance, the unsteady conduction (diffusion) heat transfer problem in immobile media with an internal heat source, q''' [W/m^3], is described by the first law of thermodynamics (the energy equation; Bejan, 1993)

$$\rho c \frac{\partial T}{\partial t} = \nabla(k \nabla T) + q''', \tag{1.17}$$

where ρ [kg/m^3] is the mass density, c [J/(kg·K)] is the specific heat at constant pressure, k [W/(m·K)] is the thermal conductivity, and T [K] is the temperature. Assuming that the thermal size of the system, that is, the size (length) of the diffusion path is L [m], then Eq. (1.17) indicates the following order of magnitude balance between its constitutive terms

$$\rho c \frac{\Delta T}{\tau_{diffusion}} \sim k \frac{\Delta T}{L^2}, \quad q_0''', \tag{1.18}$$

where ΔT is the order of magnitude of the temperature temporal and spatial "excursion," q_0''' is the order of magnitude of the heat source. The scaling relation Eq. (1.18) shows off the *diffusion time constant*, $\tau_{diffusion}$ [s], which is a measure of the time needed for the thermal transient to vanish hence for the system to reach a final steady state. As the material properties and the space scales are known in direct problems then Eq. (1.12) yields

$$\tau_{diffusion} = \frac{\rho c}{k} \frac{1}{L^2} = \frac{L^2}{\alpha}, \tag{1.19}$$

where α [m^2/s] is called *diffusivity*, here *thermal diffusivity*—a quantity found in the material properties data sheets.

If the heat source is dynamic, then $\tau_{diffusion}$ has to be compared with the source (excitation) time scale to decide the adequate form of the mathematical model to be used. For instance, if the heat source is the Joule power produced by a harmonic conduction current [e.g., the electroablation, Chapter 8: Hyperthermia and Ablation (Thermotherapy Methods)], $q_0''' = \rho_{el} J_0^2$, where ρ_{el} [Ω m] is the electrical conductivity and J_0 [A/m^2] is the order of magnitude of the electrical current density, then the time scale of the excitation is 20 ms for an electrical current source operating at 50 Hz. Assuming that the ROI is a spherical volume of liver tissue with the radius ~1 cm, and its properties are $k = 0.502$ W/m·K, $\rho = 1060$ kg/m^3, $c = 3600$ J/(kg·K) (Valvano, 2010), then the relation Eq. (1.19) yields $\tau_{diffusion}$ ~760 s, which predicts roughly the time to reach some desired steady state temperature.

However, because $\tau_{diffusion} << 20$ ms it will be reasonable to consider that the Joule power is produced by an equivalent (r.m.s.) DC current density, J_{DC} [A/m^2], whose distribution is governed by an equivalent potential (Laplace) problem, which yields

$$\rho c_p \frac{\Delta T}{\Delta t} \sim k \frac{\Delta T}{L^2}, \ \rho_{el} J_{DC}^2. \tag{1.20}$$

This approach avoids integrating, in the same time, a quasistationary EMF problem and the associated heat transfer problem—a considerably more laborious and, overall, less relevant path to follow.

It is worth noting that any unsteady diffusion process governed by a parabolic PDE is driven by such a diffusion time constant, which depends on the material properties and the size of the physical system only, and the length of the path in the diffusion direction—the largest of the diffusion paths if the process is multidirectional. For instance, in the magnetic field analysis in the unsteady diffusion problem of electroconductive media, known also as the eddy currents working conditions, the parabolic PDE to solve may be presented as

$$\frac{\partial \mathbf{H}}{\partial t} = \frac{1}{\mu\sigma}\Delta\mathbf{H}, \tag{1.21}$$

where μ [H/m] is the magnetic permeability, σ [S/m] is the electrical conductivity, and $\mathbf{H}$ [A/m] is the magnetic field strength. An analysis similar to Eq. (1.20) yields the magnetic diffusion time constant (scale)

$$\tau_{mag} = \frac{L^2}{\mu\sigma} = \frac{L^2}{\alpha_{mg}}, \tag{1.22}$$

which introduces the *magnetic diffusivity*, $\alpha_{mag} = (\mu\sigma)^{-1}\left[\mathrm{m}^2/\mathrm{s}\right]$, or the *magnetic Reynolds number*.

Transport time scale characterizes, for instance, the unsteady diffusion–advection heat flow of a system without internal heat sources (Bejan, 1984) presented through the energy equation

$$\rho c\left[\frac{\partial T}{\partial t} + (\mathbf{u}\cdot\nabla)T\right] = \nabla(k\nabla T), \tag{1.23}$$

and the momentum equation (Navier–Stokes)

$$\rho\left[\frac{\partial \mathbf{u}}{\partial t} + (\mathbf{u}\cdot\nabla)\mathbf{u}\right] = -\nabla p + \eta\nabla^2\mathbf{u}, \tag{1.24}$$

for a Newtonian fluid in incompressible, laminar flow ($\nabla\cdot\mathbf{u} = 0$). Here p [Pa] is the pressure, $\mathbf{u}$ [m/s] is the velocity field, and η [Pa · s] is the dynamic viscosity.

The momentum Eq. (1.24) may be scaled to indicate the order of magnitude balance

$$\frac{L}{U_0\tau_{transport}},\ 1\sim\frac{P_0}{\rho U_0^2},\ \frac{1}{\mathrm{Re}_L}\left(=\frac{\eta}{\rho}\frac{1}{U_0 L}\right),\ \text{or } 1,\ 1\sim 1,\ \frac{1}{\mathrm{Re}_L}, \tag{1.25}$$

where U_0 is a reference velocity [same in the heat transfer problem Eq. (1.23), if the two problems are coupled], P_0 is a reference pressure, L is the size of the physical domain [same in Eq. (1.20), if the two problems are coupled], and Re_L is the nondimensional Reynolds group based on the length scale L. Provided that U_0 is known (e.g., out of the boundary conditions) one may chose $P_0 = \rho U_0^2$, and then define $\tau_{transport} = L/U_0$, which is a transport (velocity) time scale.

Consequently the dynamics of the unsteady diffusion–advection energy Eq. (1.23) is led by either the diffusion time scale, $\tau_{diffusion}$ Eq. (1.19), or by the velocity time scale, $\tau_{transport}$. As only one time-scale may prevail in numerical simulations of the coupled convection heat transfer problem, either the largest (diffusion usually) or the smallest (transport usually) prevails. If the transport time scale is dominant (smallest) then $\tau_{transport}$ is selected, and the energy equation has then to be rescaled accordingly.

As an example, in modeling the heat transfer in localized hyperthermia it is common to use a homogenization technique in lieu of the general heat transfer model, for example, the bioheat equation (Pennes, 1948) (discussed later in this chapter, and in : Magnetic Stimulation). This approach assumes that the ROI is a homogeneous, continuum medium with distributed heat sink/source that accounts for the hemodynamic heat transfer

$$\rho c\frac{\partial T}{\partial t}+\rho_b C_b\omega(T-T_a)=\nabla(k\nabla T)+p_{Joule}, \tag{1.26}$$

which is a particular form of the energy equation. Here T_a is the arterial blood temperature (37°C), ω [1/s] is the blood perfusion rate, ρ and ρ_b [kg/m^3] are mass densities of tissue and blood, respectively, C and C_b [J/(kg·K)] are specific heat capacities of tissue and blood, respectively, and p_{Joule} [W/m^3] is the heat source, for example, the Joule effect. The scaling of Eq. (1.26) yields the order of magnitude relation

$$\tau \sim \underbrace{\frac{L^2}{\alpha}}_{\tau_{diffusion}},\ \underbrace{\frac{\rho C}{\rho_b C_b\omega}}_{\tau_{transport}}, \tag{1.27}$$

which shows off two concurring time scales: a diffusion time scale and a transport time scale. If the ROI is a spherical volume of liver tissue of diameter 1 cm, using $\rho = 1000\ \mathrm{kg/m^3}$, $k = 0.512\ \mathrm{W/(m\cdot K)}$, $C = 3600\ \mathrm{J/(kg\cdot K)}$, $\rho_b = 1000\ \mathrm{kg/m^3}$, $C_b = 4180\ \mathrm{J/(kg\cdot K)}$, $\omega = 6.4\times10^{-3}\ \mathrm{s^{-1}}$ Morega (EHB), the scaling relation Eq. (1.27) yields $\tau_{diffusion} \sim 800$ s and $\tau_{transport} \sim 140$ s. Therefore hyperthermia procedure may be successful provided that the heat source is intense enough to raise the tissue temperature at the required hyperthermia level, and to compensate for the heat loss through hemodynamic flow for the duration of the procedure.

If the general heat transfer model, Eqs. (1.23), (1.24) is recognized for larger vessels (arteries and veins) then, in general, the velocity time constant prevails. Furthermore the pace of the pulsating arterial blood flow (e.g., 60–80 bpm) may be much smaller than the diffusion time constant. To avoid a lengthy and cumbersome numerical simulation, it may be more convenient to replace the pulsatile flow with an equivalent stationary flow, with an average flow rate (r.m.s.). The stationary velocity field is then used in the transient heat transfer problem solution to avoid solving for a coupled pulsating hemodynamic flow and the associated transient heat transfer problem simultaneously [Chapter 8: Hyperthermia and Ablation (Thermotherapy Methods)]. Finer details (e.g., the oscillations in the arterial blood temperature due to the pulsating flow) may be lost but the dynamics of the tissue temperature raise is well characterized [Chapter 7: Magnetic Stimulation and Chapter 8: Hyperthermia and Ablation (Thermotherapy Methods)].

Propagation time scales may occur in EMF radiation problems. For a system without internal EMF sources, placed in an external EMF, e.g., produced by a radiant EMF external power source, the time scales for EMF interactions may be those of the external excitation. Assuming linear media, Maxwell's laws for immobile media yield the general diffusion–propagation PDE for the EMF presented in **H**

$$\Delta \mathbf{H} = \mu\sigma \frac{\partial \mathbf{H}}{\partial t} + \mu\varepsilon \frac{\partial^2 \mathbf{H}}{\partial t^2}, \tag{1.28}$$

which may lead to the following order of magnitude relation

$$\frac{1}{L^2} \sim \mu\sigma \frac{1}{\tau_{diffusion}}, \ \mu\varepsilon \frac{1}{\tau^2_{propagation}}. \tag{1.29}$$

Two time-scales are seen to emerge: an EMF diffusion time scale, $\tau_{diffusion} = L^2/\mu\sigma$, and an EMF propagation time scale, $\tau_{propagation} = L/\sqrt{\mu\varepsilon}$. Depending of the material properties and the length scale, the smallest of two is usually observed to follow in detail the dynamics of the process.

The scaling relation Eq. (1.29) conveys also valuable information on the penetration depth for the two EMF transmission mechanisms. For instance, for a harmonic EMF power source operating at $\omega = 2\pi f$, the time constant is $\tau_{source} = 1/f$. Relation Eq. (1.29) indicates two penetration depth scales: a *diffusion penetration* depth, $\delta_{diffusion} \sim \frac{1}{\omega\sqrt{\mu\varepsilon}}$, and a *propagation penetration depth*, $\delta_{propagation} \sim \frac{1}{\omega\sqrt{\mu\varepsilon}}$.

Correlating the two power sources, Ohmic and dielectric, pointed out by Eq. (1.13) for contributing to the first law energy balance with the penetration depths outlined by Eq. (1.29), it may be inferred that the hot spots inside the system that correspond to the two power sources may have different localizations, and their cumulative effect may actually show off a different thermal image to be the object of interest. This aspect will be analyzed in Chapter 8: Hyperthermia and Ablation (Thermotherapy Methods).

1.9 Properties of anatomic media

Electrical properties

Numerical modeling of living tissue is not possible without proper knowledge of their physical properties, since any interaction phenomenon (electromagnetic, thermal, or mechanical) is based on the response given by a substance to a certain form of stress. The elemental such response is quantified through adequate materials properties. In most biomedical applications, numerical modeling is performed at a macroscopic scale. The numerical analysis is based on equations of classical physics, applied to components of the human body up to the cellular level; characteristic dimensions might be as small as the typical cells, that is, down to the micron range magnitudes and the movement is very slow, that is, low speeds, like for Newton mechanics. The frequencies of the electric and magnetic fields are within the nonionizing (Hertzian) range, that is, lower than 300 GHz.

Specific macroscopic impact phenomena generated by the interactions of EMFs and biological matter (tissues) could be classified in two large groups: stimulation of excitable tissue, for low and medium frequency range, and heating, for medium and high frequency domain.

Stimulation represents the electrical activation of excitable cells membranes (local membrane depolarization, by the opening of transmembrane active channels for the transfer of selected ion flows) (Chapter 4: Electrical Activity of the Heart). The stimulus is an electrical signal (current), induced either by an applied external electric field (electrical stimulation) or by a variable magnetic field through electromagnetic induction effect (magnetic stimulation). Biophysical effects of stimulation target the activation of nerves, muscles or sensitive tissue and it is macroscopically quantified by the local distribution of the induced electric field or current density. Induced currents could also produce interferences with natural electrophysiological phenomena and disturbance of normal generation and transmission of biocurrents from various body sources (heart, brain, peripheral nerves, sensor analyzers), which are commonly used in medical diagnosis. Simulation of such phenomena requires specification of the tissue dielectric properties: conductivity and permittivity that are highly dependent on the electric field frequency. For the strengths of the electric field commonly associated to nondestructive tissue applications (bioelectrical phenomena, medical procedures, or environmental body exposure), the tissues show linear behavior to both dielectric properties (Chapter 8: Hyperthermia and Ablation). Some tissues (like bone and muscle) are anisotropic concerning dielectric properties in the low-medium frequency spectrum, up to 100 kHz. The anisotropy is acknowledged in field equations by the tensor representation of the dielectric property; however, in most applications an average value is considered as a constant dielectric property.

Heating occurs in conductive materials, which absorb the electromagnetic radiation [Chapter 7: Magnetic Stimulation and Chapter 8: Hyperthermia and Ablation (Thermotherapy Methods)]. This process is effective for high frequencies (radio waves and microwaves); the energy transferred by the incident electromagnetic waves to the target tissues is converted into heat. The higher the frequency, the lower is the penetration depth of the radiation and the heating is more superficial, but heat is further transported inside the body by thermal conduction; thermal conductivity is thus an important physical property to quantify the process. Dielectric properties of tissues do influence both the penetration depth and the distribution of the absorbed energy, while the thermal properties govern the evolution of the heat distribution in tissues. An important role in the regulation of temperature distribution is played by the cooling effect of the blood flow; the energy absorbed through metabolic intake could be considered too in the completion of the energy balance (Chapter 7: Magnetic Stimulation).

Literature presents several studies on the electrical characteristics of tissues and true collections of values associated to the electrical conductivity and permittivity of most significant tissues involved in the description of anatomical models for various frequency ranges were offered by Stuchly and Stuchly (1980) or Durney et al. (1986), according to the growth of computational resources, in the context of the increasing interest for numerical modeling. However, a systematic study dedicated to the dielectric tissues properties was not properly conducted until the 1990s, when the group led by Camelia Gabriel presented its comprehensive database, including tens of tissues, both human and animal, investigated over considerably large frequency ranges, from very low (10 Hz) to microwaves (20 GHz). They have also the merit to discuss their results, compare them with previous literature and show details on the measurement methods and on the mathematical approach for finding proper parametric expressions of continuous functions, useful to cover all needed frequencies, as shown by Gabriel et al. (1996a,b,c) and Gabriel and Gabriel (1996). The database founded by Gabriel's team is maintained and continuously enriched at the Institute of Applied Physics, Italian Research Council of Florence, as an internet public resource presented by Andreuccetti et al. (1997), available online for the benefit of researchers all over the world.

Since most applications operate in, or could be reduced to time-harmonic working conditions, the complex form of EMF equations represents a significant computational facility, adopted by a large section of software packages for the numerical analysis of AC and RF EMFs. In that context, the dielectric properties are included in one single theoretical quantity—the *complex conductivity* $\underline{\sigma} = \sigma + j\omega\varepsilon$, or its correspondent—the *complex permittivity* $\underline{\varepsilon} = \frac{\underline{\sigma}}{j\omega} = \varepsilon - j\frac{\sigma}{\omega}$, where $\sigma\ [\mathrm{S/m}]$ is the true electrical conductivity, $\varepsilon\ [\mathrm{F/m}]$ is the true dielectric permittivity $j = \sqrt{-1}$.

Biological materials (tissues) are generally nonmagnetic materials for the whole Hertzian frequency range, and the magnetic permeability is commonly considered identical with the permittivity in free space $\mu_0 = 4\pi \times 10^{-7}\,\mathrm{H/m}$. The presence of magnetic materials inclusions in biological tissues—like magnetite associated with injected chemicals used in magnetic drug targeting procedures, or the excess of ferrous oxides absorbed in various tissues due to pollution or other causes—might be considered by the increase of the magnetic permeability, according to the proper concentration.

When exposed to low frequency EMF human tissues behave as conductive materials, with specific electrical conductivity for different tissues. The blood has a major influence in the variability of the electrical conductivity (Chapter 5: Bioimpedance

Methods and Chapter 6: Magnetic Drug Targeting). Changes in the electrical conductivity of the aortic blood during a cardiac cycle are marked by a switch in orientation and shape of the red blood cells (RBC) due to the hemodynamic flow, which is the reason for change in the electrical conductivity: during the diastole when RBCs are randomly distributed the electrical resistance is high (low conductance) whereas during the systolic period the RBCs are aligned streamwise, and change their shape to favor flow, which leads to low electrical resistance (high conductance) (Hoetink et al., 2004; Gaw et al., 2008; Visser, 1992). Thus blood electrical conductivity is

$$\sigma_b = \sigma_{pl}\frac{1-H}{1+(C-1)H}, \tag{1.30}$$

where σ_b and σ_{pl} [S/m] are the electric conductivities of blood and plasma, H represents the hematocrit, and C is defined as nondimensional geometric factor for the RBC. In the round tube theory, C is a function of the tube radius, the geometry of the RBC approximated as prolate ellipsoid, the local the shear rate and an empirical time constant for cell orientation—cells changing from random to aligned orientation; Gaw et al., 2008. More details are found in Chapter 5: Bioimpedance Methods.

In electrophysiology and bioelectromagnetism, the electrically conductive medium extends continuously; it is three-dimensional, and referred to as a volume conductor (Malmivuo and Plonsey, 1995). Capacitance is distributed too because capacitive effects are related to cellular membranes, which extend continuously throughout a three-dimensional region. It may be noted that usually the electrical conductivity is seen as a constant, subject to the regular variability due to the different tissues and individuals (Gabriel and Gabriel, 1996; Valvano, 2010), and less concern is given to its hemodynamic flow dependence. However, for modeling electrical fields in anatomic regions with arterial flows it should be ascertained whether the constancy of electrical conductivity is acceptable. Thus for instance, a group of impedance methods aimed at the evaluation of the hemodynamic flow parameters, the Impedance Cardiography (Miller and Horvath, 1978) and the Electrical Cardiometry (Osypka, 2009) (Chapter 5: Bioimpedance Methods) are based on, sense, and use it.

Rheological properties of blood

The rheology of blood is of concern in the arterial hemodynamic flow numerical simulation, which implements momentum equation, in either Navier–Stokes, for "clear" fluid, or Brinkmann, Forchheimer, and Darcy, in porous media formulations, for which specific models of fluid are presumed (Chapter 5: Bioimpedance methods, and Chapter 8: Hyperthermia and ablation (Thermotherapy methods). Depending on the group of blood vessels the hemodynamic analysis may require different rheological models for blood. Fig. 1.3 renders, qualitatively, the pressure levels for the types of flow in relation to the specific vascular segment of concern (Feijóo, 2000; Morega et al., 2010).

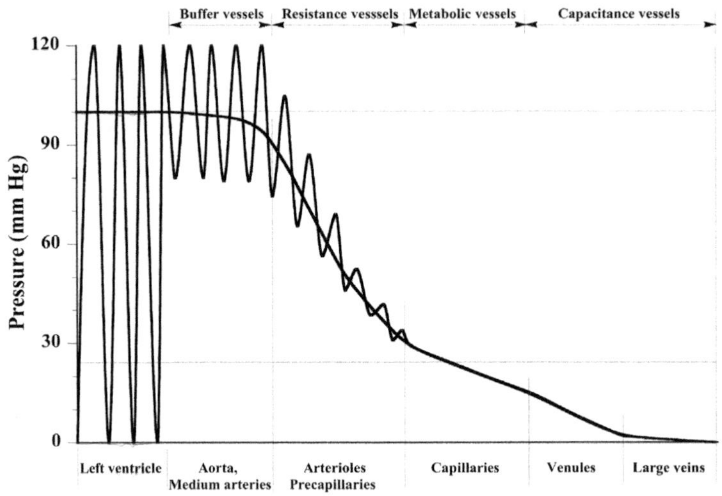

Figure 1.3 The vascular tree with different hemodynamic flows and vessel. *After Feijóo, 2000. Computational methods in biology. In: Proceedings of the 2nd Summer School LNCC/MCT, Petrópolis, Brazil.*

For instance, for larger arteries of "resistive type," the constant properties (Newtonian) fluid (Elert, 2018) with the dynamic viscosity

$$\eta = 0.005\ \text{Pa}\cdot\text{s}, \tag{1.31}$$

may be satisfactory. For medium-sized vessels other models are preferred, for example, *Carreau model* (Gambaruto et al., 2011):

$$\eta = \eta_\infty + (\eta - \eta_0)\left(1 + (\lambda\gamma')^2\right)^{\frac{n-1}{2}}, \tag{1.32}$$

where η_∞ [Pa · s] is the infinite viscosity shear rate, η_0 [Pa · s] is the zero viscosity shear rate, γ' [s^{-1}] is the shear rate tensor, λ [s] is the relaxation time, and n is a model parameter, or *Power law* model (Shibeshi and Collins, 2005):

$$\eta = m\gamma'^{n-1}, \tag{1.33}$$

where m [$\text{Pa}\cdot\text{s}^n$] and n are model parameters. The power-law model solves the discrepancies among published values of the viscosity measured using different techniques.

For smaller vessels (arterioles, venules, capillaries) porous media models may be an option (Chapter 5: Bioimpedance methods, and Chapter 8: Hyperthermia and ablation (Thermotherapy methods), provided empirical properties for the vascularized tissues are available.

1.9.3 Bioheat models, homogenization methods

The models presented in this book, many of them, are concerned with heat transfer interactions. The difficulties that occur are related to the special thermal behavior of the anatomical tissues with complex structures, heat transfer properties, and heat flow mechanisms. Heat transfer in living tissues is a multiscale process involving metabolic heat production, thermal conduction in tissues, heat convection in larger blood vessels, and heat carried by perfused tissues, capillaries, microvasculature. Mathematical models based on continuum media homogenization hypotheses are used for heat transfer analysis of the body heat balance, in thermoregulation, heat transfer in muscle and tissues, skin burns, surgical procedures, for example, hyperthermia and hypothermia cancer treatment, laser surgery, cryosurgery, cryopreservation of organs for transplant, resuscitation, thermal comfort, and extracorporeal equipment.

Pennes introduced for the first time a simple model, the bioheat equation (Pennes, 1948), by adopting the energy equation to yield

$$\rho_{tissue} C_{tissue} \frac{\partial T_{tissue}}{\partial t} = k\nabla^2 T_{tissue} - \rho_b C_b \omega_b (T_b - T_{tissue}) + q'''_m. \quad (1.34)$$

Here ω_b [s^{-1}] is the blood perfusion rate, T_b, ρ_b, and C_b are the temperature, mass density, and specific heat of the arterial blood, T_{tissue}, ρ_{tissue}, and C_{tissue} are the temperature, mass density and specific heat of the of the tissue, and q'''_m [W/m^3] is the metabolic heat rate.

This model was and still is widely used to describe the heat transfer within living tissues. Although attractive for its simplicity, it bears a number of shortcomings, which prone its predictions to error. For instance, it assumes only the venous blood flow as the fluid stream equilibrated with the tissue—tissue–blood local thermal equilibrium. Moreover it assumes uniform perfusion, neglecting the directionality of the blood flow and the important anatomical features of the circulatory network system such as countercurrent arrangement of the vascular system, the different sizes of the vessels, from micrometers to millimeters, and the pending different rheological models for blood, vascular geometry, the transvascular heat and mass transfer, the sharp spatial variations of the material properties, the necrosis that might accompany the heat generation, to name some.

Aiming to overcome these difficulties several continuum models were proposed. Wulff (1974) and Nakayama and Kuwahara (2008) assumed that the blood temperature is that of the tissue temperature within a tissue control volume and not only the local tissue temperature gradient. Klinger (1974) and later Cho (1992) assumed that the heat transfer between the hemodynamic flowing and the irrigated tissue is proportional to the temperature difference between these two media with nonuniform velocity field in space and time, and metabolic heat source (Zolfaghari, and Maerefat, 2010).

Chen and Holmes (1980) (CH–BHT) showed that the major heat transfer processes occur in the 50 μm to 500 μm diameter vessels and proposed that larger vessels be modeled separately from smaller vessels and the embedding tissue. Their model subdivides the tissue control volume into solid and bloodstream subvolumes, introduce a modified relationship for calculating the blood perfusion term, and accounts for the "eddy" conduction due to the random flow of blood, by introducing a modified thermal conductivity. However, CH–BHT model is difficult to implement because it requires detailed knowledge of the vascular network and blood perfusion. Chato (1980) introduced a countercurrent, two-temperature model. He obtained the temperature profiles along the arterial and venous vessels and evidenced the heat transfer enhancement by the perfusion "bleed-off" between the vessels as compared with constant mass flow rates models (Nakayama and Kuwahara, 2008). To account for the perfusion heat sources, which might be important particularly in the extremities (Roetzel and Xuan, 1998) replaced the convection–perfusion parameters with interfacial convection heat transfer coefficients.

Three-temperature energy equations bioheat model proposed by Weinbaum et al. (1984) account for the countercurrent blood flow effect, mainly applicable for the intermediate tissue of the skin (Minkowycz et al., 2009) where small arteries and veins are parallel, and their flow directions are countercurrent. Weinbaum and Jiji (1985) evidenced three vascular layers—deep, intermediate, and cutaneous—in the outer 1 cm tissue layer in a study performed on rabbit limbs.

Bioheat transfer models, such as in the studies by Pennes (1948), Wulff (1974), Klinger (1974), Chen and Holmes (1980), Weinbaum et al. (1984), and Nakayama and Kuwahara (2008) identified the blood perfusion with a heat source term in the energy equation, and use homogenization techniques to replace the complex anatomic media with equivalent continuum media.

As seen, progressively, one-, two- and three-energy equations (temperatures), isotropic and anisotropic models, were introduced, each of them with merits and shortcomings. Pennes one-temperature model is the simplest of them and despites its limitations it is still used intensively. Two-temperature models are used to investigate the countercurrent heat exchange between the arterial and venous blood vessels in the circulatory system for idealized one-dimensional cases. Three-energy equations may be more general since in its multidimensional and anisotropic form can be applied to all regions peripheral heat transfer from the extremity to the surroundings (Nakayama and Kuwahara, 2008). This model provides control volume-based recipes to calculate the mechanical permeability, volume fractions, interfacial heat transfer coefficients, and perfusion rates. Coupled with the continuity and Darcy's laws, it results in a mathematical model that may be solved to find both velocity and temperature fields. However, despite these progresses, it is yet much to do (Nakayama and Kuwahara, 2008; Roetzel and Xuan, 1998). For instance, there is a need for the clarification of some vasoconstrictor and vasodilator mechanisms, model "constants," physiological parameters (porosity, specific surface area), which depend on factors such as the body temperature and its interaction with the environment, and have to be determined experimentally.

Summing up, a bioheat transfer problem requires an analysis that is specific to that anatomic ROI—vascularization, tissue, heat source, and surrounding, boundary conditions—before selecting one model or another. In general most existing bioheat models reside in the theory of porous media specifically the heat and fluid flow in a fluid-saturated porous medium. Therefore to distinguish between the heat flow by larger vessels from smaller vessels, represented through participating, by diffusion and convection, heat transfer continuum media with homogenized macroscopic properties, we use models of general heat transfer through hemodynamic conduction and convection for the larger vessels, and Brinkmann-type model for the heat transfer through the participating tissue and embedded smaller vasculature.

This approach enables connecting the two coexisting types of flows, larger vessel flow and porous media flow, and is used in Chapter 6: Magnetic drug targeting, Chapter 7: Magnetic drug targeting, and Chapter 8: Hyperthermia and ablation (Thermotherapy methods).

1.10 The computational domain

The system of concern is usually outlined, drawn by a qualitative analysis aimed to explain "what and why is working" has to be materialized, to embody its abstract concept of physical domain to which the related physics and the associated interactions apply. In the end, it becomes the computational domain or simply the "geometry," drawn with the pencil on the paper, properly sized and presented such as to enable the quantitative analysis of the underlying physics—mathematical model, analytical, or numerical solution.

For engineered systems, the contemporary natural representation of the computational domain is the CAD produced "geometry." Its complexity and level of detail is then a tradeoff between the required physical detail and the available resources—hardware, software, etc. More recently, medical image–based construction techniques–fused geometries (CAD and construction based) are used to provide patient-tailored computational domains. Chapter 3: Computational Domains, provides some details and examples of the design techniques that are commonly utilized in this book.

Numerical modeling for medical engineering is about systems and anatomic structures. The sketch of the computational domain has then to be inspired and must represent the insights of the design and evolution in nature. Design phenomena are not covered through the existing laws of physics though (Bejan, 2000). Instead scientists propose empirical or theoretical skeletons to build upon: general models, fractal geometry, network theories, chaos theory (Bergé et al., 1984), optimality statements (optimum, maximum, minimum) (Bejan and Zane, 2012), allometric scaling rules (power laws) (Darwin, 1845; Huxley, 1972; D'Arcy, 1992), and so on. All these methods are descriptive rather than predictive.

Allometric laws, fractal geometry, and constructal law

The scaling method and the scaling rules were introduced and used here to provide for order of magnitude, predictive relations that are useful in constructing consistent and computationally efficient physical models. The scaling rules have much more potential though as they reflect the underlying physical principles and the generic features of the system: size, structure, and properties. In fact they may define the skeleton, predict the quantities that are related to each other, for quantitative assertions to be found through experimental statistics. Along this path comes the allometry[3] or the statistical shape analysis.

Allometry is also used in biology, where it studies the relationship of the body size to its shape, anatomy, physiology, and finally behavior. Examples of allometric rules are (Snell, 1982) *body size–mass law*, in power law form, $Y = Y_0 \cdot M^{\alpha}$, [Y is a dependent parameter, Y_0 an integration constant, M the body mass, α the scaling exponent ($\alpha > 0$ positive allometry, $\alpha < 0$ negative, allometry, $\alpha = 1$ isometry)], or in log arithmic form, $\log Y = \alpha \log M + \log Y_0$; (Murray, 1926a,b) work–uptake law for transfer processes in mass conservative networks [the cube of vessels diameter at each generation is preserved, i.e., $\Sigma D^3 \sim$ constant, in animal circulatory systems, plant vascular systems, ecosystems (e.g., forests), intracellular networks, etc., but also porous materials, *a.s.o.*]; Huo–Kassab model shows the Murray's law exponent is equal to 7/3 (rather than 3) in coronary branching (Huo and Kassab, 2012; Kleiber, 1932) metabolic rates law for of mammals and birds, $B \propto M^b$ [B is the metabolic rate, M is the body mass, b is an allometric exponent (3/4 metabolic, −1/4 heart rate, 1/4 life span, 3/8 aorta/tree trunk diameters, 1/4 genome length, −3/4 population densities in forests)], to name a few.

Allometric relationships between two measured quantities, e.g., mass and flow rate, are often expressed as power law equations. However, it should be noted that the existence per se of the power law does not indicate that the object is necessarily a fractal.

The geometric and functional complexity of the tree structures in the human body—hemodynamic system, respiratory tract, renal system, etc.—raises significant difficulties when passing from mathematical modeling to numerical simulation, and Euclidean geometry that applies to smooth and regular shapes of integer geometric dimension—zero for a point, one for a line, two for a plane, and three for a volume—fails to resolve such problems. *Fractal*[4] geometry was then introduced to describe such "mathematical monsters" (Mandelbrot, 1975), self-similar objects that have the same details in different scales, of noninteger geometric, fractal dimension Order in Chaos (2013) and Bergé et al. (1984).

[3] Greek: *allos* means different and *metrie* means to measure.

[4] Latin: *fract* means broken.

For instance, fractals are applied in modeling the diabetic retinopathy, a very common complication of diabetes disease that produces changes in the morphology of blood vessels in the diseased retina. The fractal properties of blood vessel patterns of the retina involves manual segmentation of the blood vessel patterns and the analysis of the retinal vasculature as a fractal (Cheng and Huang, 2003; Cheung et al., 2009; Uahabi and Atounti, 2015), and singles out the fractal dimension and its lacunarity—fractals that have the same fractal dimension but with different appearances—to probe of the disease progression.

Fractal design tools may be utilized to mimic, to some degree of accuracy, naturally made systems, for example, anatomic structures through fairly relevant presentations. Even complex flow, tree-like structures such as the air passage of lungs (a flow system for air), the capillaries (a flow system for blood), and the neuronal dendrites (a flow of electrical signals) may be rendered using the fractal geometry (Bejan and Zane, 2012). Although fractal geometry may render convincingly tree-like structures it neither relies nor provides on the physical meaning of the system. Elaborated to represent a natural, imperfect construction rather than the demiurgic perfect vision of the world, these mathematical "monsters" are intensively and successfully used in computer rendering, virtual reality nowadays. The apparent shape and structure of natural phenomena (electrical discharges, river basins, vascular trees, etc.) may be quite convincingly rendered but no clue on the underlying physics or the reasons of the shape is offered.

For instance, a current line of mathematical modeling in medicine assumes the tree structures of the human body as "real" fractal objects, which allows the analysis of the underlying physics fractal analysis (Uahabi and Atounti, 2015; Barnsley, 1988).

On the other hand, the Constructal law (Chapter 2: Shape and Structure Morphing of Systems with Internal Flows) explains the design seen in nature, tree-like, "monstrous," structures including the anatomic entities (Bejan, 2000, 2012). This law can be used to understand why design emerges and may predict how they will evolve. After all, it is consistent with the Aristotele's synthesis that "Nature makes all things with a purpose" and "does nothing in vain" (D'Arcy, 1992). Constructal law analyzes and explains that geometric complexity arises from the functions of the tree structures in the human body—hemodynamic system, respiratory tract, renal system, etc.

CAD-designed models are produced using mathematical algorithms implemented in software tools that no have particular physical insights in the drawn object or its functions. The shape, as close to reality as possible, is the objective and not the driving forces (origins, constraints) that morph the system in the shape and form in which it appears to the analyst. CAD made computational domains are used to modeling the problems in Chapters 5–8 (Morega et al., 2013a,b).

CAD, fractal, or hybrid CAD–fractal constructions have to represent realistically the natural systems that are subject to analysis: proper sizes, proportions, that is, realistic rate of growth, shape, structure. At this point, allometry may enter scene and join the design group as a design criterion. Moreover, the geometry of the computational domain should be consistent with the constructal law predictions and allometric laws, if available. Chapter 2: Shape and Structure Morphing of Systems with Internal Flows is devoted to this topic, and Chapters 5–8 present models that utilize this approach to construct the computational domains.

Medical image-based construction, CAD and fused computational domains

Realistic computational models envisage the real shape and structure of the body and its anatomic content, the different tissues and their properties. It is an important asset for getting meaningful results to medical problems, from the numerical simulation perspective. Image acquisition scanners used in Computing Tomography (CT), Magnetic Resonance Imaging (MRI), Ultrasound Imagery (USI) provide personalized image sets of 2D slices, average images of 3D thin volumes in the DICOM (Digital Imaging and Communications in Medicine) specific medical format. The image source dataset that is used, in general, for preoperatory reasons, provides here the basis for constructing the computational domain (Morega et al., 2010, 2016). Specialized software uses them as input data in the construction of the computational domain with the internal anatomy connections. Personalized construction of the anatomical structure is key to the patient centered therapeutic approach and the accurate representation of the anatomical organs and tissues morphology is crucial for the utility numerical modeling. Chapter 2: Shape and Structure Morphing of Systems with Internal Flows is devoted to this topic, and Chapters 4, 5, 7, 8 present models that utilize this approach to construct the computational domains.

When the computational domains combine anatomic structures with CAD elements—electrodes, transducers, prostheses, devices, sources, etc.—it is necessary to combine CAD blocks with MIR structures into a complex, numerically consistent model. This fusion may raise difficulties and it is a matter of the designer to opt for the best, more often, available solution. Concerns and results regarding this design approach are presented in Chapter 2: Shape and Structure Morphing of Systems with Internal Flows. Several computational domains introduced in Chapters 5–8 are constructed using image fusion techniques.

1.11 Diffusion–convection problems: heatfunction and massfunction

Scalar fields, which are solution to diffusion (Laplace–Poisson) problems, are commonly visualized using surfaces of constant scalar value and field lines of the gradient of the scalar, or its conjugated flux. Vector fields, solution to Lapalce–Poisson and

Helomoltz problems, are depicted usually using field lines (streamlines in fluid mechanics), and arrows. Further on the scalar fields (temperature, species, etc.) in coupled diffusion–convection (Morega and Nishimura, 1996) are visualized, in general, through isosurfaces of those scalars and the field lines of their conjugated fluxes. For instance, the solution to the stationary convection–diffusion heat transfer problem for incompressible flow, described by the following energy equation:

$$\rho c\left[\frac{\partial T}{\partial t}+(\mathbf{u}\cdot\nabla)T\right]=\nabla(k\nabla T)+q''', \tag{1.35}$$

where q''' is the local heat source, is presented through isothermal surfaces, T is a constant, and heat flux $-k\nabla T$ vector field lines, as the solution of the regular conduction (diffusion) problem.

In fact this diffusion-type visualization is less relevant and rather confusing here because the convective transport of energy that adds to the conduction is not evidenced in this way. To solve this difficulty, Eq. (1.35) may be rewritten as follows:

$$\rho c\nabla\cdot\underbrace{(\mathbf{u}T-k\nabla T)}_{\mathbf{H}_f}=\underbrace{q'''}_{\text{heat source / sink}}, \tag{1.36}$$

where a new quantity, the heatfunction (Bejan, 1984) emerges

$$\mathbf{H}_f=\underbrace{\mathbf{u}T}_{\textit{transport}}-\underbrace{k\nabla T}_{\textit{diffusion}}. \tag{1.37}$$

A vast body of work is devoted to present the energy paths unveiled through this vector field that sums up the two underlying heat transfer mechanisms (Morega, 1988; Morega and Bejan, 1993, 1994). This new function is much more than a merely mathematical consequence of diffusion–convection. It has a strong physical meaning too by relating the transport and diffusion mechanisms in a single quantity, which makes $\mathbf{H}_f$ a valuable aid in evidencing the enthalpy corridors that lead the convection–diffusion heat transfer.

Similarly all convection–diffusion scalar fields may be visualized using such companion vector fields. For example, in convection–diffusion mass transfer problems, the mass function, $\mathbf{M}_f$, may be introduced through $\mathbf{M}_f=\mathbf{u}c-D\nabla c$ (Bejan, 1984), where c [mol/m^3] is the species concentration, and D [m^2/s] is the mass diffusivity of the species.

Consider the stationary forced convection mass transfer problem in a parallel plate channel, in laminar, Newtonian, incompressible, fully developed (Hagen–Poiseuille) flow. The channel is 70-μm wide and the maximum velocity is 1 mm/s. The fluid is a methanol–water mixture with 0–90 wt.% CH_3OH, $\rho=800-1000$ kg/m^3, $\eta=0.891-1.033$ poises. The species is salicylic acid, and its diffusivity is $D=\{1, 0.5, 1\}\times 10^{-5}$ cm^2/s (Chaaraoui et al., 2017). A patch placed on the upper wall is the mass source $\frac{\partial c}{\partial n}=0$. The horizontal walls are impermeable, the inlet has a homogeneous Dirichlet condition, and the outlethomogeneous convection flux condition. The mathematical model of this mass transfer problem, in nondimensional form is as follows:

$$(\tilde{\mathbf{u}}\cdot\nabla)\tilde{\mathbf{u}} = -\nabla\tilde{p} + \frac{1}{\mathrm{Re}_d}\nabla^2\tilde{\mathbf{u}},\ \nabla\cdot\tilde{\mathbf{u}} = 0, \tag{1.38}$$

$$\left(\tilde{\mathbf{u}}\cdot\nabla\right)\tilde{c} = \frac{1}{\mathrm{Pe}_d}\nabla^2\tilde{c}, \tag{1.39}$$

where the following definitions are used: $\tilde{\mathbf{u}} = \frac{\mathbf{u}}{U_0}$, $\mathrm{P}_0 = \rho U_0^2$, $\tilde{c} = \frac{c - c_0}{c_{\max} - c_0}$, $\mathrm{Re}_d = \frac{U_0 d}{\eta}$, $\mathrm{Pe}_d = \frac{U_0 d}{D} = \mathrm{Re}_d Sc$, and $\mathrm{Sc} = \frac{\nu}{D}$ is Schmidt number: viscous diffusion rate/molecular (mass) diffusion rate (Bejan, 1984). Here $\mathrm{Re}_d = 200$ and $\mathrm{Pe}_d = \{0.08, 8, 80\}$.

Fig. 1.4 shows the mass transfer process seen through isoconcentration lines (thin, black curves) and mass function, $\tilde{\mathbf{M}}_f = \tilde{\mathbf{u}}\tilde{c} - \mathrm{Pe}_d^{-1}\nabla\tilde{c}$, field lines (thicker black lines).

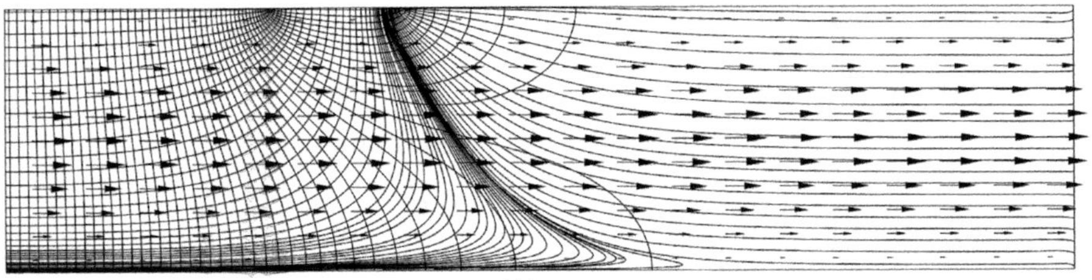

(A)

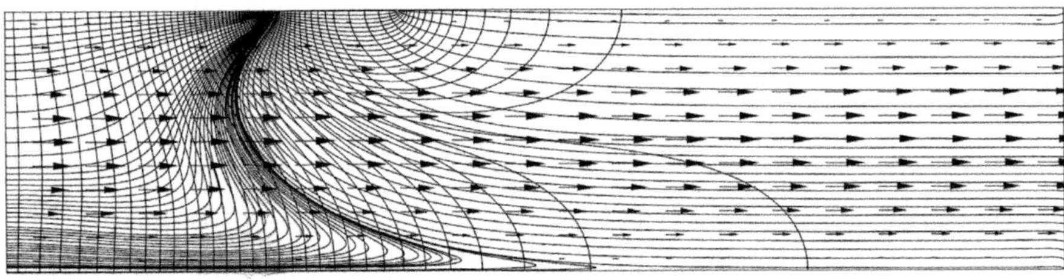

(B)

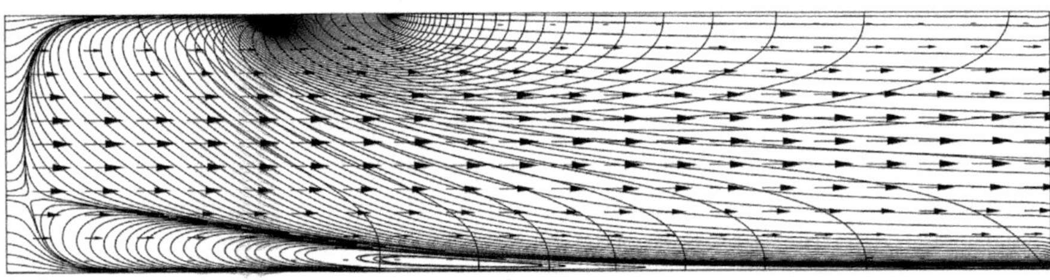

(C)

Figure 1.4 Isoconcentration lines and mass function field lines in forced convection mass transfer in laminar, stationary, incompressible, fully developed channel flow of a Newtonian fluid: (A) $\mathrm{Pe}_d = 0.08$; (B) $\mathrm{Pe}_d = 8$; and (C) $\mathrm{Pe}_d = 80$.

Apparently although the species field was solved considering the mass convection process, Eq. (1.39), the image provided by the isoconcentration representation is less relevant in depicting the mass transfer mechanism, which is fully unveiled by the mass function field lines.

1.12 A roadmap to a well-posed, direct problem and its solution

In view of the complexity posed by multiphysics problems of medical engineering, some strategy has to be envisaged. In general the evolution of a system of finite size, under internal and external constrains, is presented by the laws of Physics. In a direct problem, the system hence its boundary, structure, properties, internal sources, boundary constraints, and initial state are known input data, and the concern is to find the states that the system "visits" during its evolution. A practical path to tackle the problem is presented sequentially, but the steps to follow may be reiterated to define, in the end, a consistent problem with a realistic, robust solution.

The keyword in the first initial stage of the solution is the qualitative analysis. It may provide order of magnitude solutions and hints useful in reducing the complexity of the problem while representing the outlining physics. A qualitative sketch of the system and its boundary is drawn, its structure (materials, sources), the internal and external constraints are presented, and the underlying physical laws are reviewed. The physical (thermodynamic), mathematical, and numerical concepts of system and boundary have to be considered. Then an order of magnitude analysis is performed with the aim to find the characteristic space and time scales, and to contain, "bracket" the solution. The outcome of this analysis is a well-defined system, and (hopefully) a simpler; however, consistent physical model that accounts for the underlying mechanisms, couplings that need to be observed.

In medical physics problems, it is more likely that multidisciplinary or "multiphysics" models with some irreducible, occurring couplings, for example, properties, sources, body forces. Different time scales (diffusion, transport, propagation) of different phenomena (flow heat transfer, mass transfer EMF) and dynamics for the EMF may intervene, for example, stationary, harmonic, pulse, and PWM.

For example, as discussed earlier, in analyzing the heating and heat transfer processes in MW hyperthermia and ablation the local EMF heat source may be introduced through the r.m.s. value of the electric field strength in the SAR term (Chapter 8: Magnetic Stimulation). This assumption has physical grounds—the same power level—and avoids the simultaneous integration of a dynamic EMF problem and the transient HT problem, which happen at considerably different time scales. Moreover if arterial hemodynamic contributes to the HT then the characteristic pulsating flow may be replaced with an equivalent per mass flow rate stationary flow because the blood flow rate happens faster than the HT rate within the surrounding medium.

Within the limits of the continuum media hypothesis, small to large space scales phenomena may act simultaneously. For instance, the hemodynamic flow occurs through a complex network of interconnected arterial, capillary, and venous vessel trees whose sizes vary considerably, under different flow mechanisms, and with different local, rheological characteristics for the blood. Homogenization techniques are then used to avoid the difficulty of analyzing small and large, slow and fast flows altogether, while keeping the "big picture" of the underlying physics: the small size vessels (slow flow) and the embedding tissue are seen as an equivalent continuous medium that has their combined functionality. For instance, in heat transfer analysis, the bioheat equation (Pennes, 1948) presents a perfused tissue as an equivalent, continuum medium that acts as a heat sink/source that models the convective contribution to the HT of capillaries and small size vessels.

In selecting the leading phenomenon in a model with coupled physics, a qualitative power level analysis may be useful. For instance, in modeling the transcutaneous electrical stimulation, the external power source exceeds that of the electrical activity of the heart. It is then reasonable to neglect the latter, and model the body as a volume conductor to find the electrical current density distribution inside the thorax (Chapter 4: Electrical Activity of the Heart).

Moreover an order of magnitude analysis of the physical model may provide the arguments to simplify it, by neglecting the small and or slow (in the order of magnitude sense) terms, which provide negligibly small information while introducing, perhaps, undesirable numerical complexities.

These are only some considerations. Other valuable conclusions that may be useful in simplifying the numerical problem to solve may be drawn through the qualitative analysis as will be see in the Chapters that follow.

Next the route marked by the following steps ought to be traveled mark the route to solve the model:

1. Construct the system and its boundary, called the computational domain in what follows. CAD tools, or medical image–based construction techniques, or fractal algorithms, or a combination of these may be used. When using CAD or fractal geometry, the resulting computational domain has to represent realistically the physical system under analysis.
2. Build the physical model:
   a. list the laws of physics that describe the system state, evolution, the material laws that apply and the internal sources;
   b. define the boundary conditions, that is, the interactions with the surroundings (heat transfer and, or mechanical work equivalents); and
   c. define the initial conditions, that is, the initial state of the system.
3. Build the mathematical model out of the physical model. The set of equations that make the physical model is reduce to a smaller set of higher order PDE(s) and/or ODE(s).

4. Solve the mathematical model. Here analytic methods (if existent) and numerical methods may be used. The FEM technique is mostly used because of its capacity to approach complex computational domains, to model-coupled physical modes and produce numerical solutions within prescribed numerical accuracy limits.
5. Check the numerical accuracy of the result. This step should go without mentioning; however, more than seldom the first numerical solution is taken as accurate enough without checking it.
6. Postprocess the results.

This approach is consistent with Hadamard's definition of a well-posed direct problem.

References

Andreuccetti, D., Fossi, R., Petrucci, C., 1997. An internet resource for the calculation of the dielectric properties of body tissues in the frequency range 10 Hz–100 GHz, IFAC–CNR, Florence, Italy. Based on data published by C. Gabriel et al. in 1996 [online]. Available from: <http://niremf.ifac.cnr.it/tissprop/>.

D'Arcy, T.W., 1992. On Growth and Form, Canto ed. Cambridge University Press.

Barnsley, M.F., 1988. Fractals Everywhere. Academic Press, Inc., San Diego, CA.

Bathe, K.J., Wilson, E.L., 1976. Numerical Methods in Finite Element Analysis. Prentice-Hall Inc, Enhglewood Cliffs, NJ.

Bejan, A., 1984. Convection Heat Transfer. Wiley.

Bejan, A., 1988. Advanced Engineering Thermodynamics. Wiley & Sons, New York.

Bejan, A., 1993. Heat Transfer. Wiley, New York.

Bejan, A., 2000. Shape and Structure, from Engineering to Nature. University Press, Cambridge.

Bejan, A., Zane, J.P., 2012. Design in Nature. How Constructal Law Governs Evolution in Biology, Physics, Technology, and Social Organization. Anchor Books, Random House, Inc, New York.

Bergé, P., Pomeau, Y., Vidal, C.H., 1984. Hermann publishers in arts and science, Paris Order within Chaos. Towards a Deterministic Approach to Chaos. Wiley & Sons, New York, Toronto, Chichester, Brisbane, Singapore.

Brebia, C.A., Telles, J.C.F., Wrobel, L., 1984. Boundary Element Techniques. Springer Verlag, Berlin.

Brown, R.J., 2019. Thought Experiments. Stanford Encyclopedia of Philosophy, https://plato.stanford.edu/entries/thought-experiment/.

Canuto, C., Quateroni, A., 1984. Preconditioned minimal residual methods for Chebyshev spectral calculations. Comput. Phys. 60, 315–337.

Chaaraoui, Z., Al–Taiar, A.H., Othman, A.A., 2017. Comparative conductimetric studies of salicylic acid in methanol–water mixtures at 25°C. Arab. J. Chem. 10, S2004–S2008.

Chato, J.C., 1980. Heat transfer to blood vessels. ASME J. Biomech. Eng. 102, 110–118.

Chen, M.M., Holmes, K.R., 1980. Microvascular contributions in tissue heat transfer. Ann. N.Y. Acad. Sci. 335, 137–150.

Cheng, S.C., Huang, Y.M., 2003. A novel approach to diagnose diabetes based on the fractal characteristics of retinal images. IEEE Trans. Inf. Technol. Biomed. 7 (3), 163–170.

Cheung, N., Donaghue, K.C., Liew, G., Rogers, S.L., Wang, J.J., Lim, S.W., et al., 2009. Quantitative assessment of early diabetic retinopathy using fractal analysis. Diabetes Care 32 (1), 106–110. Available from: https://doi.org/10.2337/dc08-1233.

Coherence, 2018. Available from: <https://en.wikipedia.org/wiki/Coherence_(units_of_measurement)>.

Cho Y.I, Ed., 1992. Bioengineering heat transfer. In Advances in Heat Transfer, Academic Press, 22.

Darwin, C.R., 1845. Journal of Researches into the Natural History and Geology of the Countries Visited During the Voyage of H.M.S. Beagle Round the World, Under the Command of Capt. Fitz Roy, R.N., second ed. John Murray, London. Cited in Steven, C.F., 2009. Darwin and Huxley revisited: the origin of allometry. J. Biol. 8, 14. Available from: <https://doi.org/10.1186/jbiol119>.

Delaunay, B., 1934. Sur la sphère vide. Bull. de. l'Académie des. Sci. de l'URSS, Cl. des Sci. mathématiques et. naturelles 6, 793–800.

Durney, Ch, Massoudi, H., Iskander, M.F., 1986. Radio Frequency Radiation. Dosimetry Handbook, fourth ed. USAF/SAM, Brooks AFB, TX.

Ehrenstein, U., Peyret, R., 1989. A Chebyshev collocation method for the Navier–Stokes equations with applications to double–diffusive convection. Int. J. Numer. Methods Fluids 9, 427–452.

Elert, G., 2018. The Physics Hypertextbook—Viscosity. Physics.info. Available from: https://physics.info/viscosity/.

Falconer, K.J., 1990. Fractal Geometry: Mathematical Foundations and Applications. Wiley, London.

Feijóo, 2000. Computational methods in biology. In: Proceedings of the 2nd Summer School LNCC/MCT, Petrópolis, Brazil.

Fletcher, C.A.J., 1984. Computational Galerkin Methods. Springer-Verlag, NY.

Fox, L., Parker, I.B., 1968. Chebyshev Polynomials in Numerical Analysis. Oxford Press, London.

Gabriel, C., Gabriel, S., 1996. Compilation of the dielectric properties of body tissues at RF and microwave frequencies. Report N.AL/OE–TR–1996–0037. Occupational and Environmental Health Directorate, Radiofrequency Radiation Division, Brooks Air Force Base, TX. Available from: <http://www.dtic.mil/dtic/tr/fulltext/u2/a305826.pdf> (authorized mirror at <http://niremf.ifac.cnr.it/docs/DIELECTRIC/home.html>).

Gabriel, C., Gabriel, S., Corthout, E., 1996a. The dielectric properties of biological tissues. I. Literature survey. Phys. Med. Biol. 41, 2231–2249.

Gabriel, C., Gabriel, S., Corthout, E., 1996b. The dielectric properties of biological tissues. II. Measurements in the frequency range 10 Hz to 20 GHz. Phys. Med. Biol. 41, 2251–2269.

Gabriel, C., Gabriel, S., Corthout, E., 1996c. The dielectric properties of biological tissues. III. Parametric models for the dielectric spectrum of tissues 41, 2271–2293.

Gambaruto, A., Janela, J., Moura, A., Sequeira, A., 2011. Sensitivity of hemodynamics in a patient specific cerebral aneurysm to vascular geometry and blood rheology. Math. Biosci. Eng. 8 (2), 409–423.

Gaw, R.L., Cornish, B.H., Thomas, B.J., 2008. The electrical impedance of pulsatile blood flowing through rigid tubes: a theoretical investigation. IEEE Trans. Biomed. Eng. 55 (2), 721–727. Available from: https://doi.org/10.1109/TBME.2007.903531.

Gottlieb, D., Orszag, S.A., 1977. Numerical analysis of spectral methods: theory and applications. In: SIAM, Regional Conference Series in Applied Mathematics, Philadelphia, PA.

Hadamard, J., 1902. On the partial differential equations and their physical significance. Princeton Univ. Bull. 49–52 (in French).

Hadamard, J., 1923. Lectures on Cauchy's Problem in Linear Partial Differential Equations. Yale University Press, New Haven, CT, in the *Bulletin of the American Mathematical Society* ISSN 1088-9485 (online) ISSN 0273-0979 (print). Available from: <http://www.ams.org/journals/bull/1987-17-01/S0273-0979-1987-15552-2/>.

Hartung, T., 2014. Comparative Analysis of the Revised Directive 2010/63/EU for the Protection of Laboratory Animals with its Predecessor 86/609/EEC—A T4 Report. Johns Hopkins University, CAAT, US, and CAAT–Europe, University of Konstanz, Germany.

Hoetink, A.E., Faes, T.J.C., Visser, K.R., Heethaar, R.M., 2004. On the flow dependency of the electrical conductivity of blood. IEEE Trans. Biomed. Eng. 51 (7), 1251–1261.

Huo, Y., Kassab, G.S., 2012. Intraspecific scaling laws of vascular trees. J. R. Soc. Interface 9, 190–200. Published online 15 June 2011.

Huxley, J.S., 1972. Problems of Relative Growth, second ed. Dover, New York, ISBN 0-486-61114-0.

Kleiber, M., 1932. Body size and metabolic rate. Physiological Rev. 27 (4), 511–547.

Klinger, H.G., 1974. Heat transfer in perfused biological tissue. I. General theory. Bull. Math. Biol. 36, 403–415.

Malmivuo, J., Plonsey, R., 1995. Bioelectromagnetism. Principles and Applications of Bioelectric and Biomagnetic Fields. Oxford University Press, New York, Oxford.

Mandelbrot, B.B., 1975. Les objets fractals: forme, hasard et dimension. Flammarion, Paris.

Michel, A.N., Wang, K., 1995. Qualitative Theory of Dynamic Systems. The Role of Stability Preserving Mappings, Pure and Applied Mathematics. A Series of Monographs and Textbooks, Vol. 186. Marcel Dekker, NY.

Miller, J.C., Horvath, S.M., 1978. Impedance cardiography. Psychophysiology 15 (1), 80−91.

Minkowycz, W.J., Sparrow, E.M., Abraham, J.P., 2009. Advances in Numerical Heat Transfer, Vol. 3. CRC Press, Boca Raton, FL, ISBN 978-1-4200-9521-0.

Mocanu, C.I., 1981. The Theory of the Electromagnetic Field. Ed. Didactică şi Pedagocіă. Bucharest, Romania (in Romanian).

Mocanu, C.I., 1979. The Theory of Electrical Circuits. Ed. Didactică şi Pedagogiă. Bucharest, Romania (in Romanian).

Mocanu, D., 2002. Electromagnetic Models in Biomedical Engineering (Doctoral Thesis). Faculty of Electrical Engineering, University Politehnica of Bucharest, Bucharest, Romania (in Romanian).

Morega, A.M., 1988. The heatfunction approach to the thermo−magnetic convection of electro-conductive melts. Rev. Roumaine Sci. Techn. Electrotech. et. Energ. 34 (4), 359−368.

Morega, A.M., 1998. Numerical Modeling for Boundary Value Problems in Engineering. MatrixRom, Bucharest, Romania (in Romanian). ISBN 973-9390-05-6.

Morega, A.M., Bejan, A., 1993. Heatline visualization of the forced convection laminar boundary layers. Int. J. Heat. Mass. Transf. 30, 3957−3967.

Morega, A.M., Bejan, A., 1994. Heatline visualization of forced convection in porous media. Int. J. Heat. Fluid Flow. 15 (1), 42−47.

Morega, A.M., Nishimura, T., 1996. Double diffusive convection by a Chebyshev collocation method. Technol. Rep. Yamaguchi Univ. 5 (5), 259−276.

Morega, A.M., Dobre, A.A., Morega, M., 2010. Numerical simulation of magnetic drug targeting with flow−structural interaction in an arterial branching region of interest. In: Comsol Conference, Versailles, France, 17−19 November 2010.

Morega, A.M.C., Savastru, C., Morega, M., 2013a. Numerical simulation of an adaptive magnetic field source concept in magnetic drug targeted transport. In: Proceedings of the 8th International Symposium on Advanced Topics in Electrical Engineering, ATEE, 23−24 May 2013, Bucharest, Romania; pp. 1−4. Available from: <https://doi.org/10.1109/ATEE.2013.6563479>.

Morega, A.M., Savastru, C., Morega, M., 2013b. Numerical simulation of flow dynamics in the bra-chial−ulnar−radial arterial system. In: The 4th IEEE International Conference on E-Health and Bioengineering, EHB 2013, Grigore T. Popa University of Medicine and Pharmacy, Iaşi, Romania, 21−23 November 2013.

Morega, A.M., Dobre, A.A., Morrega, M., 2016. Electrical cardiometry simulation for the assessment of circulatory parameters. Proc. Roman. Acad. Series A. 17 (3), 259−206.

Mur, G., 1998. The fallacy of edge elements. IEEE Trans. Magnetics 34 (5), 3244−3247.

Murray, C.D., 1926a. The physiological principle of minimum work: I. The vascular system and the cost of blood volume. Proc. Natl. Acad. Sci. U.S.A. 12 (3), 207−214.

Murray, C.D., 1926b. The physiological principle of minimum work: II. Oxygen exchange in capillaries. Proc. Natl. Acad. Sci. U.S.A. 12 (5), 299−304.

Nakayama, A., Kuwahara, F., 2008. A general bioheat transfer model based on the theory of porous media. Int. J. Heat. Mass. Transf. 51, 3190−3199.

Order in Chaos, 2013. Available from: <https://orderinchoas.wordpress.com/2013/02/15/mathematical-monsters/>.

Orszag,, S.A., Gotlieb,, D., 1980. Lecture Notes in Mathematics, vol. 177. Springer Verlag, NY, pp. 381−398.

Osypka, M., 2009. An Introduction to Electrical Cardiometry™. Electrical Cardiometry™, pp. 1−10.

Paradigm Available from: <https://en.wikipedia.org/wiki/Paradigm> (retrieved October 2018).

Pennes, H.H., 1948. Analysis of tissue and arterial blood temperature in the resting human forearm. J. Appl. Physiol. 1, 93−122.

Perkowitz, S., 2010. Gedankenexperiment. Encyclopædia Britannica Online (retrieved 27.03.2017).

Peyret, R., 2002. Spectral Methods for Incompressible Flow. Applied Mathematical Sciences, Vol. 148. Springer, NY.

Peyret, R., Taylor, T.D., 1983. Computational Methods for Fluid Flow. Springer—Verlag, New York.
Purcell, E.M., 1984. 2nd ed. Electricity and Magnetism, Berkeley Physics Course, Vol. 2. McGraw-Hill.
Roetzel, W., Xuan, Y., 1998. Transient response of the human limb to an external stimulus. Int. J. Heat. Mass. Transf. 41, 229—239.
Ruhdel, I., 2007. *Überarbeitung der EU—Richtlinie 86/609/EWG Ergebnisse der Internetbefragungen der Europäischen Kommission*, Akademie für Tierschutz des Deutschen Tierschutzbundes, Neubiberg, Deutschland (in German).
Schroeder, W., Wolff, I., 1994. The origin of spurious modes in numerical solutions of electromagnetic field eigenvalue problems. IEEE Trans. Microw. Theory Tech. 42 (4), 644—653.
Shibeshi, S.S., Collins, W.E., 2005. The rheology of blood flow in a branched arterial system. Appl. Rheol. 15 (6), 398—405.
Snell, O., 1982. Die Abhängigkeit des Hirngewichts von dem Körpergewicht und den geistigen Fähigkeiten. Arch. Psychiatr. 23 (2), 436—446.
Stuchly, M.A., Stuchly, S.S., 1980. Dielectric properties of biological substances—tabulated. J. Microw. Power 15 (1), 19—25.
Takahashi, N., Nakata, T., Fujiwara, K., Imai, T., 1992. Investigation of effectiveness of edge elements. IEEE Trans. Magnetics 28 (2), 1619—1622.
Tikhonov, A.N., Samarskii, A.A., 1963. Equations of Mathematical Physics. Pergamon Press Ltd, Oxford, England.
Uahabi, K.L., Atounti, M., 2015. Applications of fractals in medicine. Ann. Univ. Craiova Maths. Comput. Sci. Ser. 42 (1), 167—174.
Valvano, J.W., 2010. Tissue thermal properties and perfusion. In: Welch, A., van Gemert, M. (Eds.), *Optical—Thermal Response of Laser—Irradiated Tissue*. Springer, Dordrecht.
Vetterling, W.T., Teukolsky, S.A., Press, W.H., Flannery, B.P., 1997. Numerical Recipes in Fortran 90, 2nd ed. Cambridge University Press.
Visser, K.R., 1992. Electric conductivity of stationary and flowing human blood at low frequencies. Med. Biol. Eng. Comput 30, 636—640.
Warburg, E.G., 1881. Magnetic investigations. IEEE Ann. Phys. 13, 141—164.
Webb, J.P., 1993. Edge elements and what they can do for you. IEEE Trans. Magnetics 29 (2), 1460—1465.
Weinbaum, S., Jiji, L.M., 1985. A new simplified bioheat equation for the effect of blood flow on local average tissue temperature. ASME J. Biomech. Eng. 107, 131—139.
Weinbaum, S., Jiji, L.M., Lemons, D.E., 1984. Theory and experiment for the effect of vascular micro-structure on surface tissue heat transfer. Part I. Anatomical foundation and model conceptualization. ASME J. Biomech. Eng. 106, 321—330.
Wulff, W., 1974. The energy conservation equation for living tissue. IEEE Trans. Biomed., Eng. BME 21, 494—495.
Zolfaghari, A., Maerefat, M., 2010. Bioheat transfer. In: Dos Santos Bernardes, M.A. (Ed.). Developments in Heat Transfer. IntechOpen, pp. 153—170 (Chapter 9). ISBN: 978-953-307-569-3. Available from: <www.intechopen.com>, <https://doi.org/10.5772/22616>.

A.1 Scalar and Vector Fields

The laws of Physics, the boundary conditions, and the initial conditions that make the physical model yield the mathematical model that has to be solved to find the unknowns of the problem—the physical quantities of interest (e.g., temperature, velocity, pressure, electric potential, magnetic potential, species concentration). Qualitatively the physical quantities may be scalar and vector, and the mathematical physics is concerned with the solution of the problems where they occur.

Scalar fields

The internal thermodynamic disequilibrium of a system may be accompanied by gradients and fluxes of the scalar, state quantities (e.g., temperature, pressure, mass density, concentration, potential), and state, vector fields (e.g., EMF, flow field, stress field).

Scalar fields are presented (visualized) using surfaces (lines in 2D models), which are loci of constant value scalars. For instance the electrical potential field, $V(P)$, is visualized using equipotential surfaces of $V(P) = \text{const}$. The regional distribution of the equipotential surfaces, or how fast the scalar varies along a specific direction, is presented by the directional derivatives. A particular directional derivative, oriented in the direction of scalar growth, and orthogonal to the local equipotential surface, is the gradient

$$\nabla V = \mathbf{n}\frac{\partial V}{\partial n}, \tag{A1.1}$$

where $\nabla = \frac{\partial}{\partial x}i + \frac{\partial}{\partial x}j + \frac{\partial}{\partial x}\mathbf{k}$ (in Cartesian coordinates) is Hamilton's vector operator (nabla), and $\mathbf{n}$ is the unit vector orthogonal to the local equipotential surface pointing in the direction of $V(P)$ increase.

The gradient is an invariant with respect to the system of coordinates. It may also be introduced using Gauss-divergence theorem for the volume integral of a gradient field (Purcell, 1984; Mocanu, 1981)

$$\nabla V = \lim_{\Delta v_\Sigma \to 0} \frac{\oint_\Sigma V\mathbf{n}dA}{\Delta v_\Sigma}, \tag{1.2}$$

where v_Σ is the volume and Σ is its boundary.

Gradients to drive fluxes are postulated by the *fundamental law of thermodynamics* for the systems outside the internal equilibrium (Bejan, 1988). Onsager relations present the particular analytical forms of these relations, which turn to be laws that relate gradients and fluxes (e.g., Fourier, Fick, Ohm, Peltier, Seebeck). These relations may provide for the basis of analysis of the systems with internal multiple conjugated and coupled gradients and fluxes.

Vector fields

The *fundamental theorem of vector fields* states that a vector field, $\mathbf{F}(\mathbf{r})$, is uniquely determined everywhere in the ROI of volume v_Σ, bounded by the closed surface Σ, if and only if its divergence, $\rho(\mathbf{r})$, and curl (rotor), $\mathbf{R}(\mathbf{r})$, are known as follows (Mocanu, 1981):

$$\nabla \times \mathbf{F} = \mathbf{R}(\mathbf{r}),\ \nabla \cdot \mathbf{F} = \rho(\mathbf{r}), \tag{A1.3}$$

and if on the surface Σ either its normal component or its tangential component are specified

$$\mathbf{n}\cdot\mathbf{F}(\mathbf{n}) = \mathbf{F}_n,\ \mathbf{n}\times\mathbf{F}(\mathbf{n}) = \mathbf{F}_t. \tag{A1.4}$$

The curl and divergence are local physical quantities and the mathematical models, to which they lead, may be higher order partial differential equations (PDEs) of different types—Laplace, Helmholtz, and so on.

The integral form of the theorem (A.1.3) requires that the circulation, $\mathbf{R}(\mathbf{r})$, and the flux, $Q(\mathbf{r})$, of the vector field $\mathbf{F}(\mathbf{r})$ are specified

$$\oint_{\Gamma}\mathbf{F}\cdot\mathbf{ds} = \int_{A_\Gamma}(\nabla\times\mathbf{F})\cdot\mathbf{dA} = R,\ \oint_{\Sigma}\mathbf{F}d\mathbf{A} = \int_{v_\Sigma}(\nabla\cdot\mathbf{F})dv = Q. \tag{A1.5}$$

The two integral relations, Stokes and Gauss, that relate the circulation and flux integrals to surface and volume integrals, respectively, are used as discussed earlier.

The PDEs are solved numerically by using domain (interior) methods, such as finite differences and finite element (Bathe and Wilson, 1976; Peyret and Taylor, 1983; Morega, 1998). The mathematical models derived using the integral forms may be solved numerically using boundary numerical methods (Brebia et al., 1984).

Remarkably, the integral forms (3) are used to define the divergence and the curl, which are invariant with respect to the system of coordinates. The curl is given by

$$\nabla\times\mathbf{F} = \mathbf{n}\lim_{\Delta A_\Gamma\to 0}\frac{\oint_\Gamma\mathbf{F}\mathbf{ds}}{\Delta A_\Gamma},\ \text{or}\ \nabla\times\mathbf{F} = \mathbf{n}\lim_{\Delta v_\Sigma\to 0}\frac{\oint_\Sigma(\mathbf{n}\times\mathbf{F})dA}{\Delta v_\Sigma}, \tag{A.1.6}$$

where $\mathbf{n}$ is the unit vector of the direction along the which $|\nabla\times\mathbf{F}|$ is maximum, A_Γ is the area of an open surface A that is bounded by the closed curve Γ, and v_A is the volume bounded by A. The divergence is defined by

$$\nabla\cdot\mathbf{F} = \lim_{\Delta v_\Sigma\to 0}\frac{\oint_\Sigma\mathbf{F}\cdot\mathbf{n}dA}{\Delta v_\Sigma} \tag{A1.7}$$

The special forms of the curl and divergence for interfaces between media with different material properties are as follows:

$$\nabla_S\times\mathbf{F} = \mathbf{n}_{12}\times(\mathbf{F}_2 - \mathbf{F}_1),\ \nabla_S\cdot\mathbf{F} = \mathbf{n}_{12}\cdot(\mathbf{F}_2 - \mathbf{F}_1), \tag{A1.8}$$

where $(\cdot)_{1,2}$ denote the two different media separated by the interface, and $\mathbf{n}_{12}$ is the normal to the interface, oriented from $(\cdot)_1$ to $(\cdot)_2$. If a superficial source resides on the interface then the finite size "jump" of the components thus selected equals that source—for example, sheet current for the tangent and surface charge for the normal.

These forms are related either to continuity conditions at the interfaces separating regions with different properties or surfaces, curves, points with field sources—charges and currents (Mocanu, 1981).

The vector fields are presented through force or field lines (called *streamlines* in fluid mechanics), which are the *geometric loci* of the lines to which the vector field is tangent. From this perspective, a vector field with a divergence-type of source has open lines that start from positive sources and end on negative (sink) sources. In turn, a vector filed produced by a curl-type source shows off closed lines (eddies). Of course, should the sources be situated outside the ROI (divergence-free or curl-free vector fields) and if the ROI is under the action of that vector field then the field lines inside the ROI are open—start and land on the boundary.

The physical model of a vector field problem has to consider the laws that provide for the two possible types of the vector field sources—a vector source, $\mathbf{R}(\mathbf{r})$, and a scalar source, $\rho(\mathbf{r})$. For instance, for the EMF problems, Maxwell–Hertz laws apply. Consider the particular case of the electrostatic field problem (in immobile media). The electric field quantities are the electric field strength, $\mathbf{E}$, and the electric flux density, $\mathbf{D}$. Maxwell laws that present the divergence and the curl of the electric field are the electromagnetic induction (Faraday) law

$$\nabla \times \mathbf{E} = 0. \tag{A1.9}$$

and the electric flux (Gauss) law, which in the absence of volume electrical charge density is as follows:

$$\nabla \cdot \mathbf{D} = 0. \tag{A.1.10}$$

Apparently Eqs. (A1.7) and (A1.8) imply two different vector fields ($\mathbf{E}$ and $\mathbf{D}$); therefore a law that provides for a supplementary relation between them is needed. The substance inside the system enters scene through a material (continuity) law. Assuming a homogeneous, linear, isotropic medium without polarization (Mocanu, 1981), the constitutive relation is as follows:

$$\mathbf{D} = \varepsilon \mathbf{E}, \tag{A.1.11}$$

where ε [F/m] is the electric permittivity, a material property. At this point, it is worth to note that there are as many constitutive laws as many different media may exist, and from this perspective Maxwell's system of laws (equations) is open.

CHAPTER 2

Shape and structure morphing of systems with internal flows

2.1 Natural form and organization—quandary, observation, and rationale

The quest for a principle that could rule the shape or "geometric" organization of the macroscopic entities that regenerate everywhere in nature is prevalent in the rational, deterministic understanding of the natural self-organization.

More complex systems may host fluxes of different nature. Convection-diffusion processes are such examples, but the list may continue: coupled gradient-flux systems (piezoelectric, piezomagnetic, thermoelectric), electric and magnetic fluxes, etc. For instance, the spatiotemporal structure of "convection–diffusion wave" made by a natural heat and mass convection system may show off (in experiment and numerical simulation, Nishimura et al., 2000a,b) similarity with the reaction-diffusion wave, presented by the mathematical Turing's model (Turing, 1952). This resemblance gives perception of the existence of a spatial pattern, which appears in convection–diffusion system similar to that of the Turing model, and has an eigen dimension, independent of the scale of experimental apparatus resembling the Turing reaction-diffusion wave.

An optimal weaving of interlaced internal details for compact heat exchangers—fins, hydraulic diameters, channels, etc.—is observed in the biological organization too. The volume of the system (device) is limited to perform its objective, which may be to minimize the global, internal thermal resistance between the solid parts "packed" in the system and the fluid flow that bathes the system. Moreover, in many natural volume-to-point flows the high permeability paths are empty spaces: cracks (fissures, channels) in bidimensional systems and vessels (tubes, ducts) in tridimensional systems.

Certain shapes that are observed in natural, "alive" systems with flows and leading gradients (pressure, temperature, etc.) are manifestations of disequilibrium, but the number of identifiable *flow shapes* that are seen inside and outside the animated and unanimated natural entities is limited to just three: *tree*-like arborescences, *round*, and *slice* (lemon-slice like) cross-sections.

Natural systems may be similar in shape but never identical (as the fingerprints). For instance, vascular trees are always different, as their vessels cross-sections are round shapes, but never a perfect circle.

Computational Modeling in Biomedical Engineering and Medical Physics
DOI: https://doi.org/10.1016/B978-0-12-817897-3.00002-6

A tree is not a net (mesh) and there are no loops in its structure, such that nodes are connected through unique branches (unique paths), which render a certain tree topologically equivalent with other similar trees. They may be the most difficult to describe but when we see it we recognize it, and call it *tree*. Fractal geometry seems dedicated to their representation as "mathematical monsters" (Mandelbrot, 1975, 2020; Falconer, 1990; Fractal, 2020).

Trees are seen everywhere and at all scales, in trees, plants, leaves, roots, lungs, vascularized tissues, circulatory system, neuronal dendrites,[1] bacterial colonies, electrical discharges, cracks, hydrographic basins, deltas, and urban growths, etc. (Bejan, 2000a, b, 2016; Bejan, Lorente, 2010). They are flow systems that connect a node, called root, with an irrigated territory (an infinity of points) of finite size.

It may be inferred that if a single principle may explain all these forms (tree, round, slice) then that principle acts everywhere. It may fill-in the gap between physics and biology, between two ways of thinking, reasoning, two different perspectives of perceiving the surrounding reality.

The natural macroscopic structure occurs in space *and* it evolves in time. In physiology, it is known that the heartbeats and the respiratory frequency are related (Petrescu et al., 2018), and specific to each animal. Their values decrease for increasing bodies, and allometric relations (Chapter 1: Physical, Mathematical, and Numerical Modeling) document such connections empirically. Yet again, which is that principle that generates structure and shape in such a vast and diverse spectrum? Does it exist?

In the animate realm, the breathing frequencies, the pulse rate, the river morphology, etc., their recurrent geometric or time particularities have been measured in detail and correlated successfully (Murray, 1926a,b; Schmidt-Nielsen, 1972; Peters, 1983; Huo and Kassab, 2012). However, these accurate and valuable correlations offer no hint on what (if any) physical law they might epitomize.

More recently, arborescent macroscopic organizations have been quantified mathematically with fractal geometry methods, which propose iterative constructs, which mimic shapes sized in abstract spaces of fractional order, but resembling natural tree structures, when represented in the natural (2D or 3D) space representation (Mandelbrot, 1975, 2020; Falconer, 1990; Cheng and Huang, 2003; Order in Chaos, 2013; Uahabi and Atounti, 2015). This approach is descriptive rather than predictive—it has no stemming physics skeleton. However, quoting "future progress depends on establishing a basis much more consistent on which the geometrical organization is deduced out of the mechanism which is producing it" (Kadanoff, 1986; Bejan, 2000a,b).

The *Constructal law* answers precisely this desideratum and it spells out the principle that explains and relates the system shape with its morphology dynamics (Bejan, 2000a,b). It is a completely deterministic principle that allows anticipating, predicting

[1] *dendron* (Δενδρον) means tree in Greek.

the shape and structure of systems that emerge naturally under known constraints, and it points out physics out to extend over all naturally organized flow systems as well as in biology.

The constructal law, in its early exposure states that "For a finite-size open system to persist in time (to survive) it must evolve in such a way that is provides easier access to the currents (fluxes) that flow through it" (Bejan and Errera, 1997). It emerged from the optimization work performed on engineered heat transfer systems with a purpose (optimal heat transfer rates), with minimum entropy generation (or finite time, or endoreversible thermodynamics): compact and efficient shapes and structures (Novikov, 1958; Hoffman et al., 1997; Bejan, 1980, 1996, 2001a,b; Ledezma et al., 1996; Morega and Bejan, 2005; Hoffman, 2008) aimed to enhance the thermodynamic efficiency of an engineered structure subject to global constraints by physically modifying its design. These studies led to arborescent heat transfer networks where each feature is a natural consequence and not an axiom. Relating this result with the tree networks in nature is a natural reflection of the invocated principle of global maximization in volume (territory) to point flows. The same principle should act everywhere where trees occur.

2.2 Biomimetics, bionics, fractal geometry, constructal theory

Nature provides answers to all problems encountered throughout our existence, and new technologies are more than often inspired by biological realizations that evolved through natural selection into too well adapted, developed structures and materials. Engineering objectives such as self-healing abilities, tolerance to environmental exposure, hydrophobicity, self-assembly, solar energy usage, have thus been addressed. Closely related to *bionics*,[2] or *biologically inspired engineering*, *biomimetics*[3] or *biomimicry* terms the replication of elements, functionalities, models, and systems of nature with the aim to solve convoluted human problems, which means the application of biological means and systems found in nature to the study and planning of engineering systems and modern technology (Vincent et al., 2006). For instance, the design of the aircraft wing profile and the flying techniques, physiology and methods of locomotion transferred to biorobots, etc. The Schmitt trigger device (Schmitt, 1938, 1969) replicates the squid nerve propagation in "a thermionic trigger" that allows a constant electronic signal to be changed to an on/off state, and makes the basis of analog to digital

[2] Biology & electronics; "the use of electronically operated artificial body parts"; the science of systems which have some function copied from nature, or which represent characteristics of natural systems or their analogs—largely abandoned in English speaking countries (Vincent, 2009).

[3] Ancient Greek: βίος (bios), life, and μίμησις (mīmēsis), imitation.

conversion in electronics. On its venue, biophysics aims to solve problems of biology using the laws, theory, and technology of the physical sciences.

But these theories and technologies do not purpose to explain the shape of a system as its outcome, manifestation, and struggle to adapt to constraints either embodied as optimization criteria in engineered systems, or as result of interactions with the environment, in natural animated and inanimate systems—shapes of systems with purpose and under certain constraints.

Heat transfer principles (Chapter 1: Physical, Mathematical, and Numerical Modeling) explain the geometric structure of systems, their spatial arrangement, and the amounts of their parts. If a system is purposed to convert, as well as possible, heat crossing it into work, subjected to constraints (the available heat input, the size, and costs), then the factors responsible to reducing the pending irreversibility are optimization degrees of freedom. The heat flows through the system, from the input (hot source) to the output (cold source, the ambient) and to the ambient (leaked heat), are driven by temperature gradients across thermal conductance paths (power plant parts), which are related to the entropy that is generated (Bejan, 1996). The first law analysis shows off that when the power plant is subject to size constraint, the irreversibility (entropy generation) of its functioning may be minimized by adjusting the sizes, or optimally allocating and morphing these conductances, the spatial conveyors of heat fluxes: in essence, provide good conductivity paths ("paving" materials) to "ease", facilitate the currents flows.

Spatial allotment comes with shaping the system and its parts. This, geographically means allocation of hardware is such a way that imperfection (resistance to flow) is distributed around, and its constituent parts all "work" under same stress. This holds for engineered systems, which, subjected to optimization, progressively resemble to (shape like), and function like natural systems. For instance, the animals' structures are presented through accurate power laws that relate their body sizes and other flow and performance parameters (Murray, 1926a,b; Schmidt-Nielsen, 1972); Peters, 1983; Bejan, 2000a,b). The unifying feature is that the metabolism rate (the *exergy*[4] consumption rate) is fairly, parsimoniuously shared between organs (system parts) in proportions that are reasonably insensible to animal size.

From the perspective of the *constructal theory*, which means to treat the living systems as energy systems (power plants) with flows, constraints, shaped with purpose and capable of evolution, the forecast of the metabolic rate for the body and its parts equates the prediction of the body structure and shape, its parts, their sizes, and irreversibly in relation with each other.

[4] The *exergy* (maximum available energy) of a closed system equals the maximum possible useful work that brings the system into equilibrium with a heat reservoir, reaching maximum entropy.

2.3 Shape and structure

The fundamental problem of volume to point flow and the constructal growth

The roots of the constructal law origins could be traced back to the performance optimization of engineered, artificial systems such as heat sinks, fluid systems, electrical windings, etc.—systems of finite size, with purpose, with internal fluxes, under internal and external constraints, whose structures and shapes are adapted and morphed to optimally function (Bejan and Errera, 1997). The fundamental structural problem is the minimization of the volume (territory) to point resistance of single flows (streams, currents, fluxes), which also explains the morphology of natural flow systems, such as lungs, capillary beds, river basins, etc. (Bejan, 2000a,b). The elemental system, or the unit cell out of which the growth (in size) starts is the seed for higher order tree ensembles that perform the same function (volume-to-point discharge), optimally—with minimum flow resistance.

Fig. 2.1 (Morega and Proca, 2004) shows the conduction heat transfer implementation of the constructal principle (Bejan and Errera, 1997).[5] The designer allots a small quantity of material of higher conductivity to pave a better conduction path (dark) for the heat generated in cell to reach a port on the boundary, which is otherwise insulated. In the limit $k_0/k_p << \phi_0 << 1$, $\phi_0 = D_0/H_0$ the analytic solution of the stationary diffusion problem may be used to calculate the resistance of M_0, a design parameter, defined as the ratio of the maximum internal temperature drop through the heat current that leaves the cell through the port on the boundary

$$\frac{\Delta T_0}{q''' H_0 L_0 / k_0} = \frac{1}{8} \times \frac{H_0}{L_0} + \frac{k_0}{2\phi_0 k_p} \times \frac{L_0}{H_0}, \tag{2.1}$$

that shows off two optimization parameters: the cell aspect ratio, $AR_0 = H_0/L_0$, which is the cell shape factor, and the composition factor, $(k_0/k_p)/\phi_0$, which presents the material properties and the composition of the cell. The design resistance is minimum for the slenderness factor

$$\left.\frac{H_0}{L_0}\right|_{opt} = 2\left(\frac{k_0/k_p}{\phi_0}\right)^{1/2}, \tag{2.2}$$

that, interesting enough, spells out the optimal shape of M_0.

The design may continue, with the first order ensemble, M_1, Fig. 2.1b, which is in fact a tree. An additional design parameter, assumed by the designer, intervenes here: the size of the allotted high conductivity lane, D_1. The design resistance of this

[5] The analog problem of volume-to-point current flow was addressed in the photovoltaic power generation structures (Morega and Bejan, 2005).

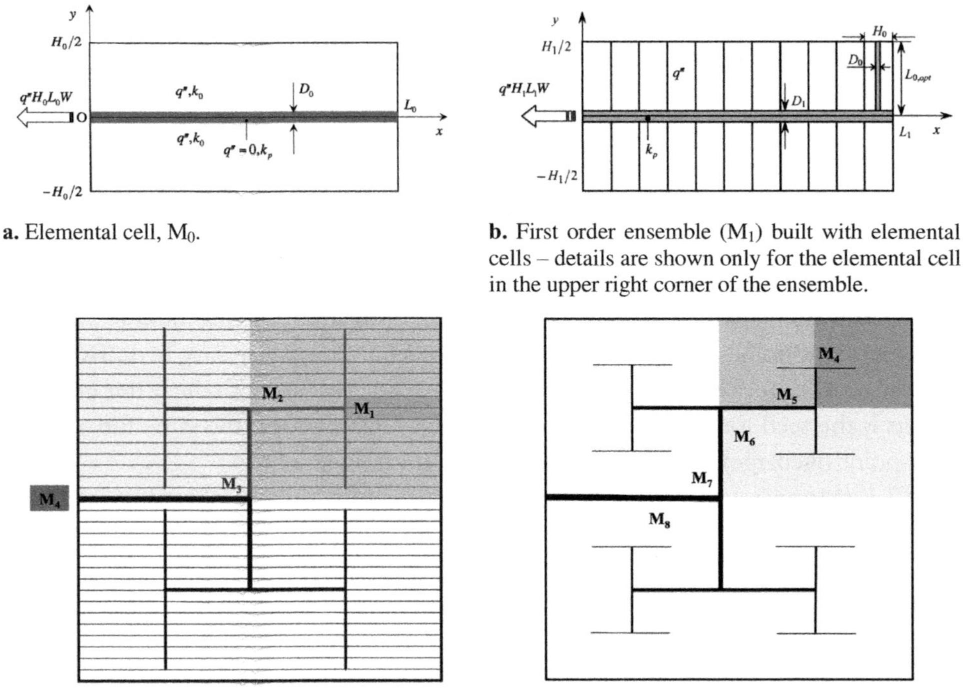

a. Elemental cell, M_0.

b. First order ensemble (M_1) built with elemental cells – details are shown only for the elemental cell in the upper right corner of the ensemble.

c. Fourth order ensemble (M_4) and its constituent lower order ensembles (M_1,M_2,M_3).

d. Eighth order ensemble (M_8) and its constituents, lower order ensembles (M_4,…,M_7).

Figure 2.1 The fundamental problem of volume-to-point flow; k_p and k_0 are conductivities, and q' is the heat generation rate. The quantity of higher conductivity material ($A_0 = D_0 \times L_0$), hence the composition of this system, and its size ($A_0 = H_0 \times L_0$) are fixed Morega and Bejan (2005)[6].

ensemble shows off a minimum with respect to $AR_1 = H_1/L_1$. The composition factor $k_1 = k_p\phi_0$ is actually an effective conductivity. Unlike AR_0, which is a continuous variable, AR_1 is quantized: the design recipe sets $H_1 = 2L_0$, and L_1 size may vary as multiple of $2H_0$, and the increment is, in fact a pair of mirrored elemental cells. The next higher order *already optimized* ensembles are constructed by recurrently mirroring the previously optimized construct and adding the high conductivity lane that connects its mass center with the discharge port, on the boundary. The higher order ensembles inherit the shape of the elemental cell, here a sequence of rectangles (even order constructs) and squares (odd order constructs).

The sketchy, less natural appearance of the constructs in Fig. 2.1 is due to the pending simplifying assumptions aimed to keep the number of optimization parameters to a minimum, which makes it possible to find analytical solutions to the

[6] The analogous problem of volume-to-point current flow was addressed in the photovoltaic power generation structures Morega and Bejan (2005).

conduction heat transfer problem (a Poisson problem, Chapter 1: Physical, Mathematical, and Numerical Modeling). This constructal design embodies and is the objective, the deterministic result of the physics that governs the system: gradient driven diffusion, or conduction (heat, mass, electrical current). It is *scalable* (self-similar) and *robust*—higher order ensembles have the same volume-to-point resistance. This principle may suggest that the core of the tree construction of many animate and inanimate systems is one single design principle: the volume-to-point constrained minimization of the overall flow resistance between one port on the boundary and the finite, contained volume (territory), which is an infinite number of points—one for all and all for one, topologically possible in a continuous medium. A design principle so general confers predictability to the tree network structure, with its main properties.

Fluid trees

The constructal optimized ensemble, Fig. 2.1 (Morega and Proca, 2004), was introduced later in physiology, for fluid trees as a three-dimensional convenient, representative model for the vascular arborescence (Cohn, 1954, 1955). It was known, experimentally, that every mother vessel splits into two smaller, daughter vessels, as it was known and demonstrated for tubes, based on flow resistance minimization, that the diameter must decrease by a constant factor ($2^{-1/3}$) at each bifurcation: this result had been derived in Thompson (1942). And the geometric description of these constructions that is presented without theory in the fractal geometry, as a heuristic model of the lung bronchial tree (Mandelbrot, 2020) is, in fact, a two-dimensional rendering of Cohn's (1954) ramified fluid system (Bejan and Zane, 2012; Morega, 2013).

To solve the essential problem of the volume-to-point flow, Fig. 2.2 Morega and Bejan (2005), with minimum resistance (Bejan, 1988, 1997a,b) a two-dimensional representation of area A was assumed.

The flow through the port on the boundary is connected to each point of the territory, and the mass, $\dot{m}'$, and volumetric, $\dot{m}'''$, flow rates are related through $\dot{m}' = \dot{m}'''A$. The volume is a single fluid saturated porous medium with constant properties. Extending Darcy model (Chapter 1: Physical, Mathematical, and Numerical Modeling) to a more general case, A is filled here by an inhomogeneous porous medium made of a low permeability material (K) and a small volume fraction of inserts (cracks, open, or filled, etc.) of significantly higher permeability (K_1, K_2,...), of unspecified thicknesses (D_1, D_2,...) and lengths (L_1, L_2,...). Thus the elemental cell (A_0) is made of the low permeability material and high permeability strip (K_0, D_0). Each successive higher order ensemble of volume (A_i) is a set of previous order ensembles of size (A_{i-1}), which are tributaries for the collector layer (K_i, D_i, L_i).

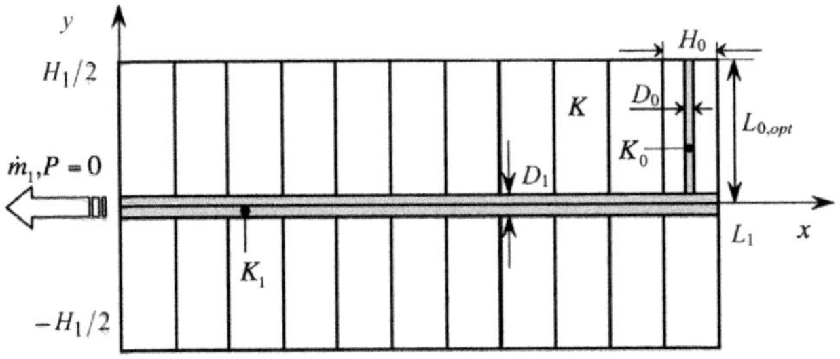

Figure 2.2 Volume-to-point pressure driven flow. The boundary is impermeable, except for the output port.

The analysis is much simplified for $K << K_0$ and $D_0 << H_0$, when the flows in the K and K_0 regions are unidirectional, yielding

$$v = \frac{K}{\mu}\left(-\frac{\partial P}{\partial y}\right), \quad u = \frac{K_0}{\mu}\left(-\frac{\partial P}{\partial x}\right), \tag{2.3}$$

where P is pressure and μ is the kinematic viscosity. The peak pressure drop on the cell (left upper and lower corners of the cell), $P_{peak,1}$ divided through the volumetric flow rate that crosses the port M, $\dot{m}'''$, yields the design flow resistance of the cell

$$\frac{P_{peak,0}}{\dot{m}'''_0 A_0 \nu} = \frac{1}{8} \times \frac{H_0}{L_0} + \frac{K_0}{2\phi_0 K} \times \frac{L_0}{H_0}, \tag{2.4}$$

where $\phi_0 = D_0/H_0$. As for the conduction problem discussed above, two factors are evidenced as optimization parameters: the aspect ratio of the cell, $AR_0 = H_0/L_0$, and the composition factor ϕ_0. The optimization (minimization) of Eq. (2.4) yields

$$\left.\frac{H_0}{L_0}\right|_{opt} = 2\left(\phi_0 \tilde{K}_0\right)^{-1/2}, \quad \Delta\tilde{P}_0 = \frac{1}{2}\left(\phi_0 \tilde{K}_0\right)^{-1/2}, \tag{2.5}$$

where the definitions $\left(\tilde{H}_i, \tilde{L}_i\right) = \frac{(H_i, L_i)}{A_0^{1/2}}$, $\tilde{K}_i = \frac{K_i}{K}$, $\phi_i = \frac{D_i}{H_i}$ are used. As expected, this flux-gradient flow is analogous to the heat conduction problem above, hence the heat transfer results Eqs. (2.1) and (2.2) may translated into their Darcy flow counterparts. The constructal design sequence, toward higher order ensembles and its outcomes are the same.

Moreover, for channels with clear fluid Hagen–Poiseuille flow (Chapter 1: Physical, Mathematical, and Numerical Modeling), if D_i is small enough then $K_i = D_i^2/12$ $(i = 1,2,\ldots)$ (Leopold et al., 1964). This analysis may be repeated sequentially, noting that the permeabilities are no longer independent. The results of this constructal growth as well as a comprehensive discussion on more complex, three-dimensional flow structures may be found in Bejan (2000a,b).

Living trees

The constructal principle initially explained the shape of the systems that optimizes the volume-to-point flow resistance and lays the theoretical bridge to explain several known empirical allometric laws for living tree structures. In biology, where the viewer usually contemplates a tree construct of higher order, the size concerns global and external quantities such as the total volume, V_i, the total mass, $M_i = \rho V_i(1 - \phi_i)$, the cross-sectional area of the tree's root, $S_i = (\pi/4)D_i^2$. Then the total interior area of all tubes inside the constructal, A_i, are the outcome of the optimal constructal structure. In nondimensional form their definitions are

$$\hat{\mathrm{M}}_{\mathrm{i}} = \frac{M_i}{\rho L_0^3} = 2^i(1 - \phi_i),\ \hat{\mathrm{S}}_{\mathrm{i}} = \frac{S_i}{\rho L_0^2} = \frac{\pi}{\lambda^2} 2^{(2i-5)/3},\ \hat{\mathrm{A}}_{\mathrm{i}} = \frac{A_i}{L_0^2} = \frac{\pi}{\lambda} a_i, \tag{2.6}$$

where ρ is the tissue density, $\lambda = L_0/D_0$ is the shape factor of the smallest tube, and i is the construct order.

Figs. 2.3 and 2.4 (Morega and Proca, 2004) plots these quantities versus each other using the points representing each level of assembly, i, marked at the upper part of the curves, to show their origin. They synthesize the observations made from outside (measurements performed by the biologist), that is, from the perspective of an

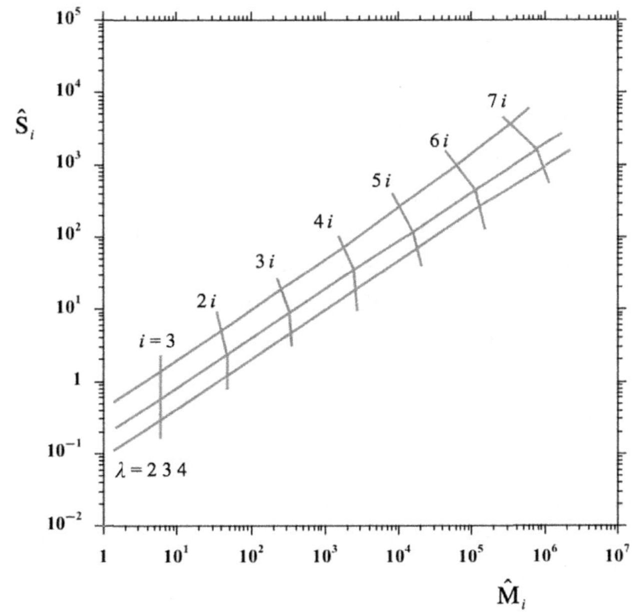

Figure 2.3 The relation between the cross-sectional area of the root and the total mass of the construct.

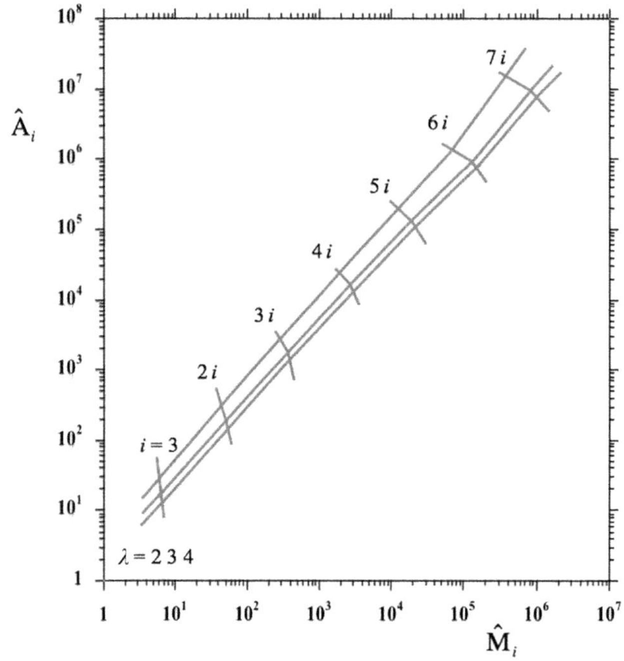

Figure 2.4 The relation between the total internal surface of the tubes and the total mass of the construct.

observer, who examines the whole organs belonging to large and small animals, unaware of the principle that generated the organ.

These representations may suggest the existence of a power law between mass, $\hat{M}_i$, the contact surface, $\hat{A}_i$, and the cross-sectional area of the root, $\hat{S}_i$. In particular, $\hat{S}_i$ increases almost proportionally with $\hat{M}_i$ risen to 0.7 power, which concurs well with the allometric law averaged over the organs of different sizes, where the exponent of M was $\sim 3/4$ (Murray, 1926a,b), and the total surface of the exposed tree of ducts is nearly proportional with the mass of the constructal

$$\hat{S}_i \sim O(10^{-1})\hat{M}_i^{0,7}, \ \hat{A}_i \sim O(1)\hat{M}_i^{1.03}. \tag{2.7}$$

The influence of the smallest tube slenderness (aspect ratio, $\lambda = L_0/D_0$) is minute, suggesting that these macroscopic relations are robust regarding the design feature represented by λ. More robustness with respect to λ is evidenced in Figs. 2.5 and 2.6 (Bejan, 2000a,b; Morega and Proca, 2004).

The volume and the mass are almost proportional but density, M_i/V_i, decreases slightly when the order of the construct (M_i, V_i) increases. Smaller and simpler constructals (small i) are denser than the higher order constructals (large i).

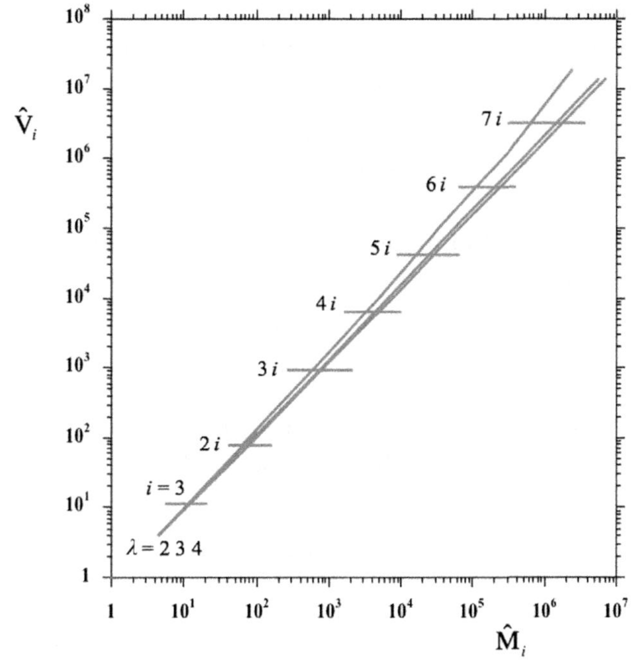

Figure 2.5 The relation between the volume and the mass of the construct.

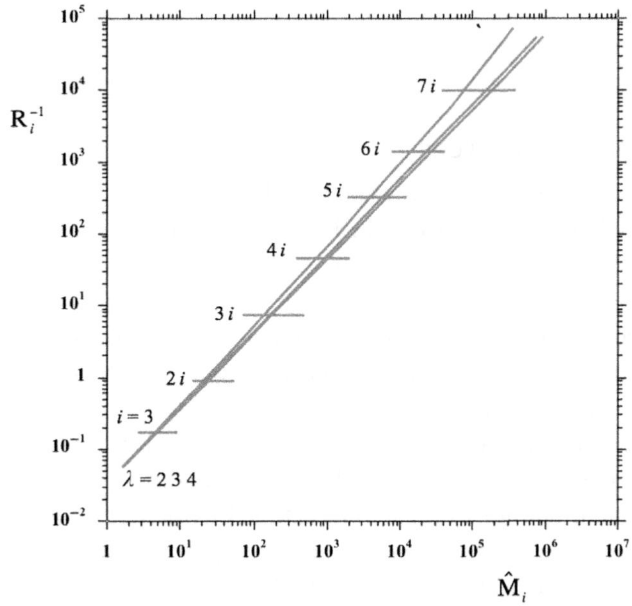

Figure 2.6 The flow conductance of each construct increases proportionally with $\hat{M}_i^{\alpha}$, where $\alpha \approx 0.9$.

These theoretical considerations concur with the allometric laws qualitatively and quantitatively. The allometric law accepted for the respiratory contact surface in mammals is (Weibel, 1972; Schmidt-Nielsen, 1984)

$$\frac{A}{1\text{m}^2} \cong 3.31\left(\frac{M}{1\text{kg}}\right)^{0.98}, \tag{2.8}$$

where A and M are presented in m^2 and in kg, respectively. In the ambit of the mammals corporal masses ($10^{-3}\ldots10^3$ kg) it may be approximated by

$$\frac{A}{\text{m}^2} \sim 3\frac{M}{\text{kg}}. \tag{2.9}$$

The theoretical relation $\hat{A}_i \sim \hat{M}_i$ suggests $A_i \sim M_i/(\rho\ \lambda\ D_0)$, and for $\rho \sim 10^3\ \text{kg/m}^3$, and $D_0 \sim 10^2\ \mu\text{m}$, for the smallest of the length scale (alveoli) yields

$$A_i \sim \frac{10\,M_i}{\lambda\ \text{kg}}\text{m}^2. \tag{2.10}$$

Because $\lambda \sim 3$, this theoretical formula is, essentially, the same empiric correlation Eq. (2.9).

Fig. 2.7 (Morega and Proa, 2004) was obtained using Figs. 2.5 and 2.6. The overall contact surface is nearly proportional with $(\phi_i\hat{V}_i)^b$ where the exponent $b \sim 0.9$, independent of λ.

The constructal theory may thus predict the allometric relations between the size of the body, the metabolic rate, the breathing frequencies, and heart beatings.

Counterflow convection trees

Living systems, tissues and animals are vascularized by pairs of trees of wider vessels (arteries, veins) interlaced in counterflow, which are also heat convection paths (Keller and Seiler, 1971; Wulff, 1974; Chato, 1980; Chen and Holmes, 1980), where the heat flux is relative to the temperature gradient of the ensemble. For instance, the inhalation/exhalation function of the lung relies on two countercurrent streams, which facilitate the heat drain to lower temperatures serving in the mean time as a thermal insulation structure—the better, the better the thermal contact between the two counterflows is. The most important convective configuration of vessels is that in which the two vessels are perfectly interlaced such that in each pair of parallel tubes, embedded in an interstitial space, the flows are in countercurrent, for example, Fig. 2.8, left (Morega and Proca, 2004). Fig. 2.8, right shows the same interlaced constructal concept applied to concurrent, redundant electric power distribution networks (Morega et al., 2008).

The ratios between the diameters of the succeeding tubes and their lengths minimize the flow resistance under the constraints of the total volume of the tubes and the

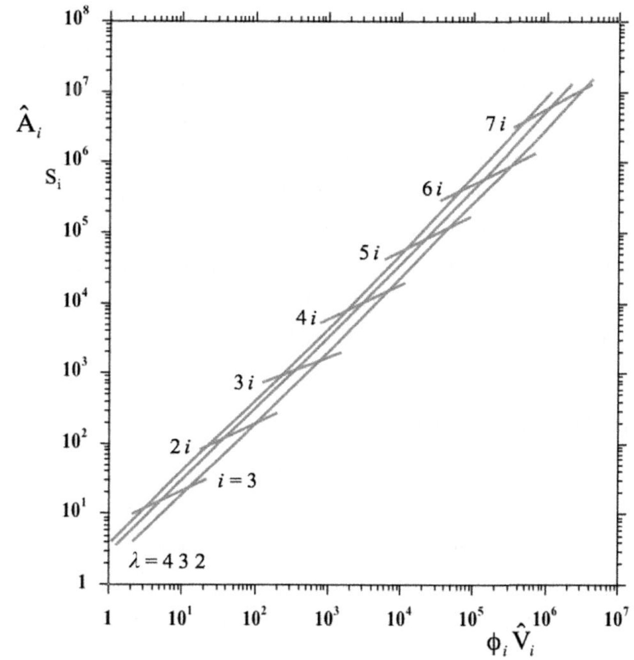

Figure 2.7 The total internal surface of the tubes versus the total volume of the tree of tubes.

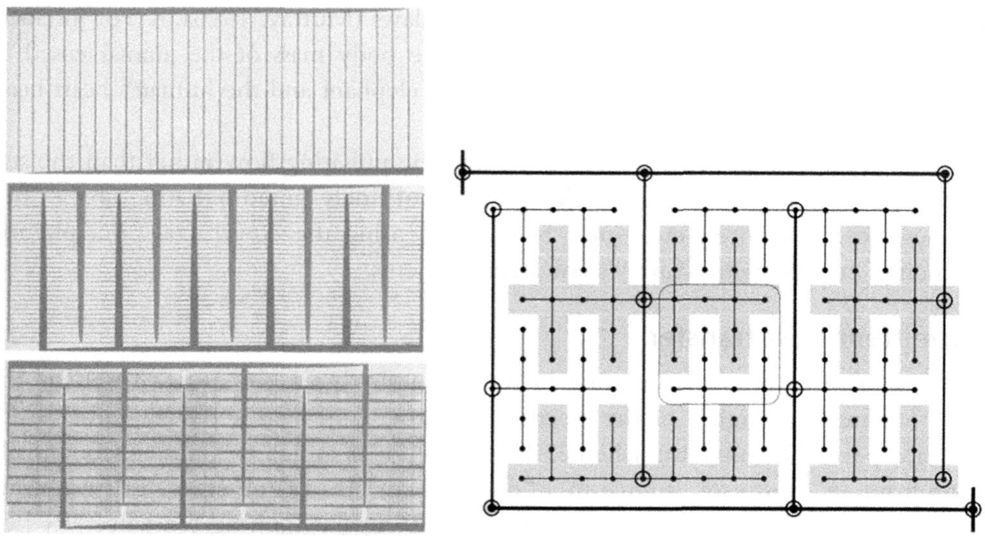

Figure 2.8 Countercurrent (*left*) and concurrent (*right*) flows.

total volume, and it explains why the metabolism rate (the total heat rate) of an animal has to be proportional to its size, to the total volume, or to the body mass. It is the resistance to the loss of body heat that relates the metabolism rate and the size of the body in all animals, with warm blood or cold blood. At the limit of the small dimensions of the bodies of warm-blooded animals, the heat is transferred mainly through the conductive tissues and the exponent used in the power law (exponential) is, as observed, of 1/3. In oysters and amphibians, the resistance is dominated by the outward convection of the body surface, and the exponent is the noticed one, of 2/3. In mammals and birds, the resistance is dominated by counterflows and the exponent is the one observed, that is, 3/4 (Bejan, 2001a,b).

2.4 Structure in time: rhythm

All flow systems have the unifying principle of the spatial, geometric shape. Nature however, exhibits not only spatial shape and structure but also temporal structure: rhythm. The same is true for the engineered systems. Breathing and respiration have many aspects in common with the engineered processes controlled by the dynamic transfer through unsteady diffusion. The lung and the vascularized tissue have similar objectives and constraints. The fluid (blood) as soon as admitted inside the smallest passages (alveoli, capillaries) initiates the mass transfer within the surrounding tissues. The effectiveness of this transport mechanisms decreases in time. The maximization of the mass transfer rate implies the removal of the old fluid batch such that a new batch produces again higher fluxes.

The most known natural flows that exhibit also rhythm are the breathing and the heartbeats of animals, surprised by the allomoteric laws (Murray, 1926a,b). For instance, the frequency of the heartbeats are relative to the body mass of the animal raised to power of ~ 0.25. A mouse breathes faster than an elephant and the human's heart beats faster than the heart of the horse.

An inherent question is why nature selected the pulsating flow rather than the stationary flow to serve these beings. The existence of finely tuned pulsating flows—frequencies, flow resistances, etc.—may be exposed using the constructal principle and the findings concurs with Murray's law ascertainments.

Intermittent heat transfer

A glimpse in the optimal structuring in time of a rhythmic process is offered by the simple loading/discharging of the electric capacitor, C [F], in a PWM (pulse width modulated) flyback convertor scheme. The capacitor is aimed to transfer sequentially power from an "upstream" source to a "downstream" load (Veli et al., 2019). During the loading phase, C is connected in series with a current limiting resistance, R_1 [Ω], to be powered by a primary voltage source, U_0 [V]—for simplicity U_0 is assumed constant, or the source is of "infinite" power, concept equivalent to the "thermostat" in

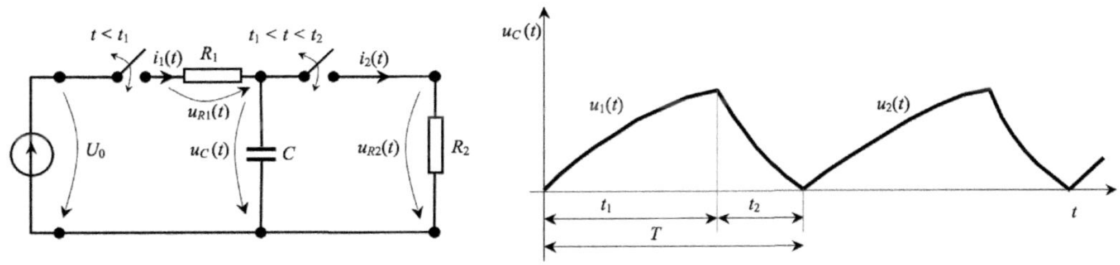

Figure 2.9 The voltage on an energy storage device (capacitor) during intermittent charge / discharge – the electric circuit (*left*) and the signal (*right*).

heat transfer. This loading circuit is active until the capacitor is fully loaded, and the voltage to its terminal reaches U_0; then the capacitor is switched to a discharging resistance, R_2, until its full discharge. The voltage at the capacitor terminals during this load/discharge sequence is, respectively

$$u_C(t) \underset{t \leq t_{on}}{=} u_1(t) = U_0 e^{-t/\tau_1}, \; u_C(t) \underset{t_{on} < t \leq T}{=} u_2(t) = U_0 e^{-t/\tau_2}, \tag{2.11}$$

where $\tau_1 = R_1 C$, $\tau_2 = R_2 C$ [s] are the time constants of the two, charge/discharge phases, t_{on} is the charging time and T is the period of cyclic charging/discharging process, Fig. 2.9.

The electric energy stored in the capacitor at any time t, with respect to a homogeneous initial state, is $W_e(t) = Cu_C^2(t)/2$. The two resistances (electric here) may be used to adjust the timings—either faster or slower—and to tune the power loss, by Joule effect, to the environment, that is, the efficiency of the flyback convertor. For the cyclic, rhythmic functioning of this element, the rhythm (period) is constrained, obviously, by the two time constants, τ_1 and τ_2. Should the optimal working regime for C be full charge followed immediately by the full discharge, than $T_{opt} = t_1 + t_2 \sim \tau_1 + \tau_2$.

Respiration

The same optimization of the power sketched in Fig. 2.9 leads to optimal frequencies of the intermittent flows (closed-opened) for models for organs (lungs, circulatory systems), which comply with irreversibilities, such as the viscous flow (with friction) and the mass transfer (Bejan and Errera, 1997).

For the inhalation process, which lasts t_1, the thoracic cage sustains a volume increase V of atmospheric air (T_0, P_0) which is driven by a lower than atmospheric pressure ($P_0 - \Delta P_1$), with the average velocity U_1, Fig. 2.10.

The pressure drop is proportional to the average inspire velocity

$$\Delta P_1 = rU_1^n, \tag{2.12}$$

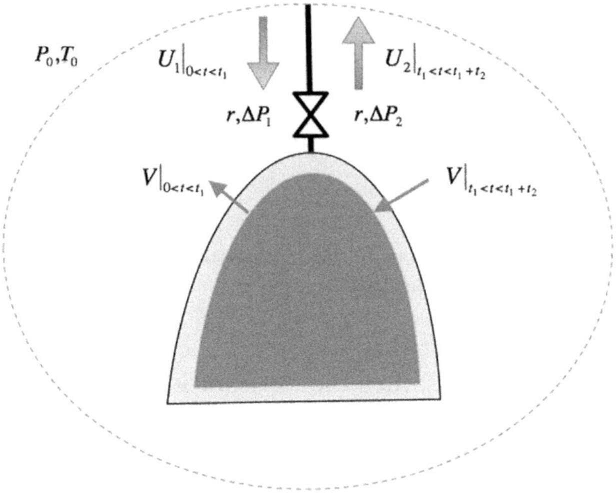

Figure 2.10 The sketch of the respiratory system and its two-stroke cycle.

where r is the flow resistance of the airways, and n is a type of flow index that varies from $n \cong 1$ for the laminar flow to $n \cong 2$ for turbulent flow. Expiration, which lasts t_2, is driven by the overpressure ΔP_2 inside the thorax

$$\Delta P_2 = rU_2^n, \tag{2.13}$$

where U_2 is the average expiration velocity.

For the duration of these processes, mass conservation yields, respectively

$$\rho_0 U_1 A_f t_1 = \rho_1 V, \ \rho_1 U_2 A_f t_2 = \rho_1 V, \tag{2.14}$$

where A_f is the effective cross-sectional area of the airflow, ρ_0 is the atmospheric air density, and ρ_1 is the density of the inspired air (at $P_0 - \Delta P_1$ and T_b). The total, per cycle ($t_1 + t_2$) mechanical work performed by the thoracic muscles is

$$W = \oint \left(P_0 = P_{cavity}\right) \mathrm{d}V_{cavity} = (\Delta P_1 + \Delta P_2)V, \tag{2.15}$$

such that the related average power is $\dot{W} = W/(t_1 + t_2)$, or

$$\dot{W} = r\frac{V^{n+1}}{A_f^n}\frac{t_1^{-n} + t_2^{-n}}{t_1 + t_2}. \tag{2.16}$$

This result shows off that the needed power drops monotonically when either t_1, or t_2, increase. It may be inferred then that the effortless respiration corresponds to the longest possible inspire and expire intervals. On the other hand, the main respiration function demands, t_1 and t_2 ought to be finite, as required to ease the O_2 transfer from the inhaled fresh air and the vascularized tissue of the pulmonary passages, through the surfaces that secede the air passages from the pulmonary vascularized tissue. Oxygen transfer occurs through diffusion, on both sides of the parting interface, due to the small sizes of both the terminal branching of the airways (alveoli) and the capillary blood vessels. In the region irrigated with blood, more restrictive to mass transfer, it is of the order $j \sim D\Delta C/\delta$, $\delta \sim (Dt_1)^{1/2}$, where δ is the mass penetration depth.

For example, humans' respiration is at ~60 bpm, that is, $t_1 \sim 1$ s. Oxygen diffusivities in the region with air and the region with blood are 10^{-5} and 10^{-9} m^2/s, respectively (Hydei et al., 1996). After 1 s, in the alveoli the penetration depth is $\delta \sim 3$ mm, two orders of magnitude larger than the alveoli scale, ~50 μm (Grotberg, 1994). In the blood irrigated region $\delta \sim 30$ μm, that is, of the same order of magnitude as the capillaries (Schmidt-Nielsen, 1972). The amount of Oxygen transferred is $m = jAt_1$

$$m \sim \underbrace{\frac{D\Delta C}{(Dt_1)^{1/2}}}_{j} At_1, \tag{2.17}$$

where A is the area of the overall contact mass transfer surface of the airways, and the per cycle averaged mass transfer rate

$$\dot{m} = m/(t_1 + t_2) \sim AD^{1/2}\Delta C \frac{t_1^{1/2}}{t_1 + t_2}, \tag{2.18}$$

is due to the lungs and their muscles, which pose a constraint on the needed inspire and expire time intervals

$$\frac{t_1^{1/2}}{t_1 + t_2} \sim \frac{\dot{m}}{AD^{1/2}\Delta C} = K\ (\text{constant}). \tag{2.19}$$

Returning to the average power requirement, $\dot{W}$, eliminating t_2 and solving $\partial\dot{W}/\partial t = 0$ for minimum power consumption for respiration, yields the optimal inspiration time (Bejan and Errera, 1997).

$$\left(1 - Kt_{1,opt}^{1/2}\right)\left[\left(\frac{1}{Kt_{1,opt}^{1/2}} - 1\right)^n + 1\right] \sim \frac{n}{2n+1}. \tag{2.20}$$

A periodic flow is then required by the minimization of the respiration mechanical power. The particular flow regime, n, has no significant influence on the order of magnitude of $t_{1,opt} \sim K^{-2}$. For instance, when the animal is small enough such that the airways are very narrow and U_1 is sufficient small for the flow to be laminar $(n = 1)$, $t_{1,opt} \sim (9/16)K^{-2}$, yielding the optimal expiration time $t_{2,opt} \sim (3/16)K^{-2}$

The theoretical high point that the respiration intervals are of the same scale

$$(t_1, t_2)_{opt} \sim K^{-2} = \left(\frac{AD^{1/2}\Delta C}{\dot{m}}\right)^2, \tag{2.21}$$

is in good concordance with a significant body of observation archived and correlated in biology literature (Schmidt-Nielsen, 1972). This theoretical time of respiration increases with the size of the animal's body as a power law with the exponent ~ 0.25, again, in excellent agreement with the data correlated in biology.

Heart beating

Relying on the minimization of the mechanical power constrained by a global mass transfer limit the constructal principle may predict the existence of an eigen heartbeats frequency, finely adjusted, that is inverse proportional with the body size. The two-chamber piston model of the circulatory system, Fig. 2.11, feeds the pulmonary

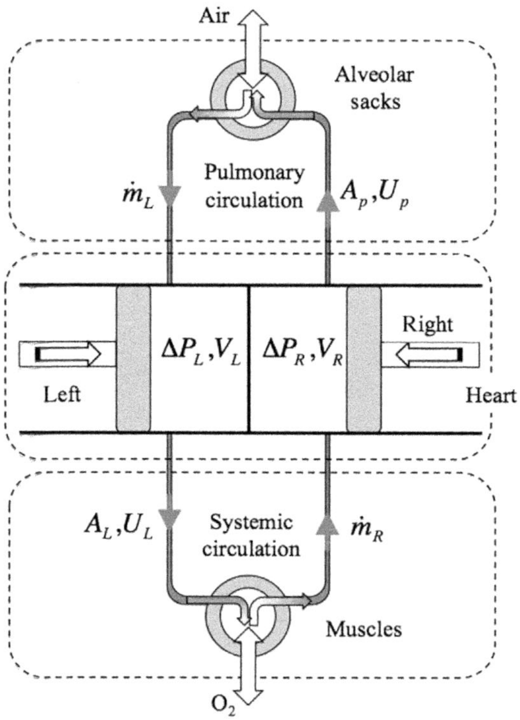

Figure 2.11 The human circulatory system model with two chambers heart model.

circulation—from the arteries that convey the deoxygenated blood to the lungs, and ends with the pulmonary veins that transport back the oxygenated blood to the left side of the heart—and the systemic circulation—that carries the oxygenated blood to the muscles, where O_2 is delivered, and the veins that return the deoxygenated blood to the heart.

The working chambers (left, L; and right, R) that pump the blood from the core to the capillaries of the pulmonary circulatory system contract simultaneously in the interval t_1. The heart muscle contracts both chambers, increasing the pressure of the blood, which returns to the heart, from P_0 (reference) to $P_0 + \Delta P$, where

$$\Delta P = r_s U_s = r_p U_p, \tag{2.22}$$

U_s and U_p are the average velocities of the blood ejected from the heart and r_s, r_p are the resistances posed by the systemic and pulmonary flow paths, dominated by the contribution of the smallest blood vessels and capillaries. Consequently, the vascularized tissues may be homogenized as porous media saturated with fluid, with laminar Darcy flow (Chapter 1: Physical, Mathematical, and Numerical Modeling).

Mass conservation law yields

$$\dot{m}_L = \dot{m}_R = \dot{m},$$

where

$$\dot{m}_{R,L} = \frac{\rho V_{R,L}}{t_1} = \rho A_{p,s} U_{p,s}, \tag{2.23}$$

and ρ is the blood density, A_s and A_p are the effective areas of the cross-sectional surfaces related to U_s and U_p. The contractions of both chambers are, hence, the same, $V_L = V_R = V$.

The per cycle mechanical average power required by the myocardium is

$$\dot{W} = \frac{2V\Delta P}{t_1 + t_2} = \frac{2r_s V^2}{A_s t_1 (t_1 + t_2)}, \tag{2.24}$$

where t_2 is the rest time between two consecutive heartbeats. In this time interval O_2 diffuses from the alveoli to the lung capillaries and from the systemic capillaries to the muscle.

These mass diffusion transfer is governed by the scale relations order $j \sim D\Delta C/\delta$, $\delta \sim (Dt_1)^{1/2}$, $m \sim \frac{D\Delta C}{(Dt_1)^{1/2}} A t_1$, with t_2 instead of t_1 where A is the area of the surface of mass transfer for all vessels capillaries, pulmonary, and systemic. The amount of O_2 transferred in the interval t_2 is proportional to $t^{1/2}$, and the average mass transfer rate is proportional to the group $K \sim t^{1/2}/(t_1 + t_2)$. The metabolic rate of the animal emerges as a constraint, as in Eq. (2.21)

$$\frac{t_2^{1/2}}{t_1 + t_2} \sim K(\text{constant}). \tag{2.25}$$

The power consumption of the heart is then proportional to the inverse of $t_2\left(1 - Kt^{1/2}\right)$. It follows that $\dot{W} \to \infty$ for $t_2 \to \{0; K-2\}$, and at the intermediate value $t_{2,opt} \sim \left(4/9\right)K^{-2}$ it exhibits a relatively sharp minimum. The constraint K on the mass transfer provides the contraction time $t_{1,opt} \sim \left(2/9\right)K^{-2}$. Interesting enough (Bejan, 2000a,b)

$$(t_1, t_2)_{opt} \sim K^{-2} = \left(\frac{AD^{-1/2}\Delta C}{\dot{m}}\right), \; t_{1,opt}/t_{2,opt} \sim 1/2, \tag{2.26}$$

which are valid for animals in a wide range of sizes (Peters, 1983; Schmidt-Nielsen, 1984).

It may be inferred the existence of an eigen frequency of the heartbeats that minimizes the mechanical power consumed by the heart, subject to the constraints of the interface and the mass transfer, or metabolic rate of O_2. This maximization with respect to the active diffusion time interval is the foundation of all optimal pulsations in the engineered and natural systems.

Coupled rhythms in the cardio-pulmonary system

Within the framework of the thermodynamics with finite speed (TFS), the cardio-pulmonary system (CPS), may be considered as an ensemble of two biological machines, naturally designed and optimized: the heart, a "naturally designed" blood (liquid) pump, and the lungs, "naturally designed" air compressors (Petrescu et al., 2018). Studies based on a large number of measurements, for stationary states related to different positions: walking, sitting, laying, repetitive exercise, etc., have shown that the two frequencies—for heart, f_H, and lung, f_L—are correlated, for a healthy person, through

$$f_H = f_L\left(2 + N/4\right). \tag{2.27}$$

Here N is an integer, called *quantum number of the interaction between the heart and lung* in a stationary state. It is thought of as the interaction parameter that links the two biological machines (heart and lungs), namely the difference in phase between them (Petrescu et al., 2018). The analogy with classical thermodynamics indicates that Eq. (2.27) relates three state parameters (f_H, f_L, N), of which obviously only two may be independent. Three particular processes are then identified: (1) *iso-pulse* (f_H = const.); (2) *iso-rhythm* (f_L = const.); and (3) *iso-quantum* (N = const.). Moreover, it has been observed that CPS works properly (healthy) only if certain coordination in the interaction between its two biological machines exists, quantified by Eq. (2.27), where N is specific to each stationary state, throughout the underlying *circadian cycle.*[7] From the TFS perspective, this CPS activity may be characterized as a thermodynamic system by computing its power, entropy, and efficiency.

In the constructal philosophy, the structure and shape of each of the two parts (heart and lungs) is the result of purpose and constraints, and in the constraints class, each of them constraints the other. Thus together with the other body's parts they work optimally (synchronized), as a whole. Both rhythmicity and synchronization go down to the elemental cell, the excitable (here, cardiac) tissues (Chapter 4: Electrical Activity of the Heart). The malfunction of one of them triggers the disordered function of the other one, as an attempt of the whole to adapt to the new situation (Zemlin et al., 2009).

2.5 The effect of body size

The respiration frequency decreases with the animal's size as $\left(\dot{m}/A\right)^2$, Eq. (2.26). To predetermine its relation with the body size two empiric or theoretical laws that relate the metabolic rate (or $\dot{m}$) and the area of the contact surface for mass transfer, A, with the body mass, M, respectively, are needed.

As said, allometric laws (Chapter 1: Physical, Mathematical, and Numerical Modeling) are power laws, largely accepted, between geometric and functional, flow parameters of the living creatures (Kleiber, 1932; Metabolic Rate and Kleiber's Law, 2020; Murray 1926a,b; Peters, 1983; Schmidt-Nielsen, 1984). Their predetermination on a purely theoretical basis was a true endeavor. One of the most difficult to confirm is the proportionality that exists between the metabolic rate and the body mass, or volume V, raised to 3/4 power.

Because the heat leak to the environment through convection is proportional to the body's surface, the metabolic rate has to be proportional to the square of the characteristic length, $V^{1/3}$, squared that is with the body mass, or volume raised to power 2/3. However, this heat transfer theoretical result is infirmed through observations on the mammals and birds, which show off an exponent much closer to 3/4 than 2/3 for bodies of small size reptiles, amphibians, and fish. The 3/4 law does not interrelate the cold blood vertebrates although these small size bodies are equally dominated by arborescent flow structures. From another standpoint, other theoretical studies are based on

[7] Circadiam comes from latin: *circa* (around), and *diem* (day). The circadian rhythm is a natural, internal process that regulates the sleep—wake cycle and repeats roughly every 24 hours (Brain Basics: Understanding Sleeping, 2020). It can refer to any biological process that displays an endogenous, entrainable oscillation of about 24 hours.

the optimization of the pumping mechanical power minimization in tree flows in physiology and river morphology (Thompson, 1942; West et al., 1997).

However, the minimum pumping power and the minimal corporal heat loss are parts of the same optimization principle: the constructal principle or how to be best fit, and it brings in the metabolic rates of all vertebrates, with warm blood (3/4) and with cold blood (2/3). Here the heat transfer is not considered individually, exclusively, but it is combined with the pumping power minimization in fluid constructal trees (Bejan, 2000a,b).

The minimization of the fully developed Hagen–Poiseuille flow resistance of a fluid tree network (each mother tube, of length L_i and diameter D_i, bifurcates into two identical daughter tubes, of lengths L_{i+1} and diameters D_{i+1}) when limiting the total volume of the tubes, unveils the optimal ratio of the successive diameters $D_{i+1}/D_i = 2^{-1/3}$. This finding, the old Murray law, is particularly robust because it is independent of the lengths (L_i, L_{i+1}) and of the relative positions of the three (mother and daughters) tubes. It is not even essential to recognize the constructal method as foundation of this structure. The heat transfer analysis may be conducted by accepting, heuristically, this tree network (Bejan, 2000a,b) or the one in (West et al., 1997)—it is as an assumption only.

The blood vessels trees and the superposition of the venous and arterial trees is so intimate that each tube of one tree is in counterflow with a tube of the other (Fig. 2.12, left), and the reality of counterflow of blood and other fluids (Fig. 2.12, right) are widely recognized in physiology (Schmidt-Nielsen, 1984; Vogel, 1988). In the analysis of the heat current between the root and the canopy in a warm-blooded animal of concern is the heat lost through the subcutaneous level, and heat is transferred laterally, between the arterial (warmer) and venous (colder) currents (Section 2.3.4).

In the volume delimited by the adiabatic boundary the enthalpy of the warmer fluid is larger than that of the colder fluid, the counterflow conveys longitudinally the

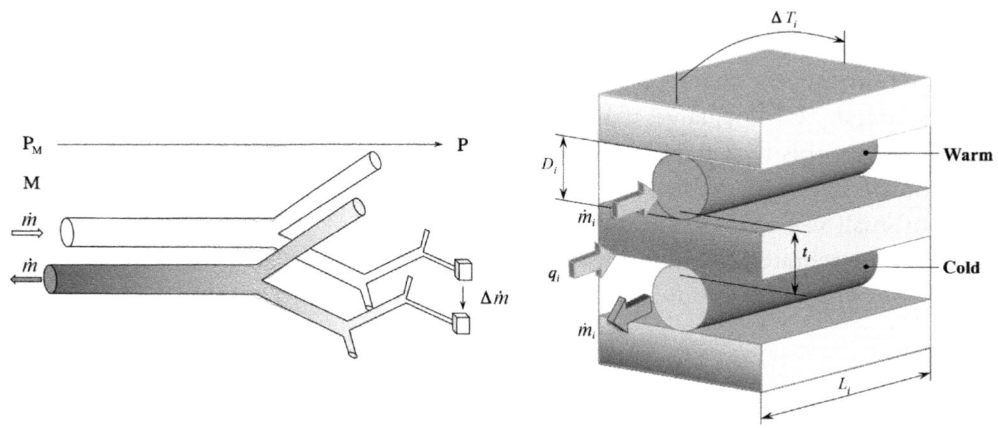

Figure 2.12 Trees of counterflow convective trees.

energy current $q_i = \dot{m}_i c_p \Delta T_{t,i}$ — c_p is the specific heat of blood and ΔT_i is the temperature difference between the two currents at level i. Such a counterflow provides a longitudinal temperature gradient $\Delta T_i/L_i$. Its companion convection heat current between the vessels and through the tissue in between is

$$q_i = \frac{(\dot{m}_i c_p)^2}{h_i p_i} \frac{\Delta T_i}{L_i}, \tag{2.28}$$

where h_i and p_i are the total heat transfer coefficient between the currents and, respectively, the contact perimeter between the two currents. An order of magnitude analysis of Eq. (2.28) yields

$$\Delta T_i \sim \frac{q_i L_i}{h_i p_i} \frac{k}{c_p^2}, \tag{2.29}$$

where $k \sim k_f$ (thermal conductivities of blood and tissue, respectively) and $h_i \sim k/D_i$. The resistance of the fluid inside the channel is of the order $\sim D_i/k_f$ and the resistance of the solid tissue in between the two tubes scales as $\sim t_i/k$, where t_i is of the same order of magnitude with D_i.

The two fluid currents make a single convective tree in counterflow, with zero net mass flux, which spans from the internal, metabolic temperature of the animal (at $i = 0$) to the skin temperature. Using the flux conservation laws, for mass flow rate, $N_i \dot{m}_i = \dot{m}_0$ (constant), and heat rate, $N^i q^i = q^0$ (constant), where N^i is number of branches, the overall temperature difference ΔT (constant) associated to the warm blood vascularization is then (Bejan, 2001a,b)

$$\Delta T = \sum_{i=0}^{n} \Delta T_i \sim \frac{q_0 k}{\dot{m}_0^2 c_p^2} \sum_{i=0}^{n} N_i L_i, \tag{2.30}$$

which, using $L_{i+1}/L_i = f$, $L_i = L_0 f_i$, $L_n = L_0 f_n$, and $N_i = 2^i$, yields an estimation of the total heat rate

$$q_0 \sim \left(\frac{q_o}{\dot{m}_0}\right)^2 \frac{k L_n f^{-n}\left[(2f)^{n+1} - 1\right]}{c_p^2 \Delta T (2f - 1)}. \tag{2.31}$$

For the reason that both q_0 and $\dot{m}_0$ are proportional with the metabolic rate it follows that their ratio, $q_0/\dot{m}_o$, does not depend on the size of the body, n. Moreover, from the standing point of the constructal method, the length of the elemental volume L_n is assumed constant. It was shown (Bejan, 2001a,b) that the order of magnitude of the hosting volume is

$$V \sim L_n^3 \left(\frac{2}{f}\right)^n \frac{1 - f^{n+1}}{1 - f}, \tag{2.32}$$

which paves the path to a theoretical (constructal) relation between the metabolic rate and the total volume. In the limit, Eq. (2.31), $(2f)^{n+1} >> 1$ and $f^{n+1} << 1$ $q_0 \sim 2n$ and $V \sim (2/f)^n$, which yields

$$\frac{\log q_0}{\log V} = \frac{3}{4}, \tag{2.33}$$

which means that $q_0 \sim V^{3/4}$, or (because $q_0/\dot{m}_o$ does not depend on the body size)

$$\dot{m} \sim M^{3/4}, \tag{2.34}$$

which is, in fact, the allometric law.

This two fold geometric optimization with the two restrictions and the pairing of tubes in constructs larger than the preset elemental volume is the essence of the constructal method. The constructal theory combines two ideas—(1) the optimized tree for minimum pumping power and subject to spatial constraints, and (2) the convection heat transfer or, better, the insulation characteristic to the two counterflow fluid trees—and reinstates geometry in the place it deserves, in physiology, river morphology, and any other domain where macroscopic structure and form define the flow system outside equilibrium.

The convection thermal resistance of the counterflow trees, R1, resides within the animal, Fig. 2.12 bottom. This resistance acts in parallel with the internal resistance, R2, which models the conductive heat loss through the solid tissue. Outside the animal, the heat current passes through the body surface that is exposed to the environment.

For the cold blood vertebrates, the temperature drop on R1 is minimal and, when changes in the environment temperature intervene, the leading resistance is R3. In consequence the heat loss rate and the metabolic rate follow closely $V^{2/3}$. For warm blood animals, the body side of the skin poses a significant thermal resistance. In larger mammals R1 $<$ R2 and the heat current is transferred by the convection tree, and the metabolic rate matches the predominant tendency $V^{3/4}$.

The lung is also a convection currents tree, the result of two overlapping air trees: the inspiration and the expiration flows. The convection tree is made of heat currents and constitutes a corridor for heat flow of the same type as the tree considered above. During the inspiration the cold air warms up progressively along the passages through which it travels. During the expiration, the warm air cools down progressively along the same passages. The tissues of passages walls act as a regenerative heat exchanger in engineering terminology. Supplementary to the convection tree for thermal insulation, and relying on the same inlet–outlet mechanism, the lung acts as a tree path for minimizing the water loss (Nield and Bejan, 1999).

The thickness of the tissue penetrated by the mass diffusion through the respiration or the heartbeat is proportional to $t^{1/2}$. The volume or mass of the tissue penetrated by the mass diffusion in this interval is $M \sim A\ t^{1/2}$. Using the scaling $t \sim \left(A/\dot{m}\right)^2$ and utilizing $\dot{m} \sim M^{3/4}$ yields

$$A \sim M^{7/8} \text{ and } t \sim M^{1/4}. \tag{2.35}$$

This allometric law is sustained by the large body of observations archived in the literature, for example, Schmidt-Nielsen (1984), Peters (1983), Vogel (1988).

The evolutionary design shaped through the laws of physics for animate or inanimate systems is now explained through the constructal law and the theories that rely on it. The power law relations produced by the allometric laws and other theoretical or empirical observations have been seen as a signs of possible fractal shapes. Fractal geometry might present so convincingly some then but it is physics that explains them—from functioning to shape. In the constructal philosophy, vascularized tissues and organs, the organism, work, adapt, morph, and survive as a optimally balanced whole, and are the result of a constrained evolution as described by the laws of physics, following the time arrow.

References

Bejan, A., 1988. Advanced Engineering Thermodynamics. Wiley & Sons, New York.

Bejan, A., 1996. Method of entropy generation minimization, or modeling and optimization based on combined heat transfer and thermodynamics. J. Appl. Phys. 79 (418−419), 1191−1218.

Bejan, A., 1997a. Theory of organization in nature: pulsating physiological processes. Int. J. Heat. Mass. Transf. 40 (9), 2097−2104.

Bejan, A., 1997b. Constructal-theory network of conducting paths for cooling a heat generating volume. Int. J. Heat. Mass. Transf. 40, 799−816.

Bejan, A., 2000a. Shape and Structure, From Engineering to Nature. Cambridge University Press, Cambridge.

Bejan, A., 2000b. From heat transfer principles to shape and structure in nature: constructal theory. 1999 Max Jakob Memorial Award Lecture, Trans. ASME 122, 430−449. August.

Bejan, A., 2001a. Entropy generation minimization: the method and its applications. J. Mech. Eng. 47 (8), 345−355.

Bejan, A., 2001b. The tree of convective heat streams: its thermal insulation function and the predicted 3/4-power relation between body heat loss and body size. Int. J. Heat. Mass. Transf. 44, 699−704.

Bejan, A., 2016. The Physics of Life. The Evolution of Everything. St. Martin's Press, New York.

Bejan, A., Errera, M.R., 1997. Deterministic tree networks for fluid flow: geometry for minimal flow resistance between a volume and one point. Fractals 5 (4), 685−695.

Bejan, A., Lorente, S., 2010. The constructal law of design and evolution in nature. Philos. Trans. R. Soc. B 365, 1335−1347.

Bejan, A., 1980. Entropy Generation Through Heat and Fluid Flow. Wiley, New York, pp. 180−182.

Bejan, A., Zane, J.P., 2012. Design in nature. How Constructal law Governs Evolution in Biology, Physics, Technology, and Social Organization. Anchor Books, Random House Inc, New York.

Brain Basics: Understanding Sleeping, 2020. <https://www.ninds.nih.gov/Disorders/Patient-Caregiver-Education/Understanding-Sleep> (accessed in June).

Chato, J.C., 1980. Heat transfer to blood vessels. ASME J. Biomech. Eng. 102, 110−118.

Chen, M.M., Holmes, K.R., 1980. Microvascular contributions in tissue heat transfer. Ann. N.Y. Acad. Sci. 335, 137−150.

Cheng, S.C., Huang, Y.M., 2003. A novel approach to diagnose diabetes based on the fractal characteristics of retinal images. IEEE Trans. Inf. Technol. Biomed. 7 (3), 163–170.

Cohn, D.L., 1954. Optimal systems: I. The vascular system. Bull. Math. Biophys. 16, 59–74.

Cohn, D.L., 1955. Optimal systems: II. The vascular system. Bull. Math. Biophys. 17, 219–227.

Falconer, K.J., 1990. Fractal Geometry: Mathematical Foundations and Applications. Wiley, London.

Fractal, Fractal dimension, 2020. *Allometry, Power law, Sierpinski triangle, Kleiber's law.* <http://en.wikipedia.org/> (accessed in June).

Grotberg, J.B., 1994. Pulmonary flow and transport phenomena. Annu. Rev. Fluid Mech. 26, 529–571.

Hydei, R.W., Forster, R.E., Power, G.G., Nairn, T., Rynes, R., 1996. Measurement of diffusing capacity of the lungs with a stable O_2 isotope. J. Clin. Invest. 45 (7), 1178–1193.

Hoffman, K.H., 2008. An introduction to endoreversible thermodynamics. Atti dell'Accademia Peloritana dei Pericolanti Cl. di Scienze Fisiche, Matematiche e Naturali LXXXVI (C1S0801011 (Suppl. 1)), 18.

Hoffman, K.H., Burzler, J.M., Schubert, S., 1997. Endoreversible thermodynamics. J. Non-Equil. Thermodyn. 22 (4), 311–355.

Huo, Y., Kassab, G.S., 2012. Intraspecific scaling laws of vascular trees. J. R. Soc. Interface 9, 190–200. Published online 15 June 2011.

Kadanoff, L.P., 1986. Fractals: where's the physics? Phys. Today 6–7. Feb.

Keller, K.H., Seiler, L., 1971. An analysis of peripheral heat transfer in man. J. Appl. Physiol. 30, 779–789.

Kleiber, M., 1932. Body size and metabolic rate. Physiol. Rev. 27 (4), 511–547.

Ledezma, G.A., Morega, A.M., Bejan, A., 1996. Optimal spacing between fins with impinging flow. ASME J. Heat. Transf. 118, 570–577.

Leopold, L.B., Wolman, M.G., Miller, J.P., 1964. Fluvial Processes in Geomorphology. Freeman, San Francisco, <https://archive.org/details/fluvialprocesses0000greg> (accessed in June 2020).

Mandelbrot, B.B., 1975. Les objets fractals: Forme, hasard et dimension. Flammarion, Paris.

Mandelbrot, B.B., 2020. *Fractals and the geometry of nature.* pp. 157–159. <https://archive.org/details/fractalgeometryo00beno> (accessed in June).

Metabolic Rate and Kleiber's Law, 2020. <http://universe-review.ca/R10-35-metabolic.htm> (accessed in June).

Morega, A.M., Bejan, A., 2005. A constructal approach to the optimal design of photovoltaic cells. Int. J. Green Energy 2 (3), 233–242.

Morega, A.M., Ordonez, J.C., Morega, M., 2008. A constructal approach to power distribution networks design. In: Int. Conf. on Renew. Energy and Power Quality, ICREPQ'08, 441, Santander, Spain, 12–14 March.

Morega A.M., Proca, A., 2004. Shape and Structure, from Engineering to Nature (in Romanian), Trans. AGIR Publishing House & Romanian Academy Publishing House, Translation of Bejan A., 2000. Shape and Structure, from Engineering to Nature, Cambridge University Press.

Morega A.M., Design in nature. How the constructal law governs evolution in biology, physics, technology, and social organization (in Romanian), Ed. AGIR, Academia de Ştiinţe Tehnice din România, 2013, 245 pp., Translation of: Bejan A., Zane J.P., 2012. Design In Nature. How The Constructal Law Governs Evolution In Biology, Physics, Technology, And Social Organization, Anchor Books, Random House Inc. New York.

Murray, C.D., 1926a. The physiological principle of minimum work: I. The vascular system and the cost of blood volume. In: Proc. of the National Academy of Sciences of the United States of America, 12 (3), 207–214.

Murray, C.D., 1926b. The physiological principle of minimum work: II. Oxygen exchange in capillaries. In: Proc. of the National Academy of Sciences of the United States of America, 12 (5), 299–304.

Nield, D., Bejan, A., 1999. Convection in Porous Media, second ed. Springer Verlag, New York.

Nishimura, T., Sakura, S., Gotoh, K., Morega, A.M., 2000a. Traveling plumes generated within a double-diffusive interface between counter shear flows. Phys. Fluids 12, 3078–3081.

Nishimura, T., Kunitsugu, K., Morega, A.M., 2000b. Direct numerical simulation of layer merging in a salt-stratified system. Numer. Heat. Transf. A 37, 323–341.

Novikov, I.I., 1958. The efficiency of atomic power stations (a review). J. Nucl. Energy II 7, 125–128.

Order in Chaos, 2013. <https://orderinchoas.wordpress.com/2013/02/15/mathematical-monsters/>.

Peters, R.H., 1983. The Ecological Implications of Body Size. Cambridge University Press, United Kingdom.

Petrescu, S., Costea, M., Borcila, B., Petrescu,V., Bolohan, R., Danes, S., et al., 2018. What is quantum biological thermodynamics with finite speed of the cardio-pulmonary system: a discovery or an invention? In: Morega, A.M., Lorente, S. (Eds.), Proc. of the Romanian Academy, Series A, pp. 249–254 (special issue).

Schmidt-Nielsen, K., 1984. Scaling (Why is Animal Size So Important?). Cambridge University Press, p. 112.

Schmidt-Nielsen, K., 1972. How Animals Work. Cambridge University Press, Cambridge.

Schmitt, O.H., 1938. A thermionic trigger. J. Sci. Instrum. 15 (1), 1 Jan.

Schmitt, O.H., 1969. Some interesting useful biomimetic transforms. In: Proc. of the Third Int. Biophysics Congress, 29 August–3 September 1969, Boston, MA, p. 297.

Thompson, D.'A.W., 1942. *On Growth and Form*, ("On the form of branching of blood-vessels". Cambridge University Press, Cambridge, pp. 948–957.

Turing, A.M., 1952. The chemical basis of morphogenesis. Phil. Trans. R. Soc. B 237, 37–52.

Uahabi, K.L., Atounti, M., 2015. Applications of fractals in medicine. Ann. Univ. Craiova, Math. Computer Sci. Ser. 42 (1), 167–174.

Vincent, J.V.F., 2009. Biomimetics—a review. In: Proc. IMechE, **223**, Part H. J. Eng. Med. JEIM561, 919–939.

Vincent, J.V.F., Bogatyreva, O.A., Bogatyrev, N.R., Bowyer, A., Pahl, A.-K., 2006. Biomimetics: its practice and theory. J. R. Soc. Interface 3, 471–482. Published online 18 April 2006.

Veli, Y., Morega, A.M., Pislaru–Danescu, L., Morega, M., Telipan, G., 2019. The design of the secondary electrical circuit of a flyback transformer with hybrid magnetic core. In: 11th Int. Symp. on Adv. Topics in Electrical Eng., IEEE-ATEE, 28–30 March, 2019, Bucharest, Romania.

Vogel, S., 1988. Life's devices. The Physical World of Animals and Plants. Princeton University Press, New Jersey.

Weibel, E.R., 1972. Morphometric estimation of pulmonary diffusion capacity. V. Comparative morphometry of alveolar lungs. Respir. Physiol. 14, 26–43.

West, G.B., Brown, J.H., Enquist, B.J., 1997. A general model for the origin of allometric scaling laws in biology. Sci. Rep. 276, 122–126. 4 April.

Wulff, W., 1974. The energy conservation equation for living tissue. IEEE Trans. Biomed., Eng. BME 21, 494–495.

Zemlin, C.W., Mitrea, B.G., Perstov, A.M., 2009. Spontaneous onset of atrial fibrillation. Physica D. 238 (11–12), 969–975. June 1.

CHAPTER 3

Computational domains

3.1 Physical domains generated using computer-aided design techniques

The use of computational power in the generation, editing, or optimization of a design is called computer-aided design (CAD) (Narayan, 2008). CAD is frequently used in the industrial design and testing phases. Engineers are using it for creating complex multipart models and to study their behavior under certain conditions: mechanical or thermal stress analysis, repeated cycle movements, wear, deflections, etc. Regardless of its applications, it is difficult to imagine that a new product release is possible without the use of CAD in the incipient phases, shortly after the main idea was sketched by pen on paper.

CAD is now also used in the medical and clinical fields, from the prosthesis drawings and patient-specific geometry optimization to virtual platforms that help the medical staff to train and improve different surgical techniques with low costs and great benefits. In the following sections, a set of CAD-generated body parts will be presented.

A CAD construct for an intervertebral disc

The intervertebral discs can be found in the spinal cord at the junction between two vertebrae. They play an essential role in the damping of the mechanical stress exerted upon the spinal cord and allowing the movement of the vertebrae. Two components can be distinguished in the intervertebral disc structure: the *annulus fibrosus*, disposed as concentric lamellas, made of collagen fibers, wrapped around a second part, the *nucleus pulposus* (Newell et al., 2017).

Next a CAD-generated model of a human intervertebral disc is presented. This solid model can be created using simple drawing elements (lines and splines) from any CAD software environment, for example, Solidworks, Autocad, Inventor, and Catia. First, the resulted contour of the nucleus (Fig. 3.1A), which represents the anatomical details (shape and size), was extruded, generating a tridimensional model (Fig. 3.1B). Then to obtain the surrounding annulus, the nucleus was duplicated and rescaled. The final model of the intervertebral disc presented in Fig. 3.1B was created. The two parts, annulus and nucleus are both homogenous in structure. This is not a bad approach when we look at the nucleus, but in some cases, it could become a drawback to consider the annulus as a whole. For example, in mechanical stress analysis, a more accurate description of the annulus internal structure as concentric lamellas made

Computational Modeling in Biomedical Engineering and Medical Physics
DOI: https://doi.org/10.1016/B978-0-12-817897-3.00003-8

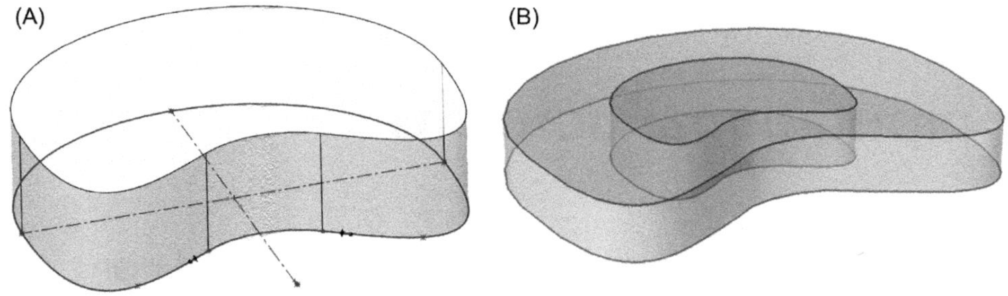

Figure 3.1 The CAD model of an intervertebral disc. The 2D sketch of the nucleus before the extrusion (A) and the final 3D solid model (B).

of collagen fibers, inclined in opposite direction in each pair, is crucial. Nevertheless, there are several situations in which such a simple geometry is enough. The model presented in Fig. 3.1 was successfully implemented as computational domain for the numerical study of the heat transfer problem in minimally invasive intradiscal electrothermal annuloplasty procedures (Tsou et al., 2010) used in the low back pain treatment. This medical procedure consists of inserting a catheter in the affected region to locally heat up the annulus and irreversibly damage the nociceptors due to the collagen fibers contraction (Thiyagarajah et al., 2018).

In this case, the homogeneous annulus that behaves as only one large lamella is a good enough assumption for numerically solving the bioheat (Pennes) or heat transfer by conduction equations (Chapter 1: Physical, Mathematical, and Numerical Modeling Essentials). The obvious advantage that this simple structure brings is given by the low number of elements resulting in the computational mesh when using the finite element method (FEM) for numerical calculus. This will translate as reduced computational effort and satisfying numerical errors.

A CAD abstraction of the kidney

Another CAD construction for biomedical applications is represented by a human kidney and the main renal artery and vein branches, which are specific to that anatomical region. The construction begins with a 3D sketch of the blood vessels paths (Fig. 3.2A) made of splines, roughly representing the real morphology. Then a circular surface representing a radial section of the blood vessel is lofted upon the imposed trajectories created in the previous step. The result is presented in Fig. 3.2B.

The solid model presented in Fig. 3.2 could play an important role as computational domain for the blood flow analysis in the main arterial and venous branches of a kidney. For an extended study, such as the blood filtering process, a solid model for the kidney should be provided. Its bean-shaped geometry is quite simple and easy to generate, starting from a revolved circular surface that defines a slightly bended cylinder, completed with two semispheres at both ends (Fig. 3.3A). Furthermore additional

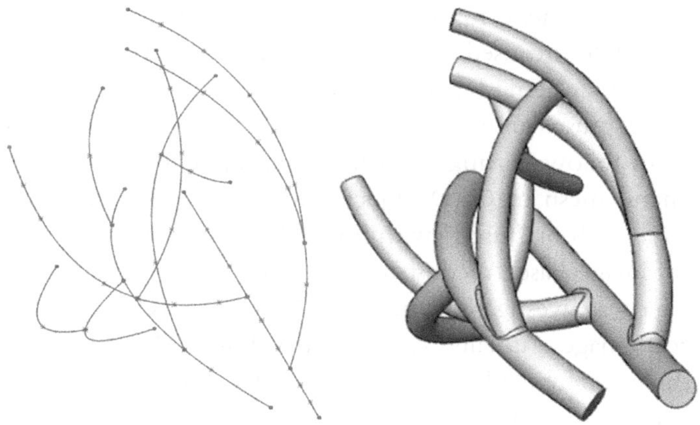

(A) (B)

Figure 3.2 The renal artery and vein construction. The 3D sketch of the blood vessels paths (A) and the solid domains (B).

CAD objects can be added to the model. For example, an umbrella-like antenna (Fig. 3.3B) could be useful in the study of tumor ablation procedures (van Sonnenberg et al., 2010; Novanta et al., 2018).

In this case besides the antenna, an ellipsoid delimiting the tumoral tissue was also added.

3.2 Image-based reconstruction of anatomically accurate computational domains

Numerical modeling in medicine and biology requires realistic computational domains to obtain accurate results. This chapter introduces several examples where computed

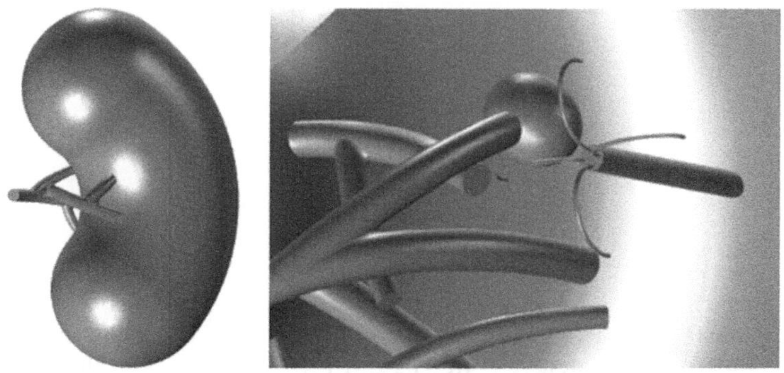

(A) (B)

Figure 3.3 The blood vessels integrated in a kidney (A) and the antenna used in tumor ablation procedures (B).

tomography (CT) or magnetic resonance imaging (MRI) image-based reconstructed solid models (Jan, 2006) are used as computational domains to numerically simulate physical processes such as hemodynamic flows, with and without interaction with the vessel walls and the muscular embedding tissue, bioelectric field mapping. Fig. 3.4 sketches the path to follow starting from raw data input to the end constructs: either FEM (finite element) mesh or CAD object. Moreover, CAD objects may be further "meshed"—either as such or "fused" with image-based constructs.

The process of image-based reconstruction begins with the data acquisition step. In this phase, a subject is investigated using a CT or an MRI scanner. A set of high-resolution medical images, usually saved in the DICOM (Pianykh, 2012) format, acquired from three different planes, sagittal, coronal (frontal), and transversal (axial), will be stored. Then a set of software tools, toolboxes, software plugins, or standalone software applications will be needed for the region of interest (ROI) segmentation. This step is crucial for the image-based reconstruction process because it will generate the first 3D solid model out of 2D data (the medical image set). Thus the segmentation algorithm and tools should be carefully chosen. The reconstructed 3D model will be postprocessed according to its final purpose. If the model is meant to become the computational domain of a numerical study solved with the FEM, a certain set of features and details should be removed. For example, it is preferable to clear the steep angles, to fillet or chamfer the edges, and to smooth whenever possible. Otherwise, it is highly probable that the meshing algorithm will be difficult and it might generate

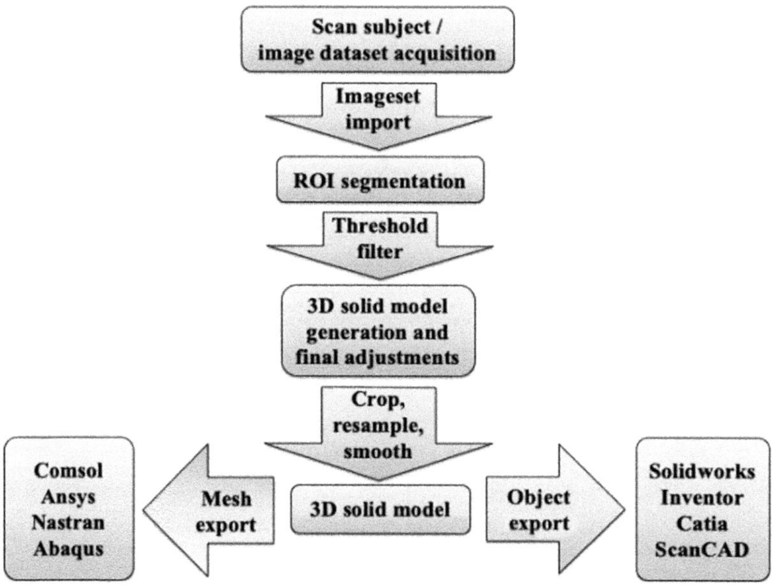

Figure 3.4 The main steps in the image-based reconstruction process.

several errors. If the model will serve as a precise 3D CAD figure of the reconstructed organ and tissue, the anatomical details become important, for example, the shape and volume of a preoperatory and a postoperatory tumor. In such cases, where no meshing process is needed and a high level of morphology detail is targeted, the important features should no longer be removed.

Rigid and elastic arterial networks

In the medical and clinical engineering fields, the interest in the concept and development of numerical models used as powerful and trustworthy tools for the investigation of the arterial hemodynamics is rapidly increasing. These models can help us understand the influence that different arterial networks, from physiologically normal to aneurysm affected or stenosed blood vessels, from our circulatory system manifest upon the transport of nutrients, oxygen, or substances with pharmacoclinical purpose, for example, medication used in chemotherapy or magnetic drug targeting procedures.

Due to the human circulatory system complexity and individualities, the image-based reconstruction for generating anatomically accurate computational domains seems to be a promising virtualization method. The models described in this chapter bear the significant advantage of representing the real morphology of the source (original) blood vessels used in the segmentation process.

The first set of models simplifies the study of hemodynamic problems, considering that the blood vessel walls are rigid. When advancing in age, the blood vessels tend to lose their elastic properties and begin an atherosclerotic, calcifying process (Sangiorgi et al., 1998). This assumption eliminates the flow—structural interaction generated by the pulsatile arterial blood flow. More complex models with elastic blood vessel walls, which allow the study of blood pulsation vessel walls embedding tissue, are also described here.

Cardiovascular disease represents the main death cause in the modern world. For example, atherosclerosis, which mostly affects elderly people, is responsible for the vessel walls stiffening due to the cholesterol deposits (Pyörälä et al., 1994). The atheroma plaque rupture can trigger myocardial or brain ischemia, causing myocardial infarction or cerebrovascular accident (Zaman et al., 2000).

The blood vessel dimensions are well fitted to the blood flow stream and to the viscous stress exerted upon the vascular endothelium (Davies, 1995). The morphology of the arteries, geometry, and bifurcation regions is one of the most influential factors in determining the blood flow, the wall viscous stress, and the mechanical forces that play a key role in the atherogenesis process (Khanafer and Vafai, 2008), which is also governed by the atherogenic macromolecules and the mass transport that occurs at the arterial walls, from the blood stream to the surrounding tissues.

Considerable effort is being made to study and understand the hemodynamic flow and the associated mass and heat transfer processes. The arterial hemodynamics is strongly related to the complexity of the blood vessels geometry, with shapes always moving and deforming due to the heart pulsations and respiratory displacement (Zeng et al., 2003; Broboană et al., 2008). Thus the computational domain generation is a very delicate and crucial step when defining a model. The software tools for image-based reconstructions are now well developed, with a high enough performance level to make them reliable when reproducing the 3D morphologies of real organs or tissues. The numerical models elaborated for the pulsatory blood flow analysis, in rigid arterial networks, bring the advantage of using computational domains segmented out of MRI datasets.

When the final 3D geometry of the blood vessel is ready, the next step in the modeling process is represented by the solid FEM discretization of the model, followed by the physical and mathematical model definition and numerical (FEM) solving of the latter one. The blood flow patterns are analyzed after postprocessing the numerical result as surface plots, color maps, streamlines, and arrows that describe the viscous stress, velocity, or pressure field dynamics. This will be detailed later in Chapter 6, Magnetic Drug Targeting.

Next the image-based reconstruction of an arterial network made of large blood vessels (main blood vessel diameter larger than 20 mm) is presented. The source is a high-resolution MRI DICOM image dataset (Fig. 3.5) acquired from a healthy subject.

The ROI segmentation, the most delicate step in the image-based reconstruction process, will significantly affect the final 3D solid model quality. The segmented object (ROI) should accurately respect the morphology of the source organ and tissues. This is granted by the overall quality of the source image set: the number of artifacts, the resolution, the sharpness, and brightness levels, etc., on the one hand; and by the optimal

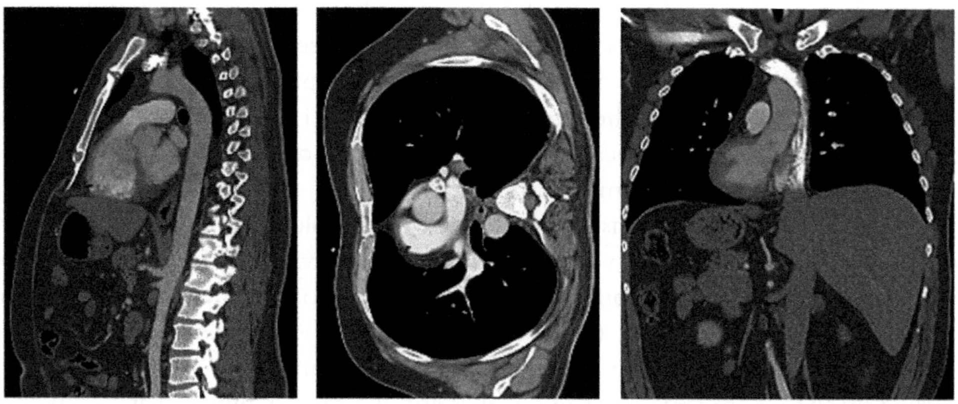

(A) (B) (C)

Figure 3.5 The source of the image-based reconstruction process of rigid arteries. Slices from the high-resolution DICOM dataset: sagittal plane (A), transversal plane (B), and coronal plane (C).

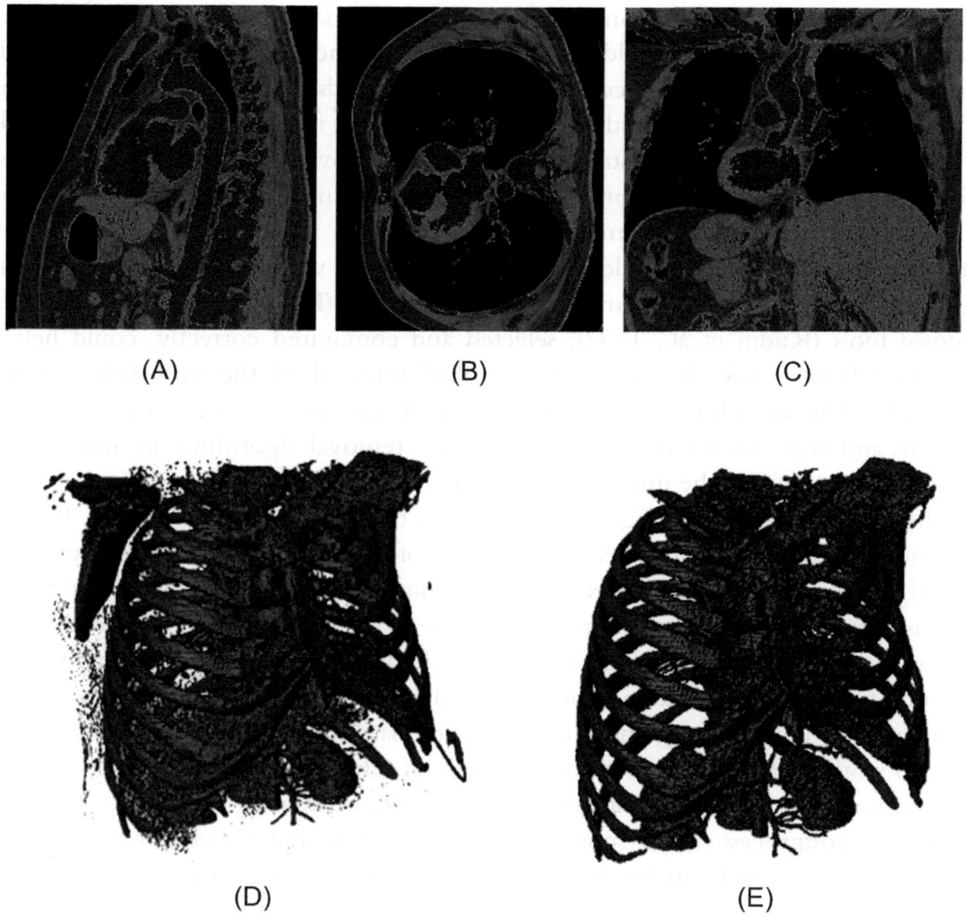

Figure 3.6 ROI segmentation process: the grayscale interval isolation (A)–(C) and the segmentation result—mask with (D) and without (E) noise and artifacts.

parameter values selected for the segmentation tool, for example, the grayscale interval of a threshold filter (Vincent, 1993), which extracts the ROI from the images, on the other.

In some cases, different organs or tissues could be represented in the image dataset by close grayscale values. For example, when using contrast-enhancing substances, the blood will be represented by grayscale values similar to the ones used to depict the bone tissue. This will become a drawback for the image-based reconstruction process: the blood vessels and the bone tissue will be both segmented out from the image dataset at the same time (Fig. 3.6). If only one of the two tissues is of interest, additional time will be spent to postprocess and remove the unwanted parts.

Although very different regarding their state or structure, the blood and bone tissue are hard to differentiate on the source images, due to their similar intense grayscales.

Before the manual postprocessing of the segmented mask, a flood-filling algorithm (Fathi and Hiltner, 1999) should be used. This will check for discontinuities in the mask and will eliminate the disconnected regions (islands). Also a cavity removal tool (Chris and Garland, 1990) should repair the mask, filling in the cavities, once applied.

The source image quality, strongly related to the overlapping artifacts and noise, affects the ROI reconstruction process. Even a well-optimized segmentation algorithm will provide a preliminary 3D model of the ROI with a set of additional unwanted features (Fig. 3.6D). These should be carefully removed, without accidentally touching the mask of the ROI. For example, smoothing filters (Kuan et al., 1985) and noise removal tools (Rudin et al., 1992), selected and configured correctly, could help at generating better masks by the user-controlled removal of the undesirable regions (Fig. 3.6E). The smoothing filters attenuates the images noise level and evens the sharp contours and edges of the ROI, while the noise removal algorithms are mainly concerned with improving the images background.

The 3D-smoothed mask (Fig. 3.6E) is made of both blood and bone tissue. Thus the latter needs to be removed to create a 3D mask of the arterial network. This process is manual and consists of carefully selecting and deleting the bone tissue straight from the 2D mask as depicted in the source images (Fig. 3.6A–C) or from the 3D solid model generated by the segmentation algorithm (Fig. 3.6E). Although the 3D elimination process of the bone tissue is easier because it brings the advantage of a more precise and effortless selection of the unwanted regions, it needs enough hardware resources, especially 3D graphics, to work well. This aspect becomes significant when dealing with complex models, comprised of several masks, for different types of organs or tissues.

A final, postprocessed, 3D solid model of a rigid arterial network is presented in Fig. 3.7B. This model can be discretized and used as computational domain in an

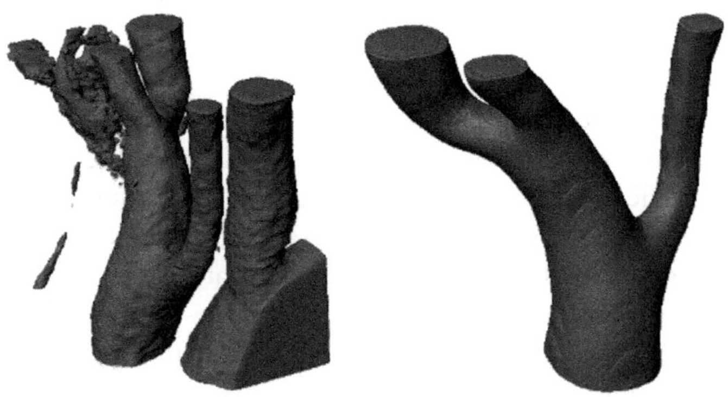

(A) (B)

Figure 3.7 Cropping and postprocessing the rigid arterial network: before (A) and after (B) applying the flood filling, smoothing, and noise removing algorithms.

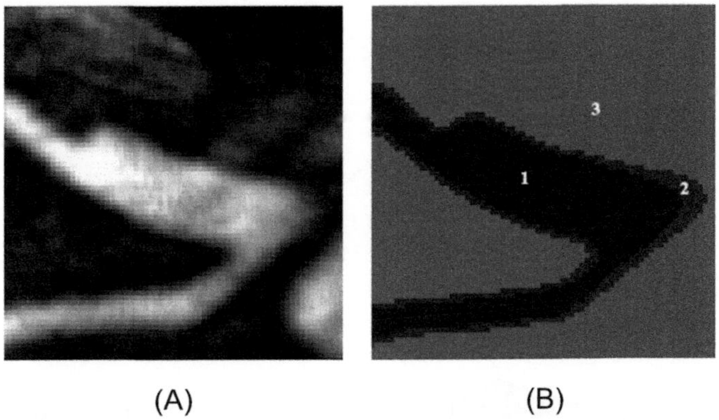

Figure 3.8 ROI segmentation for blood (B, 1), vessel walls (B, 2), and surrounding tissue (B, 3) using MRI-acquired DICOM image datasets (A).

FEM study for the pulsatile blood flow in stiffened arteries, as presented in Chapter 6, Magnetic Drug Targeting, or it can be used as solid CAD for visually investigating a certain region of interest for clinical purpose.

A more complex set of models, useful for achieving a more detailed view upon the arterial hemodynamic, is considering the mechanical interaction between the pulsatile blood flow, the vessel walls, and the embedding tissue. In this particular case, the 3D solid models are always multipart, comprised of at least two subdomains, instead of only one, the case of rigid arteries. The ROI segmentation is similar to the same process described earlier for the rigid arterial network (Fig. 3.7). The same steps for artifact and noise removal should be carefully implemented to enhance the quality of the final 3D model. Besides the mask for the blood, two additional subdomains are segmented out of the source image set, one for the arterial walls and another for the surrounding tissues (Fig. 3.8).

Since the blood vessel walls cannot be distinguished in the source dataset, once the arterial blood segmentation process is finalized by the smoothing step and the deletion of undesired regions (Fig. 3.9A), the mask for the blood will become the starting point for generating the vessel walls (Fig. 3.9B). A morphological filter (Roushdy, 2006) is used to dilate the mask of the blood. Then the original (nondilated) mask will be subtracted from the dilated one through Boolean operations (Masuda, 1993; Requicha and Voelcker, 1985), generating the blood vessel walls mask (Fig. 3.9B). The surrounding tissue is obtained after inverting the original blood model. The vessel walls are subtracted out of this negative and the result is a tissue mask (Fig. 3.9C). The final 3D solid model (Fig. 3.9D) made of three parts, the blood, vessel walls, and embedding tissues, is ready to be discretized and used as computational domain in FEM studies of pulsatile blood flow—structural interactions.

The two previously presented models, for rigid and elastic arterial networks, are both related to physiologically normal morphologies. The elastic arteries model could also be

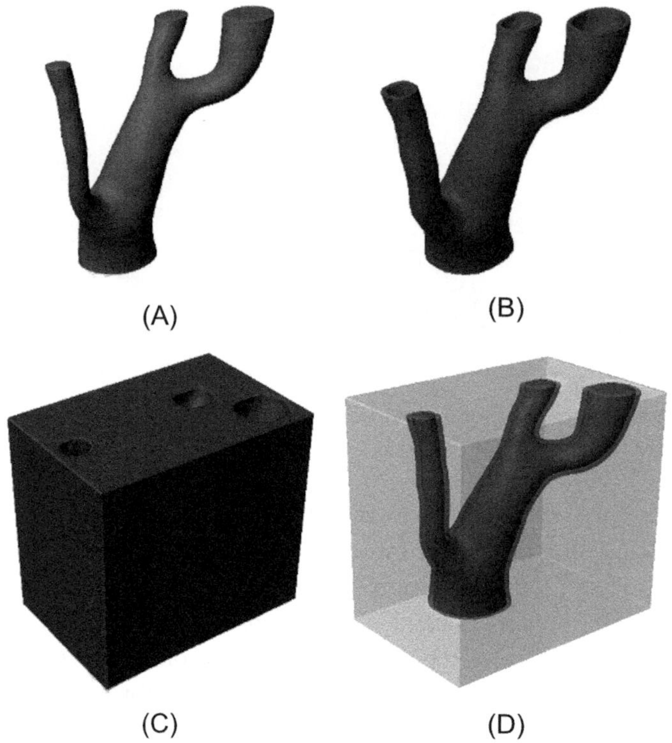

Figure 3.9 The 3D multipart solid model of an elastic arterial network (D) comprised of blood (A), vessel walls (B), and tissue (C). *After Dobre, A., 2012. Investigation Methods for the Analysis of Coupled Phenomena Specific to the Medical Engineering Field (Doctoral thesis). Faculty of Electrical Engineering, University Politehnica of Bucharest, Bucharest, Romania.*

useful in the analysis of genesis and evolution of pathological morphologies like the aneurysms (Isselbacher, 2007). These are sphere-shaped vessel wall deformations, which occur due to the wall thickening triggered by cardiovascular disease, for example, atherosclerosis, frequently developed by elderly people. While the factors that lead to these pathological formations is still the subject to debate, it is certain that the aneurysms can be congenital or developed. For the latter, the major risk factors are smoking, arterial hypertension, diabetes, and powerful mechanical shocks. There are two types of developed aneurysms, morphologically distinct: the saccular aneurysm that usually occurs in the cerebral arterial network, and the fusiform aneurysm, specific to the abdominal or thoracic regions of the aorta.

Since the aneurysms are, in general, asymptomatic, the problems start when they rupture and generate life-threatening hemorrhage. The dissection, thrombosis, or wall rupture of an aneurysm can bring clinical complications. When one or more layers of the blood vessel wall are sectioned, bleeding along the muscular fibers within the blood vessel represents a big issue.

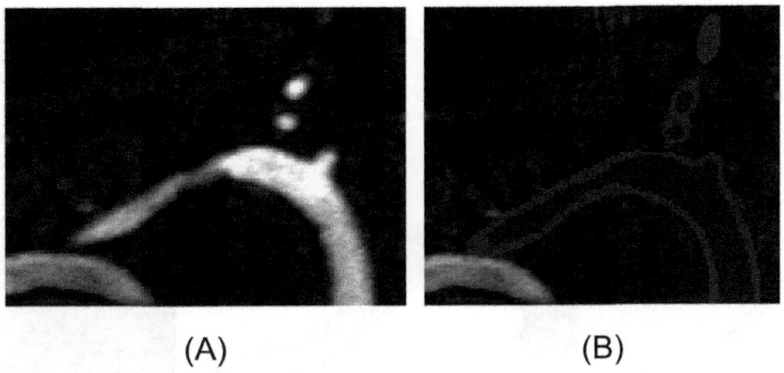

Figure 3.10 Segmentation of the left subclavian artery (B) using an MRI image set (A).

To generate a 3D solid model for an elastic arterial network affected by a saccular aneurysm, an MRI-acquired DICOM dataset (Fig. 3.10A) for the thoracic region of a healthy subject was used to segment a region of the left subclavian artery. Because the image set did not contain any pathological formations, this was created using a lucrative alternation between a flood-fill tool and a CAD primitives generation algorithm (Sun et al., 2005; Ma et al., 2011). After finishing the segmentation process of the targeted arterial network (Fig. 3.10B), a prone to aneurysm formation region of the arterial wall was searched for. This was judged after the numerical simulation results given by the model that implements the normal elastic arterial branch as computational domain (Fig. 3.11) and accounts for the presence of pulsatile blood flow—structural interaction.

The ROI reconstruction of the blood volume that flows within the subclavian artery was achieved by applying a threshold algorithm (Vincent, 1993). After removing the 3D artifacts and undesired regions, a smoothing (Kuan et al., 1985) was configured and applied several times to generate the result presented in Fig. 3.11A. The postprocessed blood volume mask was dilated by a morphological algorithm (Roushdy, 2006)

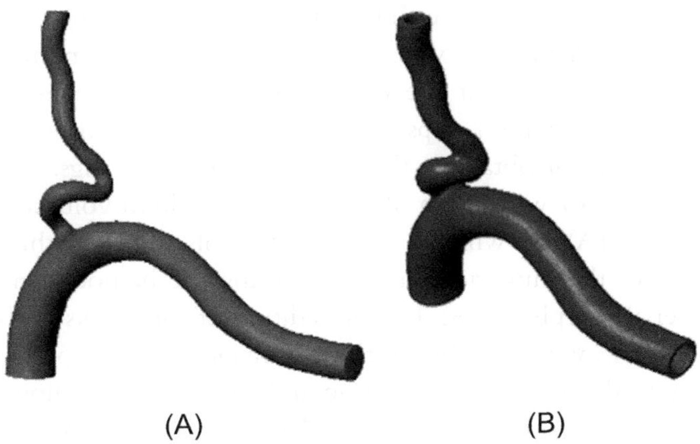

Figure 3.11 Segmented 3D masks of blood (A) and arterial wall (B) given by the MRI image datasets.

(A) (B) (C)

Figure 3.12 The 3D solid model of the elastic arterial network affected by a saccular aneurysm: selection of the region prone to aneurysm formation (A) and the postprocessed model in 3D (B) and 2D (C) views.

and the original blood mask (Fig. 3.11A) was subtracted from it, generating the vessel wall mask (Fig. 3.11B). Together the two 3D solid models (Fig. 3.11) were FEM discretized and used as computational domains for the hemodynamic problem that outlines the vessel areas where high mechanical stress is exerted due to the blood pulsation and vessel geometry. This represents the most relevant criterion for selecting a high probability region for the formation of a developed saccular aneurysm.

Starting from the original 3D model of the blood volume (Fig. 3.11A), a sphere was generated and placed on the main artery's wall (Fig. 3.12A). Then a flood-filling algorithm (Fathi and Hiltner, 1999) converted the selected spherical volume into a solid CAD object, representing the blood inside the virtually created aneurysm. This new blood mask was dilated, and after subtracting the aneurysm's and arterial blood's masks, the vessel walls were generated (Fig. 3.12B). In the end, two 3D solid models were obtained, one for the blood volume and the other for the arterial walls, affected by a saccular aneurysm (Fig. 3.12).

In the construction process of the elastic arterial network affected by aneurysm (Fig. 3.12), the masks generation steps are followed in reverse order when compared to the process presented for obtaining the arterial networks in Figs. 3.9 and 3.11. The starting point is a ready-to-work with 3D mask for the blood volume within the subclavian artery (Fig. 3.11A), on which a set of CAD tools and algorithms are applied to shape it as needed for the numerical studies regarding the hemodynamic of pathologically affected arterial morphologies. The 3D editing of the masks is changing the 2D ROI (Fig. 3.12C). This workflow brings the advantage of a high precision level when choosing the new CAD components position and, also when defining their shape and dimension. On the other side, this process is hardware consuming for a fluent, real-time execution of the desired commands and algorithms.

The 3D arteries reconstructed in this chapter are used as computational domains in the numerical models developed for the study of pulsatile blood flow patterns and impact in both rigid (pathological) and elastic (physiologically normal and aneurysm affected) arterial networks. These morphologies are next FEM discretized and linked to different physical and mathematical models that, once solved, will outline the hemodynamic specificities.

The heart

The cardiovascular system (Mohrman and Heller, 2013) is responsible for the blood oxygenation and pumping in the arterial–venous network. The mechanical (pumping) function of the heart is strongly related to the cardiac electrical activity. The electrical pulses parameters trigger the heart contraction and relaxation that generates the systemic pressure gradients, which makes the blood flow. The understanding and analysis of the mechanisms governing the complex electrical activity of the heart makes the subject of a very important topic. When the hemodynamic parameters begin to rise or fall outside of physiologically normal variation intervals, dysfunctions of the myocardial tissue that generates and diffuses the electrical signals are, usually, present.

The mathematical modeling and numerical simulation of the nonlinear activity of the heart is still a great challenge due to the complex phenomena and the internal structure of the cardiac tissue. The heart contraction is triggered by the electrical pulses generated in the sinoatrial node, which is the most important component of the heart, belonging to the intrinsic electrical conduction system. The cardiac tissue cells, called *nodal cells*, are responsible for the specific automatism of the heart. These make the heart muscle contract even when there are no links to the nervous system, as long as the nodal cells are viable.

The hemodynamic parameters mirror the electrical activity of the heart that generates them. Thus noninvasive investigation and correlation between the electrical and mechanical functions of the heart could be a good practice when setting diagnostics. Over time different methods were developed and optimized, such as the electrocardiography (ECG) (Vijay Raghawa Rao, 2017), which maps the electrical activity of the heart using various electrode configurations placed on the subject's thorax.

Recently the focus is on methods based on the evaluation of passive electrical parameters such as the electrical impedance of the thoracic region, influenced by the electrical conductivity of blood flowing within the aorta (Kubicek et al., 1970; Sherwood et al., 1990; Bernstein and Osypka, 2003; Song et al., 2014). These new noninvasive investigation techniques, based on the analysis of thoracic electrical impedance dynamics, along with the classical methods, such as ECG, provide an in-depth clear picture of the patient's condition.

The development of numerical models that use image-based reconstructed realistic morphologies as computational domains and adequate physical and mathematical models, could represent valuable tools for the analysis, understanding, and improvement of the noninvasive (bio)impedance techniques applied for the assessment of the electrical and mechanical functions of the heart. Furthermore the numerical models are not limited by

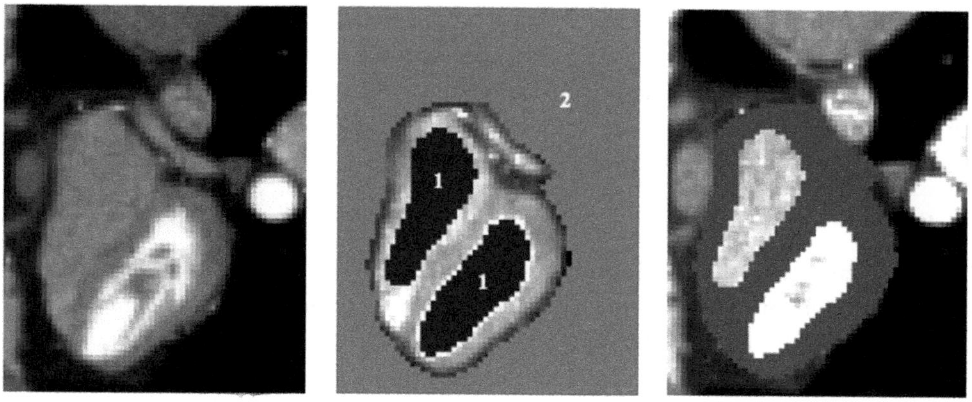

(A) (B) (C)

Figure 3.13 Segmentation of the cardiac tissue (C) using the inverse of the blood mask from atria and ventricles (B, 1) and the tissue mask surrounding the heart (B, 2) generated using an MRI DICOM image set (A).

bioethical reasons, such as the laboratory or clinical tests, or other experimental investigations. The use of anatomical morphologies is a great step forward in the modeling of cardiovascular system-related problems instead of massively idealized geometries, for example, the heart depicted as an ellipsoid, and considerably decreases the deviation errors.

Using a high-resolution MRI DICOM image set acquired from a healthy subject, the segmentation process begins with an appropriate selection of the threshold filter (Vincent, 1993) parameters that should precisely extract the cardiac tissue from the source images. In this case the reconstruction process is more laborious than the one presented for the generation of blood, vessels, and surrounding tissue masks due to the nonhomogeneous structure of the cardiac tissue.

The direct segmentation of the cardiac tissue is a complex process because of the internal nonhomogeneous structure of the heart. This would result in masks with abundant reconstruction artifacts, difficult to remove and maintain at the same time the original (real) morphology of the heart tissue. Thus a cycle of Boolean operations(Masuda, 1993; Requicha and Voelcker, 1985) was applied to masks corresponding to other ROIs. First, the threshold algorithm was configured to recognize and isolate the grayscale values specific to the blood in the atria and ventricles and to the heart's embedding tissues (Fig. 3.13). Then the blood mask (Fig. 3.13B, 1) was inverted generating its negative. The mask thus obtained was subtracted from the tissue mask (Fig. 3.13B, 2) surrounding the heart giving the first raw version of the cardiac tissue mask (Fig. 3.13C). This approach of the image-based reconstruction process of the heart eliminates the massive postprocessing steps that should have been followed in the case of a direct reconstruction and preserves the original morphology imposed by the investigated subject's anatomy.

A one-time use of the Gaussian smoothing algorithm (Taubin, 1995) along with the flood-filling filters upon the initial mask (Fig. 3.14A and B) is enough to remove the small imperfections, to guarantee spatial continuity and to obtain the desired post-processed final solid model presented in Fig. 3.14C and D. Once FEM discretized, this geometry becomes the computational domain for action potential propagation numerical studies, presented in Chapter 4, Electrical Activity of The Heart.

A vertebral column segment

Although CAD-built computational domains may offer the needed, accurate support for numerical simulations that are aimed to unveil the outlining physics that govern a certain application, however, when patient-related specificities are at a prime, then, provided they are available, image-based computational domains have to be utilized. For instance, in the analysis of the magnetic stimulation of the spinal cord [e.g., in the treatment of the spasticity (Korzhova et al., 2018)], starting with a set of images presenting CT scans in sagittal, axial, and coronal planes, CT scans (Whole Spine (Cervical—Dorsal—Lumbar—Sacral) CT image data set, 2018) a 3D model is built for the L1—L5 lumbar segment, as presented next (Baerov, 2019). Fig. 3.15 shows the data set as presented by Slicer (2018) and the (re)construction produced for medical purposes—the three main views.

The path to build a solid model usable in an FEM modeler, as part of the computational domain, for each of the L1—L5 lumbar vertebrae, is as follows. First, the whole column is visualized, and the ROI containing each of the vertebrae is defined. Next a *crop* function is used to visualize singled out vertebrae. Manual segmentation is performed using the following: (1) *threshold* tools (to select bone); (2) *save island effect* (to separate the vertebra from other entities); (3) *paint effect* (the parts that do not belong to the vertebra are erased, and the vertebra is colored to identify its tissue); and (4) *split, merge*, and *build* (to create the desired 3D volume of the vertebra). Each model is saved in the STL[1] format introduced by 3D Systems (1988, 1989, 1994). Fig. 3.16 presents the stages of construction.

Using the same path, the intervertebral discs and the spinal marrow volumes are constructed (Fig. 3.17).

Next the STL objects are discretized using the following sequence and tools (MeshLab, 2018): (1) *filtering* using RSR (*remeshing*, *simplification*, and *reconstruction*); (2) *reduction in the complexity* of the 3D model using QECD (*quadric edge collapse decimation*); (3) *filtering* using RSR; and (4) "polyface" mesh *boundaries closure and smoothing* using SPSR (*screened poisson surface reconstruction*). The resulting models (Fig. 3.18) are saved in 3DS format.

[1] The STL acronym is related to "standard triangle language," "stereolithography language," and "stereolithography tesselation language," introduced by 3D Solutions (3D Systems, Inc., 1988, 1989, 1994; Grimm, 2004).

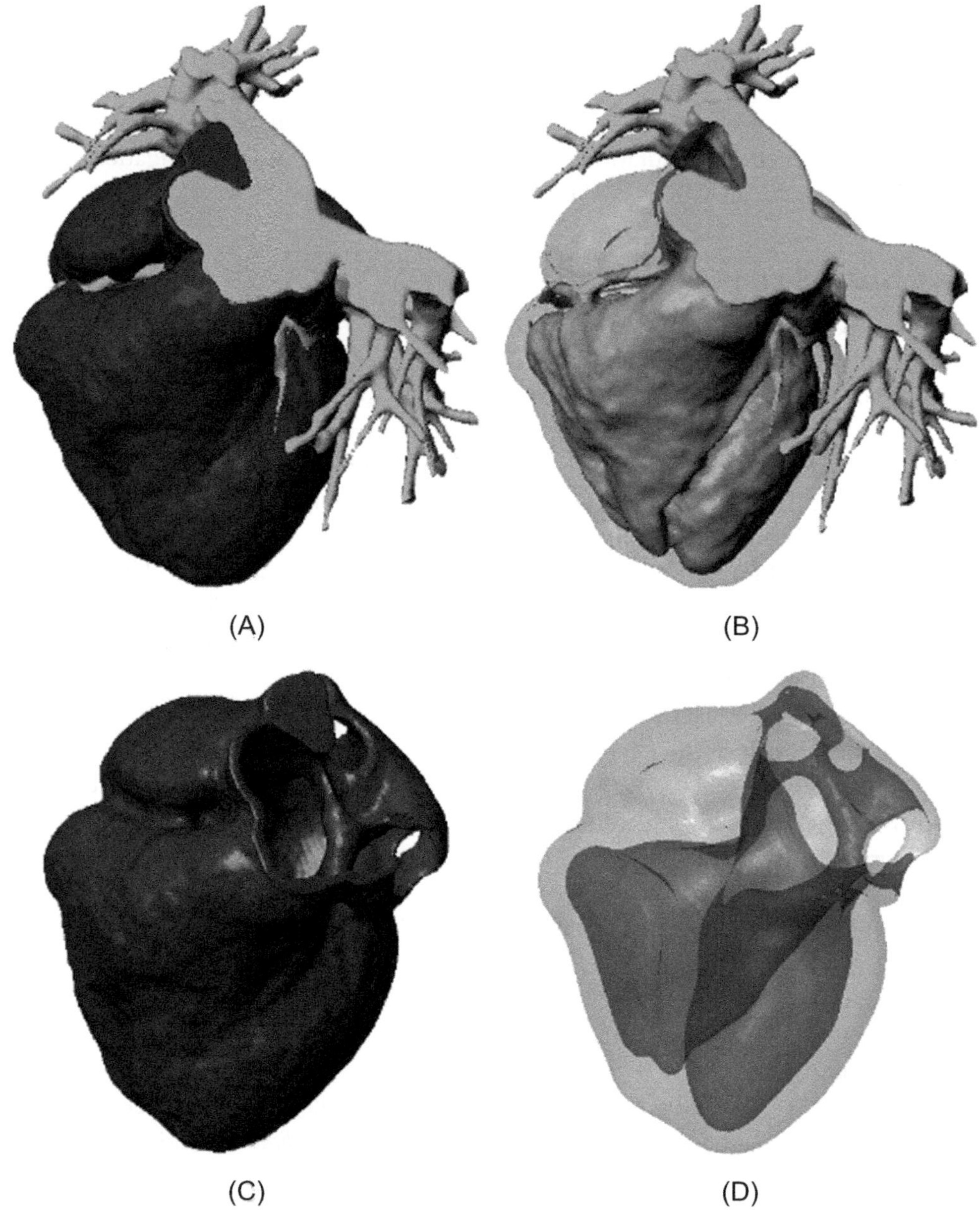

Figure 3.14 The 3D solid model of the heart filled with blood (A, opaque, and B, transparent) and empty (C, opaque and D, transparent). *After Dobre, A. (2012).*

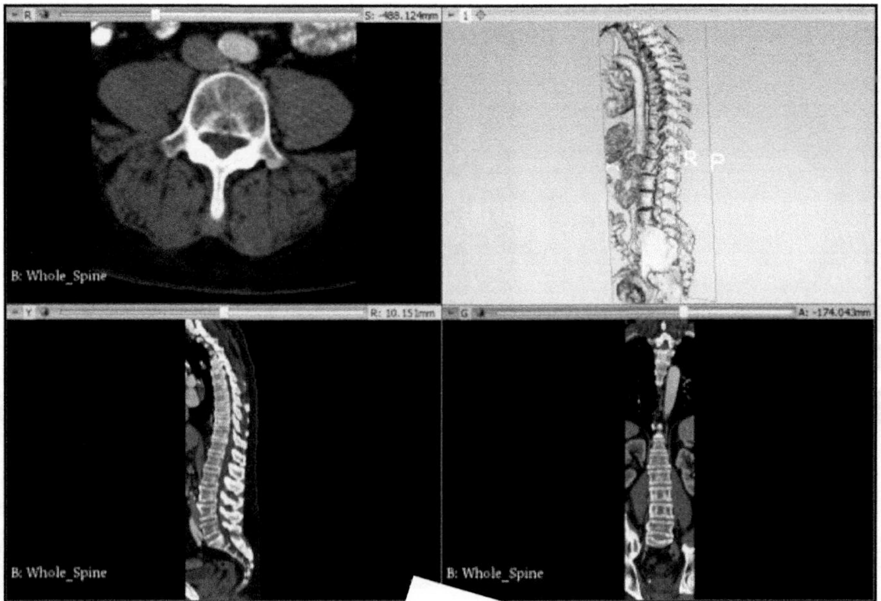

Figure 3.15 The vertebral column, input da [illegible] ise, from the top, left image: axial plane; sagittal plane; coronal, and the 3D mo [illegible]

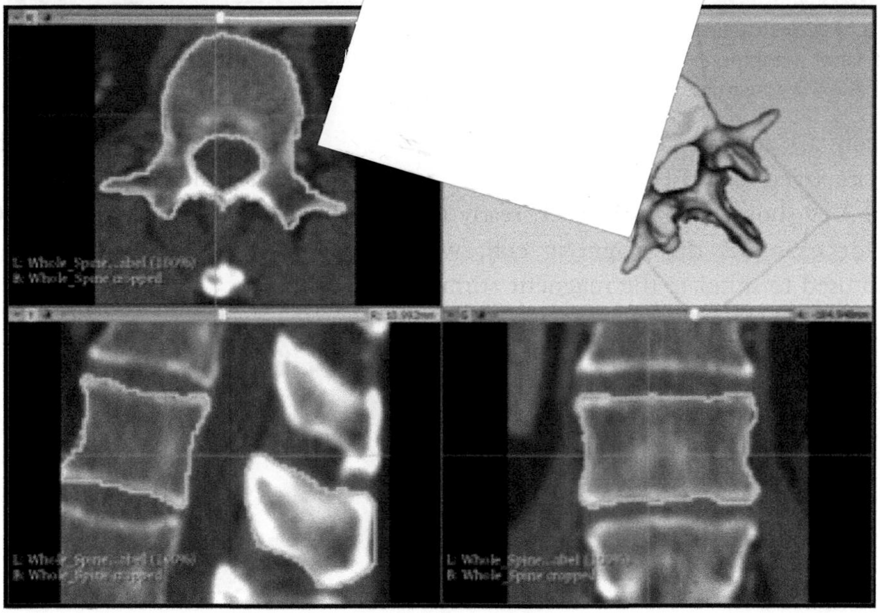

Figure 3.16 Construction of the STL models for the L1 vertebra—counter-clockwise, from the top, left image: label mapping in the axial plane; label mapping in the sagittal plane; label mapping in the coronal; the 3D construct usable for numerical modeling.

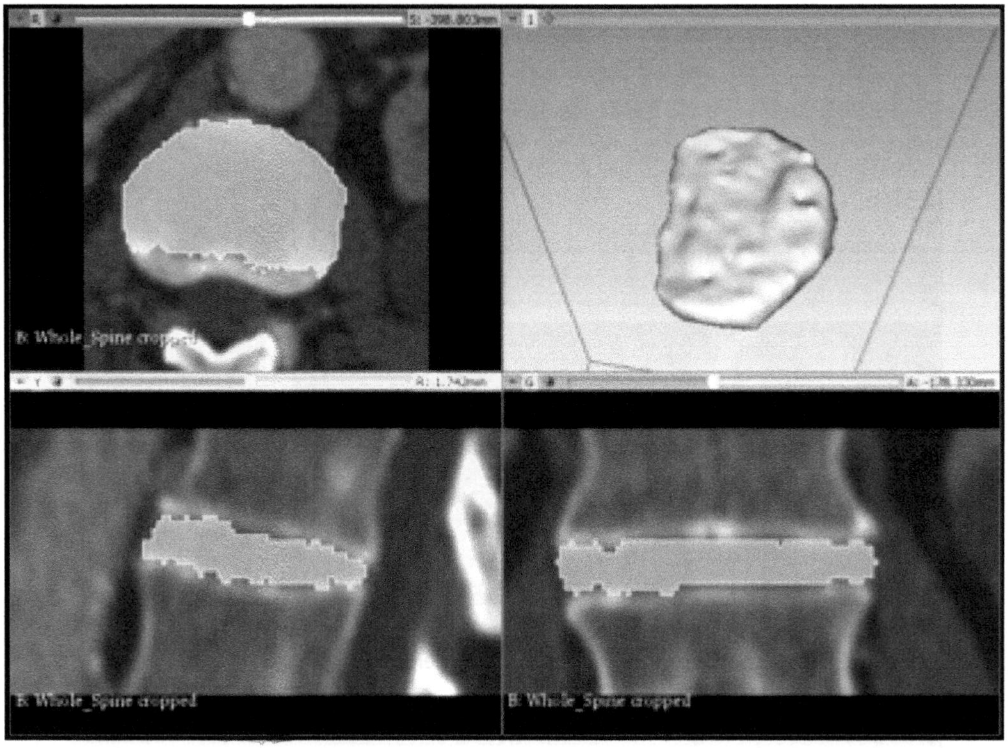

Figure 3.17 Construction of the STL models for the disc between L1 and L2—counter clock, from top/left: label mapping in the axial plane; label mapping in the sagittal plane; label mapping in the coronal; the 3D construct usable for numerical modeling.

Finally a CAD tool may be used to convert the polyface objects in solid objects, which are saved in an FEM-modeler compatible format, for example, IGES.

Fig. 3.19 shows the final object, ready to be imported in an FEM-modeler. Other parts, for example, the magnetic coil, will be added to build the computational domain used to simulate the magnetic stimulation.

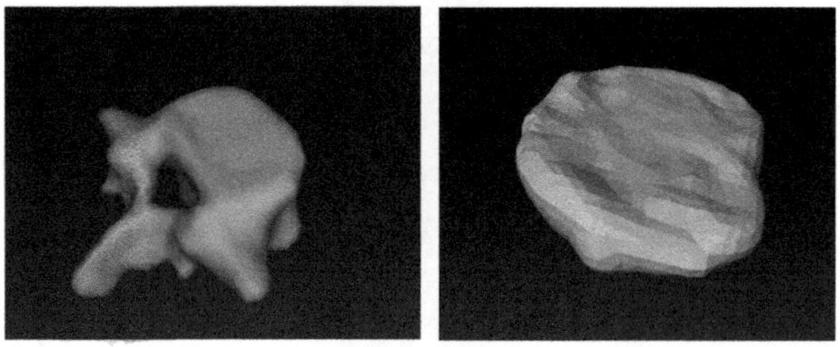

(A) (B)

Figure 3.18 L5 vertebra (A) and the L3–L4 disc (B) processed in Meshlab.

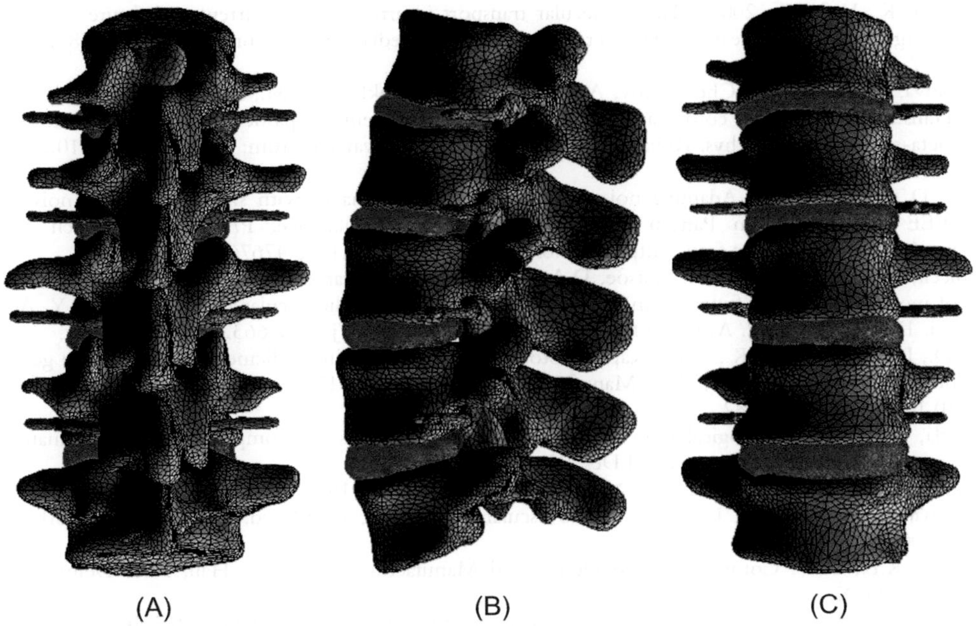

Figure 3.19 The computational domain of the lumbar column: (A) coronal view, from the front; (B) sagittal view, from the right, and (C) coronal view, from the back.

References

Baerov, R., 2019. Magnetic Stimulation of the Lumbar Column (B.S. thesis). Faculty of Medical Engineering, University "Politehnica" of Bucharest, Bucharest, Romania (in progress).

Bernstein, D.P., Osypka, M.J., 2003. Apparatus and Method for Determining an Approximation of Stroke Volume and Cardiac Output of the Heart, US Patent No. 6.511.438.

Broboană, D., Bălan, C., Morega, A.M., Gheorghe, C., Iacob, R., 2008. Modeling the vortical structures in hemodynamics of small vessels and capillaries. In: 15th International Congress on Rheology, 80th Annual Meeting of the Society-of-Rheology, Monterey, CA, 3–8 August 2008.

Chris, M.W.H., Garland, R.M., 1990. Cavity search: an algorithm for the isolation and display of cavity-like binding regions. J. Comput. Aided Mol. Des. 4 (4), 337–354. Available from: https://doi.org/10.1007/BF00117400.

Davies, P.F., 1995. Flow-mediated endothelial mechano–transduction. Physiol. Rev. 75, 519–560.

Dobre, A., 2012. Investigation Methods for the Analysis of Coupled Phenomena Specific to the Medical Engineering Field (Doctoral thesis). Faculty of Electrical Engineering, University Politehnica of Bucharest, Bucharest, Romania.

Fathi, M., Hiltner, J., 1999. A new fuzzy based flood-fill algorithm for 3D NMR brain segmentation. In: IEEE SMC'99 Conference Proceedings, IEEE International Conference on Systems, Man, and Cybernetics, 12–15 October 1999. ISBN 0–7803-5731-0. Available from: https://doi.org/10.1109/ICSMC.1999.812525.

Grimm, T., 2004. User's Guide to Rapid Prototyping. Society of Manufacturing Engineers p. 55., ISBN: 0-87263-697-6.

Isselbacher, E.M., 2007. Diseases of the aorta, Cecil Medicine, 23rd ed. Saunders Elsevier, Philadelphia.

Jan, J., 2006. Medical Image Processing, Reconstruction and Restoration. CRC Press. ISBN: 978-0-8247-5849-3.

Khanafer, K., Vafai, K., 2008. Macromolecular transport in arterial walls: current and future directions, Emerging Topics in Heat and Mass Transfer in Porous Media, vol. 22. Springer, ISBN: 978-1-4020-8177-4, pp. 219–235.

Korzhova, J., Sinitsyn, D., Chervyakov, A., Poydasheva, A., Zakharova, M., Suponeva, N., et al., 2018. Transcranial and spinal cord magnetic stimulation in treatment of spasticity: a literature review and meta-analysis. Eur. J. Phys. Rehabil. Med. 54 (1), 75–84. Available from: https://doi.org/10.23736/S1973-9087.16.04433-6.

Kuan, D.T., et al., 1985. Adaptive noise smoothing filter for images with signal-dependent noise. In: IEEE Transactions on Pattern Analysis and Machine Intelligence, PAMI–7, 2 March 1985, pp. 165–177. Available from: https://doi.org/10.1109/TPAMI.1985.4767641.

Kubicek, W.G., Patterson, R.P., Witsoe, D.A., 1970. Impedance cardiography as a noninvasive method of monitoring cardiac function and other parameters of the cardiovascular system. Ann. N. Y. Acad. Sci. 170 (2), 724–732. Available from: https://doi.org/10.1111/j.1749-6632.1970.tb17735.x.

Ma, D., Lin, F., Chua, C.K., 2011. Rapid prototyping applications in medicine. Part 2: STL file generation and case studies. Int. J. Adv. Manuf. Technol. 18 (2), 118–127. Available from: https://doi.org/10.1007/s001700170082.

Masuda, H., 1993. Topological operators and Boolean operations for complex–based non–manifold geometric objects. Comput. Aided Des. 25 (2), 119–129.

MeshLab. <http://www.meshlab.net/> (accessed in December 2018).

Mohrman, D.E., Heller, L.J., 2013. Cardiovascular Physiology, eighth ed. McGraw Hill. ISBN-13: 9780071793117.

Narayan, K.L., 2008. Computer Aided Design and Manufacturing. Prentice Hall, New Delhi. ISBN: 812033342X.

Newell, N., Little, J.P., Christou, A., Adams, M.A., Adam, C.J., Masouros, S.D., 2017. Biomechanics of the human intervertebral disc: a review of testing techniques and results. J. Mech. Behav. Biomed. Mater. 69, 420–434. Available from: https://doi.org/10.1016/j.jmbbm.2017.01.037.

Novanta, A., et al., 2018. Chart for renal tumor microwave ablation from human study. Diagn. Intervent. Imaging 99, 609–614. Available from: https://doi.org/10.1016/j.diii.2018.05.005.

Pianykh, O.S., 2012. Digital Imaging and Communications in Medicine (DICOM): A Practical Introduction and Survival Guide, second ed. Springer. ISBN: 978-3-642-10849-5.

Pyörälä, K., et al., 1994. Prevention of coronary heart disease in clinical practice: recommendations of the task force of the European Society of cardiology, European atherosclerosis Society and European society of Hypertension. Atherosclerosis 110 (2), 121–161. Available from: https://doi.org/10.1016/0021-9150(94)90200-3.

Requicha, A.A.G., Voelcker, H.B., 1985. Boolean operations in solid modeling: boundary evaluation and merging algorithms. Proc. IEEE 73 (1), 30–44. Available from: https://doi.org/10.1109/PROC.1985.13108.

Roushdy, M., 2006. Comparative study of edge detection algorithms applying on the grayscale noisy image using morphological filter. GVIP J. 6 (4).

Rudin, L.I., Osher, S., Fatemi, E., 1992. Nonlinear total variation based noise removal algorithms. Phys. D: Nonlinear Phenom. 60 (14), 259–268. Available from: https://doi.org/10.1016/0167-2789(92)90242-F.

Sangiorgi, G., et al., 1998. Arterial calcification and not lumen stenosis is highly correlated with atherosclerotic plaque burden in humans: a histologic study of 723 coronary artery segments using nondecalcifying methodology. J. Am. Coll. Cardiol. 31 (1). Available from: https://doi.org/10.1016/S0735-1097(97)00443-9.

Sherwood, A., et al., 1990. Methodological guidelines for impedance cardiography. Psychophysiology 27 (1), 1–23. Available from: https://doi.org/10.1111/j.1469-8986.1990.tb02171.x.

SLC File Specification, 1994. 3D Systems, Inc.

Slicer v.4.10, 2018. <https://www.slicer.org>.

Song, R., Rich, W., Kim, J.H., Finer, N.N., Katheria, A.C., 2014. The use of electrical cardiometry for continuous cardiac output monitoring in preterm neonates: a validation study. Am. J. Perinatol. 31 (12), 1105–1110. Available from: https://doi.org/10.1055/s-0034-1371707.

StereoLithography Interface Specification. 3D Systems, Inc., July 1988, October 1989, <https://www.loc.gov/preservation/digital/formats/fdd/fdd000504.shtml>.

Sun, W., Starly, B., Nam, J., Darling, A., 2005. Bio-CAD modeling and its applications in computer-aided tissue engineering. Comput. Aided Des. 37 (11), 1097−1114. Available from: https://doi.org/10.1016/j.cad.2005.02.002.

Taubin, G., 1995. Curve and surface smoothing without shrinkage. In: Proceedings of IEEE International Conference on Computer Vision, 20−23 June 1995, Cambridge, MA. ISBN 0−8186-7042-8. Available from: https://doi.org/10.1109/ICCV.1995.466848.

Thiyagarajah, A.R., et al., 2018. Intradiscal electrothermal therapy. Drugs & Diseases, Clinical Procedures. WebMD LLC, Medscape (updated March 2018).

Tsou, H.K., et al., 2010. Intradiscal electrothermal therapy in the treatment of chronic low back pain: experience with 93 patients. Surg. Neurol. Int. 1, 37. Available from: https://doi.org/10.4103/2152-7806.67107.

van Sonnenberg, E., McMullen, W., Solbiati, L. (Eds.), 2010. Tumor Ablation: Principles and Practice. Springer, ISBN 978-1-4419-3046-0.

Vijay Raghawa Rao, B.N., 2017. Clinical Examination in Cardiology, second ed. Elsevier. ISBN: 9788131249246.

Vincent, L., 1993. Morphological grayscale reconstruction in image analysis: applications and efficient algorithms. IEEE Trans. Image Process. 2 (2), 176−201. Available from: https://doi.org/10.1109/83.217222.

Whole Spine (Cervical−Dorsal−Lumbar−Sacral) CT image data set, 2018. <https://www.embodi3d.com/files/file/11735-whole-spine-cervical-dorsal-lumbar-sacral-ct-scan/> (retrieved in December 2018).

Zaman, A.G., et al., 2000. The role of plaque rupture and thrombosis in coronary artery disease. Atherosclerosis 149 (2), 251−266. Available from: https://doi.org/10.1016/S0021-9150(99)00479-7.

Zeng, D., Zhaohua, D., Morton, H., Friedman, M.H., Ross, C., 2003. Effects of cardiac motion on right coronary artery hemodynamics. Ann. Biomed. Eng. 31, 420−429.

CHAPTER 4

Electrical activity of the heart

4.1 Introduction

Doubtless, the cardiovascular system is one of the most complex structures in the human body responsible for vital functions. Current research interest in elucidating its mechanisms, not yet fully understood, is increasing due to the associated conditions, which are often deadly such that the cardiovascular diseases are considered the main cause of death worldwide (Thomas et al., 2004; Yancy, 2000; World Health Organization, 2017). Numerous models for studying this topic can be found in literature, treating various branches of the main cardiovascular challenges. An extensive review to the direct (forward, from cell to body) and inverse electrocardiography problems is presented by Pullan et al. (2005), who provides a glimpse in the mathematics and associated computational methods used to numerically simulate and interpret the heart's physiological function. Same topic is also detailed in the studies by Boulakia et al. (2008) and Szilagyi et al. (2002), which give numerical results on the parameters estimation, and later on, Section 4.2 provides a brief discussion on the direct and inverse electrocardiogram (ECG) problems.

Starting from the cell, the nonlinear dynamics of the ionic channels and the ionic currents driven by concentration gradients between the inner and outer cell media is described by mathematical models of the heart's pacemaker activity (Di Francesco and Noble, 1985), which may be solved numerically for the nonlinear electrical activity of the heart (Murillo and Cai, 2004).

In the beginning, the physics of the heart's sinoatrial (SA) node, the natural pacemaker, was associated with a van der Pol oscillator (Van der Pol and Van der Mark, 1928). Later on, this assumption still remains valid and is considered close enough to the real behavior of the SA node to keep developing new models (Sato et al., 1994; Zduniak et al., 2014). The SA node activity is related also with more elaborated oscillator models such as FitzHugh–Nagumo, for the normal activity of the heart, Nagumo (1962), and Landau–Ginzburg, for different pathologies and abnormal regimes (e.g., arrhythmias and fibrillations; Gong and Christini, 2004; Takembo et al., 2019). These are used to reproduce the reentrant or spiral waves that mimic the action potential (AP) propagation throughout the myocardium. The physical modeling and numerical simulation of the heart still represents a great challenge due to the complex phenomena underlying the myocardial tissue electrophysiology. The heart muscle contractions are generated through the SA pulses. The nodal cells that make the

Computational Modeling in Biomedical Engineering and Medical Physics
DOI: https://doi.org/10.1016/B978-0-12-817897-3.00004-X

myocardial tissue are responsible for the heart's fascinating automatism, which keeps the myocardium contracting as long as the nodal cells are still alive, despite the lack of any physical links with the nervous system (Royston Maloeuf, 1935; Valenzuela and Kabela, 1984). Several aspects of the nonlinear dynamics of the heart excitable tissue are touched in Section 4.3.

While the cardiovascular system is responsible for pumping blood through arteries and veins, ensuring proper oxygenation of the whole body, proper functioning of the heart, responsible for this vital function, is strongly related to the electric activity of the myocardium. Cardiac electric signal properties spark off the mechanical activity (myocardium contraction and relaxation) that produces the pressure gradients responsible for the blood flow. Showing the complex link between the electric and mechanical activities of the heart is crucial for deeply understanding its function (Capasso et al., 1983; Tournoux et al., 2007). Pathological variations of hemodynamic parameters are usually correlated with myocardial tissue dysfunctions that alter normal, physiological pathways of the electric signal propagation. Their values are a consequence of the electric activity of the heart that triggered them. Thus the noninvasive investigation of the coupled electric (ECG) and mechanical (ICG, ECM, TEB, Chapter 5: Bioimpedance Methods) activities of the heart is a way of good practice when evaluating a patient. Applanation tonometry (AT), discussed later in this chapter (Sections 4.5 and 4.6), may provide valuable information and insights concerning the pressure pulse wave, its direct and reverse components.

Physical modeling of the heart for numerical simulation still represents a great challenge due to the complex phenomena underlying the myocardial tissue electrophysiology. The heart muscle contractions are generated through a complex succession of sequences, originating with the SA electric pulses generated spontaneously. The nodal cells that comprise the myocardial tissue are responsible for the heart's impressive automatism, which keeps the myocardium contracting as long as the nodal cells are still alive (Royston Maloeuf, 1935; Valenzuela and Kabela, 1984), while the rhythm itself is under nervous control. SA autorhythmic cells depolarize spontaneously and generate peaks of their transmembrane voltage, called low action potentials (low AP).

The numerical models based on anatomically accurate computational domains, generated using medical image construction, combined with the appropriate mathematical models can become essential tools for the analysis, understanding and improvement of the cardiovascular system and its functionality. The advantage package that they carry through numerical simulation and image based construction will always include the lack of bioethical problems and restrictions, the safe, unharmful, clean, simple, usually cheap, mode of experimenting, testing, and optimization. And this concurs with European Union Directive 86/609/EEC Art. 7, Section 2 of November 24, 1986, concerning the usage of animals in experiments or other scientific purposes: "an experiment shall not be performed if another scientifically satisfactory method of obtaining the result sought, not entailing the use of an animal, is reasonably and practicably available."

Electrophysiology insights

The heart has the role of an electric generator and circuit for the transmission of cyclical voltage pulses, as local cell transmembrane voltage alternates its state between resting and active or polarized and depolarized conditions. The muscular tissue of the atria and ventricles, the conductive tissue (Hiss branches and the Purkinje network) and the specialized pacemaker cells [SA and atrioventricular (AV) nodes] are its main parts. Electrical depolarization waves initiated at the SA node diffuse through the conductive cellular network up to the muscular cells in a remarkably orderly manner and drive the cyclical contraction—relaxation of the myocardium, ensuring the blood circulation through the body.

In the cyclic functioning of a healthy heart two successive sequences are distinguishable: (1) the *systole*, or active phase, which consists in the atria contraction followed by the ventricles contraction, while the blood is forced through the atrioventricular route and circulatory pathways, and (2) the *diastole*, or passive phase, during which the myocardium relaxes and the four heart chambers are filled with blood. This happens under the electrical signals that are generated and spread throughout a complex and redundant conduction system, made as shown by Fig. 4.1 of the SA node [located between the superior vena cava (SVC) and the right atrium (RA)], the internodal atrial pathways, the AV node (located in the right posterior of the interatrial septum), and the electroconductive tissue (Ganong, 2005; Keener and Sneyd, 2009; Katz, 2011).

SA pacemaker cells cover an oval area of approximately 15-mm long and 5-mm wide, at the junction of the SVC with the RA. They are controlled by the nervous system but have the properties of automatism and rhythmicity needed to generate depolarization pulses, by self-excitation, and maintain their tact to initiate each cycle

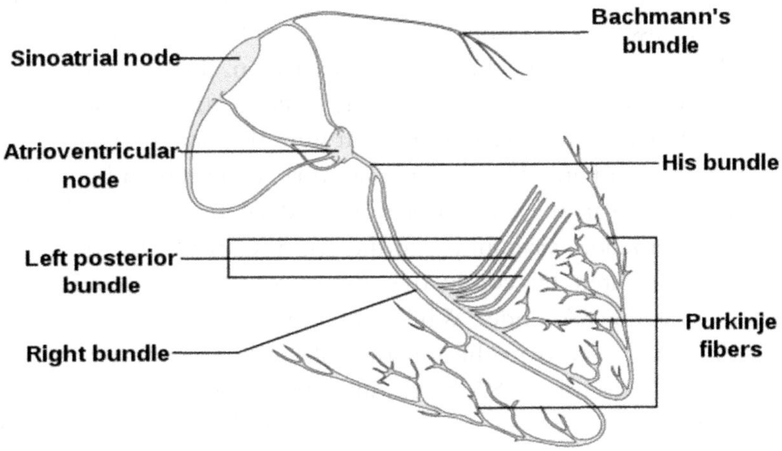

Figure 4.1 Conduction system of the heart. <*https://commons.wikimedia.org/w/index.php?curid=29922595*>.

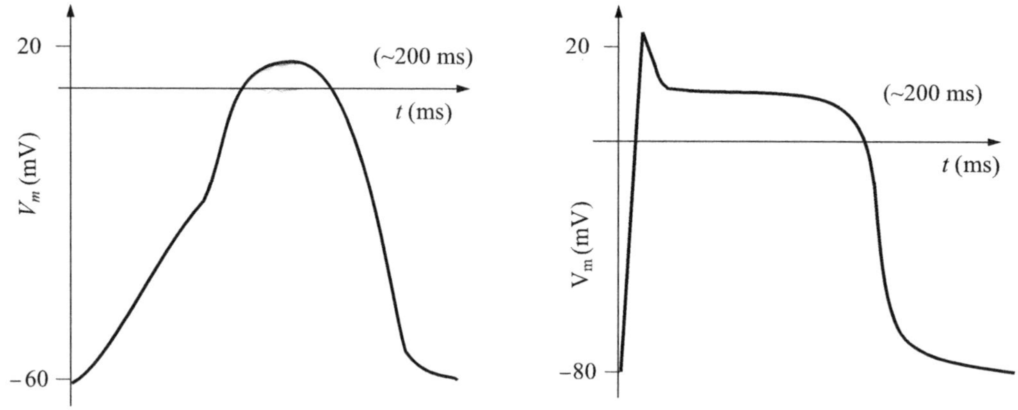

Figure 4.2 Slow AP (left), for the authorythmic cells, and fast AP (right), for the myocytes.

of depolarization of the heart with a frequency of 60–70 cycles per minute in resting state, or higher than 90 cycles per minute to exertion or state of emotion. The AP waveform, the "slow action potential" (slow AP), is as shown in Fig. 4.2, left. The electric wave generated by the SA node is transmitted through electrotonic conduction in the entire RA area and then through the interatrial beam to the left atrium contractile volumes. Depolarization of the entire atrial area occurs in about 90 ms, at ~1 m/s (Morega, 1999).

SA discharge makes the electrical signal pass through the atria pathways, AV node, Hiss bundle and the Purkinje network. This process triggers the normal heartbeat, the pulse. The radial SA node depolarization takes approximately 100 ms and heads to the AV node, which, due to low electrical conductivity, it traverses slower (0.02–0.05 m/s in ~75 ms) than the conductive tissue of the trunk and branches of Hiss (1–1.5 m/s), introducing a 100 ms delay (AV nodal delay) in the ventricles excitation. This time delay is controlled through the sympathetic nerves, the "shortening" effect, and the *vagi* stimulation, the "lengthening" effect. The wave travels in the heart from the base to the apex and from the endocardium to the epicardium. The Purkinje network, characterized by the best electrical conductivity values found in the entire electrical conduction system of the heart, rapidly spreads out the depolarization wave further to the ventricles in 80–100 ms.

The impulses generated by conductive tissues are of lower amplitude and frequency than that of the SA node. While not important in the functioning of the healthy heart, they may take over the function of local tact generator when the heart conduction pathway is disturbed (SA block, interatrial block, AV block, intraventricular block). This process affects the rhythmicity of the heart by desynchronizing the atrial contractions from the ventricular ones. Thus the AV node generates depolarization pulses with a frequency of 40–60 cycles per minute, while the Purkinje network produces pulses at the rate of 25–40 cycles per minute.

Myocardium cells (30–100 μm in length, 8–20 μm in diameter) differ significantly from nerve cells or striated muscle tissue cells. They are tightly packed and their cell-to-cell connection is intimate: the ionic channels continue from the membrane of a cell to the membrane of the neighboring cell, and the ions flow directly in between without reaching the extracellular space. Depolarization diffuses from cell to cell through the myocardium, producing peaks of transmembrane voltage, called the fast action potential (fast AP) (Fig. 4.2, right). Several types of sources that model the AP are discussed later (Morega, 1999).

The AP presents five stages: Phase 0: Na^+ channels are activated, the membrane depolarizes fast, the AP increases abruptly. Phase 1: a flux of K^+ ions exit the cell, accompanied by a temporarily decrease of AP and the Ca^{2+} channels begin to open. Phase 2: the balance between the inward Ca^{2+} and outward K^+ fluxes corresponds to the AP plateau. Phase 3: at the end of Phase 2, the interior of the cell becomes highly electropositive, and the potential approaches the electrochemical equilibrium for Na (+61 mV). Phase 4: K^+ ions flux exceeds the Ca^{2+} ions flux, accompanied by repolarization. Phase 5: return to initial state (resting potential), due mainly to the K^+ ions flux (Morega, 1999).

The electrical activity of the heart is monitored and measured using the electrocardiogram (ECG) recording (Ettinger et al., 1974; Luo and Johnston, 2010), which provides information related to the AP dynamics during the cardiac cycles. The body tissues have good electrical conductivity (the body is commonly considered a volume conductor under low frequency electric field stress; Woosley et al., 1985; Plonsey, 1963) and the electrical signal variations recorded on its surface are given by the projection of the APs of the myocardial fibers. ECG signals observe the course of the depolarization wave and they give good information to identify cardiac arrhythmias (Einthoven, 1908). In Einthoven's electrocardiographic model, the cardiac source is a current dipole with its origin inside the heart (Section 4.1), and variable orientation and magnitude (Fig. 4.3). Usually, the electrodes describing the "Einthoven triangle" are positioned on the right shoulder (UD), left (US), and abdomen (A) (Malmivuo and Plonsey, 1995; Berne et al., 1998). However, taking advantage of the good electric conductivity of the body (the volume conductor concept), it is customary to adopt one more convenient electrodes positioning system: on the wrists of the right and left hands, and on the left leg, respectively.

The morphology of the ECG signal is related to depolarization wave propagation. *P wave* is related to the RA depolarization. The pulse lasts 70–80 ms, and then the ECG signal returns to the reference level. *QRS complex* is related to the ventricular depolarization. It starts ~220 ms from the beginning of the cardiac cycle and it lasts ~80 ms, when the ECG signal returns to the reference level for ~120 ms. *T wave*, with a duration of ~200 ms, corresponds to the ventricular repolarization. *U wave*, not always present, corresponds to the repolarization of a certain type of ventricular cells with slow electrical activity, but their connection with the electrophysiological phenomena of the heart is not yet precisely established.

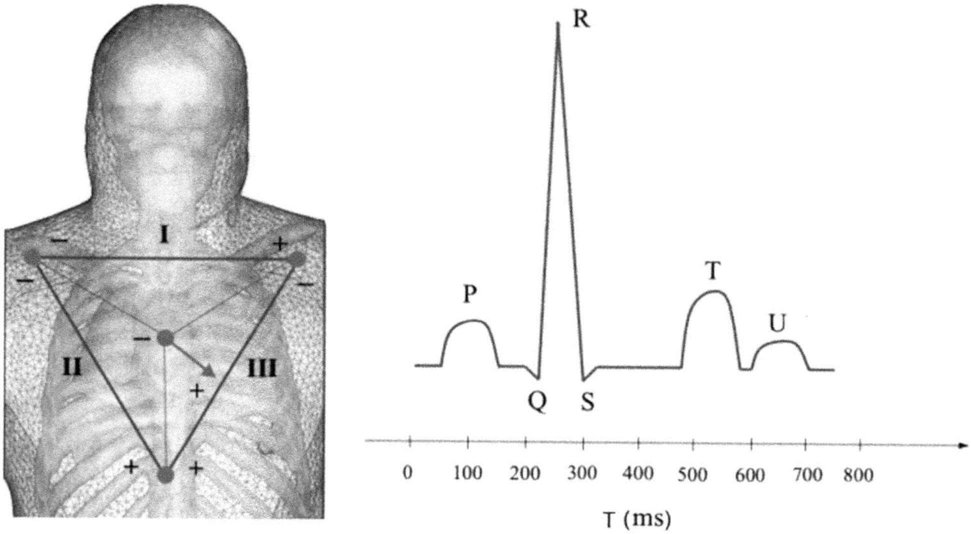

Figure 4.3 Eindhoven triangle (left) and electrocardiogram signal (right).

Bioelectric sources. The direct ECG problem

As stated by experimental results (Schwan and Kay, 1957), in the frequency range of bioelectric events, the capacitive and inductive properties of tissues are negligible as compared to their resistive properties, therefore the human body is modeled here as a continuous linear, homogeneous (by parts) medium. The bioelectrical activity of heart cells is the result of a conversion of chemical energy into electrical energy, $\mathbf{E}_i$. The electric current density is then $\mathbf{J} = \sigma(\mathbf{E}_i + \mathbf{E})$, where σ is the electrical conductivity (Chapter 1: Physical, Mathematical and Numerical Modeling). In quasistationary conditions, the electric field produced by a current dipole $\mathbf{p} = \lim_{I\to\infty, l\to 0} I\mathbf{l}$, where I is the electric current intensity and $\mathbf{l}$ is the oriented dipole length. Several representative dipole electric field sources are found in the studies by Malmivuo and Plonsey (1995) and Morega (1999).

The electric potential, V, produced by dipolar sources in a nonhomogeneous volume conductor, is the analytic solution to *the direct ECG problem* (D-ECG) (Gesselowitz, 1967)

$$\nabla\cdot(\sigma\nabla V) = \nabla\cdot\mathbf{J} = \mathbf{p}, \tag{4.1}$$

$$4\pi\,\sigma\, V(\mathbf{r}) = \underbrace{\int_v \mathbf{J}_i\cdot\nabla\left(\frac{1}{r}\right)dv}_{\text{volume source (primary source)}} + \sum_j \underbrace{\int_{S_j}\left(\sigma''_j - \sigma'_j\right)V\nabla\left(\frac{1}{r}\right)\cdot d\mathbf{S}_j}_{\text{double layer sources (secondary sources)}}, \tag{4.2}$$

where S_j are the interfaces between the homogeneous regions. The sources, conveniently placed inside the heart volume (e.g., in the myocardium), may be utilized to generate the electric potential on the epicardium, which is actually mapped on the thorax surface, accessible for voltage measurements, provided there is no epicardial electrical current outflow to the thorax.

4.2 Coupled direct and inverse ECG problems for electrical imaging

The *inverse ECG problem*, I-ECG, (or electrocardiographic imaging) aims to find the sources on the epicardium (electric potential) from voltage measurements on the chest surface. Clinically, I-ECG is important because it helps identify cardiac arrhythmias. The measurement data on the chest used as input are actually filtered images of the epicardial potential, due to the smoothing and attenuating properties of the thorax volume between the source, on epicardium, and the observation surface, the thorax outline.

As with all ill-posed problems (Chapter 1: Physical, Mathematical and Numerical Modeling), the difficulty of recovering the source here has to be alleviated by using some supplementary, consistent information on the source. Therefore a companion D-ECG problem has to be formulated and solved in the first place. The D-ECG here consists in finding the electric field inside the volume conductor of the thorax—the volume conductor between the epicardium (inner boundary) and the thorax surface (outer boundary), with known geometry and material properties—when the field sources, on the epicardium are known. The boundary, initial, and interface conditions are $V|_{\text{epicardium}} = 0, J_n|_{\text{thorax}} = 0$, $V' = V''$, $(\sigma' \nabla V)\cdot\mathbf{n}' - (\sigma'' \nabla V)\cdot\mathbf{n}'' = 0$, respectively, and there are no internal sources, $\mathbf{J}_i = 0$. Its solution is found usually using numerical methods such as the finite element method or the boundary element method (Mocanu, 2002). The analytic solution of D-ECG problem is then a Fredholm integral of first type (Yamashita, 1982)

$$V(P) = \int_{S_H} K(P, Q) V(Q) \, dS_H, \quad P \in S_T, \quad Q \in S_H, \tag{4.3}$$

where S_H is the epicardium, S_T is the thorax surface, and V is the electric potential. The cardiac sources are given through the epicardial electric potential. The kernel $K(P,Q)$ is a compact operator of L^2 class. Given $K(P,Q)$ (e.g., by the D-ECG) and the function $V(P)$, the problem is to find the function $V(Q)$.

The kernel may be seen as the normal component of the current density at a point $Q \in S_H$, produced by a unit current source at a point $P \in S_T$, when the epicardium is of zero potential. Another interpretation is that $K(P,Q)$ is the potential at $P \in S_T$ produced by a unit potential at the $Q \in S_H$ point, while the rest of the epicardium is of zero potential.

The stem of the I-ECG problem here is the transfer equation

$$\mathbf{b} = \mathbf{C}\mathbf{x}, \tag{4.4}$$

which relates the potential on the epicardium (the source) to the potential on the thorax surface (measurement, calculated potential). The outcome of the companion D-ECG is the availability (construction) of the transfer matrix, **C**, which maps the electrical potential on the epicardium, **x** ($m \times 1$) vector of the potentials, onto the electrical potential on the thorax, **b** ($n \times 1$) vector of the potentials on the chest. The D-ECG is solved numerically, by successively assigning a unit potential to each of the m nodes on the epicardium, while the rest of the nodes are of zero potential, and the chest surface is electrically insulated (Mocanu, 2002). Each column of **C** corresponds to the n potentials on the thorax calculated for a single-activated node (on the epicardium). The selection of the n nodes on the epicardium (electrodes for measuring the potential during open heart surgery) and the m on the thorax (electrodes used to measure the potential) renders a collocation approach to the solution.

Then the potential on the epicardium, **x**, is calculated by inverting **C**. Tikhonov regularization (with l_2 constraint on the energy, gradient, or laplacian of the solution) was often used to stabilize the inverse solution (Iakovidis and Martin, 1991; Velipasaoglu et al., 2000; Johnston et al., 1994; Mocanu et al., 2005), and the validation of reconstructions was performed experimentally, using electrolytic tanks (Oster et al., 1998) or by measuring epicardial electrical potentials during open heart surgery (Shahidi et al., 1994; Burnes et al., 2000). Other techniques encompass more *a priori* information about the solution: local spatial regularization (Johnson and MacLeod, 1996), constraints on the normal component of epicardial current density (Velipasaoglu et al., 2000), constraints based on supraregularized and subregularized solutions (Iakovidis and Martin, 1991). Another approach, the laplacian electrocardiography (He et al., 1997; Mocanu, 2002) uses "Laplacian" (concentric dipolar) electrodes (Carvalhaes and de Barros, 2014) to measure thoracic potentials needed in the reconstruction of the epicardial potentials. Laplacian potential electrodes are surface reference free electrodes have been shown to improve the spatial resolution of surface bioelectric signal recordings in EHG (electrohysterogram) and EEG measurements. Technically, their information is proportional to the local surface charge (laplacian). These concentric, ring electrodes, show off better immunity to noise but they are fainter and present a shorter distance coherence to the source (He et al., 1997; Li et al., 2005; Gao et al., 2017; Makeyev, 2017, 2018). Fig. 4.4 depicts an array of dipolar laplacian electrodes used to measure the electric field on the thorax surface in the I-ECG problem where the source of electric activity is a current dipole.

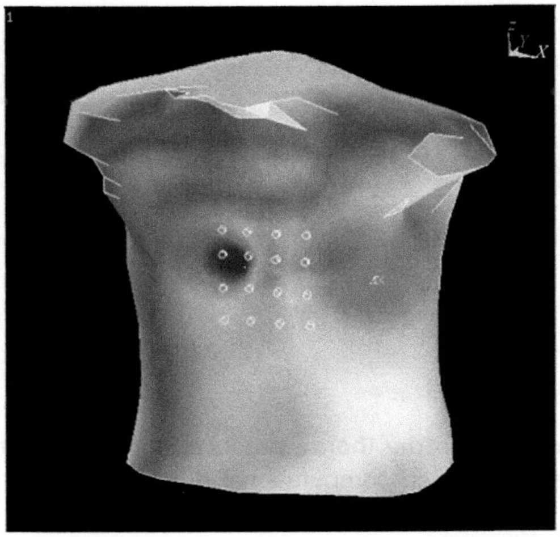

Figure 4.4 Laplacian electrodes are used to measure the thoracic electrical potential produced by an intracardiac dipole source (Fig. 4.4).

Technically, inverse solutions are obtained through the minimization of

$$\mathbf{x}_\alpha = \underbrace{\arg\min_{\mathbf{x}} \left\| \mathbf{C}\mathbf{x} - b \right\|_2^2}_{\text{least sqaures approx.}} + \underbrace{\alpha^2 \left\| \mathbf{L}\mathbf{x} \right\|_p^p}_{\text{penalty term}}, \tag{4.5}$$

where the penalty term provides a priori information on the solution. Should $\mathbf{L} = \mathbf{I}$ and $p = 2$ then Eq. (4.5) is the Tikhonov regularization of order zero, yielding

$$\mathbf{x}_{Tik} = \left(\mathbf{C}^T\mathbf{C} + \alpha^2\mathbf{I}\right)^{-1}\mathbf{C}^T b. \tag{4.6}$$

When $\mathbf{L} = \mathbf{D}$ (the gradient operator) and $p = 1$, Eq. (4.5) yields the total variation (TV) of the potential $\mathbf{x}$ (Rudin et al., 1992; Mukherjee et al., 2016)

$$\mathbf{x}_{TV} = \left(\mathbf{C}^T\mathbf{C} + \alpha^2\mathbf{D}^T\mathbf{W}_x\mathbf{D}\right)^{-1}\mathbf{C}^T b, \tag{4.7}$$

where $\mathbf{W}_x$ is the weight diagonal matrix

$$\mathbf{W}_x = \frac{1}{2}\text{diag}\left(\frac{1}{\sqrt{[\mathbf{D}\mathbf{x}]_i^2 + \beta}}\right). \tag{4.8}$$

Successive approximation (fixed point iterations)

$$\mathbf{x}^{(k+1)} = \left(\mathbf{C}^T\mathbf{C} + \alpha^2\mathbf{D}^T\mathbf{W}_{x^{(k)}}\mathbf{D}\right)^{-1}\mathbf{C}^T\mathbf{y}, \tag{4.9}$$

may be used to solve for $\mathbf{x}$.

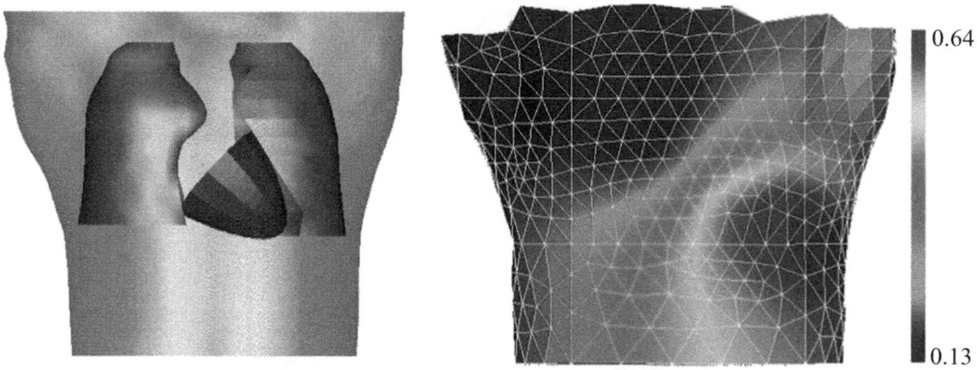

Figure 4.5 The electric potential on the thorax, produced by a step distribution of potential unit on the surface of the heart. Simplified computational domain (left) and the electric potential on the thorax produced by a step function epicardial voltage (Fig. 4.4a).

In mathematical and numerical modeling, a representative epicardial source is the step function, which characterizes the abrupt variation of the potential AP front-like shape (Fig. 4.5) and it qualifies to compare the inversion methods used in the solution of the inverse problem that aims to find the epicardial potential. The optimal parameter for Tikhonov regularization was determined with the L-curve method. For VT regularization, α was determined by visual inspection (Mocanu, 2002).

The model ought to include all anatomical structures that could significantly influence the direct and inverse solutions (Morega et al., 2000). However, the complexity of the internal geometry of the thorax may lead in the end to a very large number of degrees of freedom (grid nodes). Heart geometry is much simplified here (Fig. 4.5) to allow the smooth calculation of the gradient operator on a structured surface network. A realistic representation of the heart shape requires an unstructured grid and complicated algorithms for determining the surface gradient operator (Srinidhi, 1999). The reconstruction of this particular type of signal (step function), out of this potential distribution on the thorax using several inversion methods, is shown in Fig. 4.6.

Tikhonov regularization, which is extensively used to stabilize the inverse solution, acts as a linear filter. In an attempt to remove high frequency noise, this method also filters the high frequency components in the solution, leading to the oversmooth of the AP front. In contrast, the TV constrains the norm l_1 of the potential gradient, preserving the discontinuities in the solution. In the case of smooth potential, the two regularization methods provide similar results, but the Tikhonov method better locates the potential field extrema, while the TV overextends them (Mocanu et al., 2002, 2005).

Image-based construction of a human heart and thorax

The models presented here are developed as a numerical instrument for the study of normal and pathological heart conditions, supporting the experimental (in vivo)

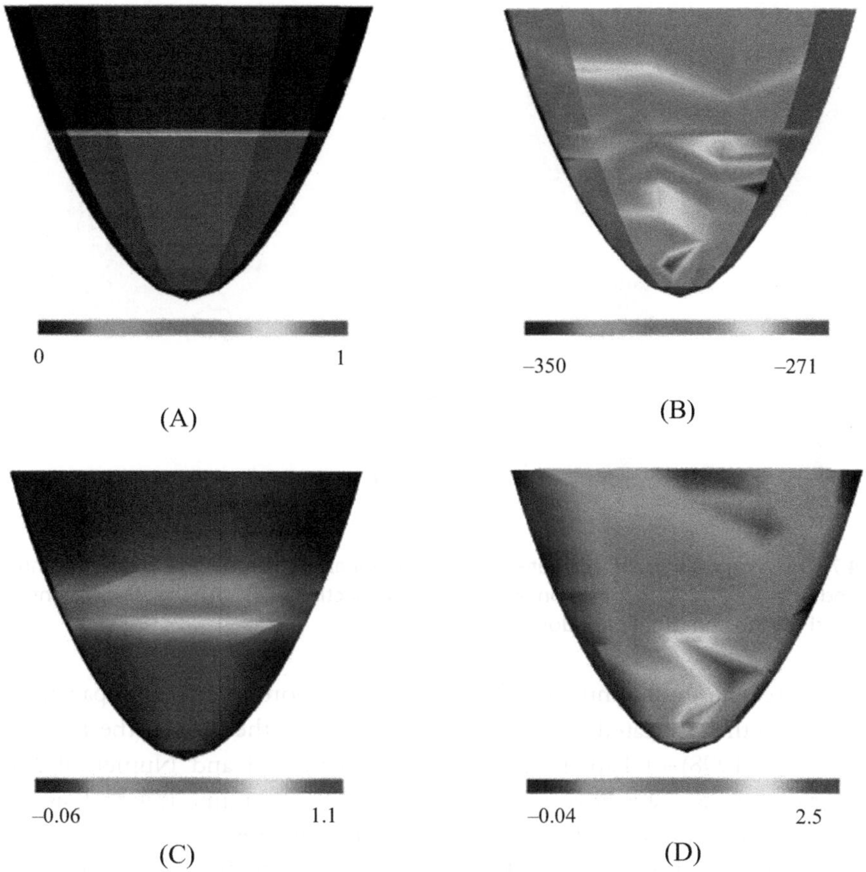

Figure 4.6 The epicardial AP spectrum (A) and its reconstructed image, (B) reconstruction using the generalized inverse, (C) reconstruction by TV, and (D) Tikhonov reconstruction.

investigations limited by bioethical reasons (Spielman, 2007; Rogozea et al., 2015). Several approaches rely on basic one- or two-dimensional models (Mocanu et al., 2007) that can be used in the numerical assessment of heart pathologies such as fibrillation and arrhythmias, while more complex models (Vigmond et al., 2009) numerically couple the electric and mechanical activities of the heart.

The modern medical investigation techniques (CT, MRI, ultrasound) are a good source for image datasets that can provide for anatomically accurate and patient-specific 3D computational domains. Recently, several models that implement image-based constructed geometries were developed to numerically simulate and study the electrophysiological behavior (Prakosa et al., 2014) or the electrical and mechanical functions (Vadakkumpadan et al., 2010) of the normal and diseased hearts. Our models implement anatomically accurate computational domains generated using CT or MRI image based

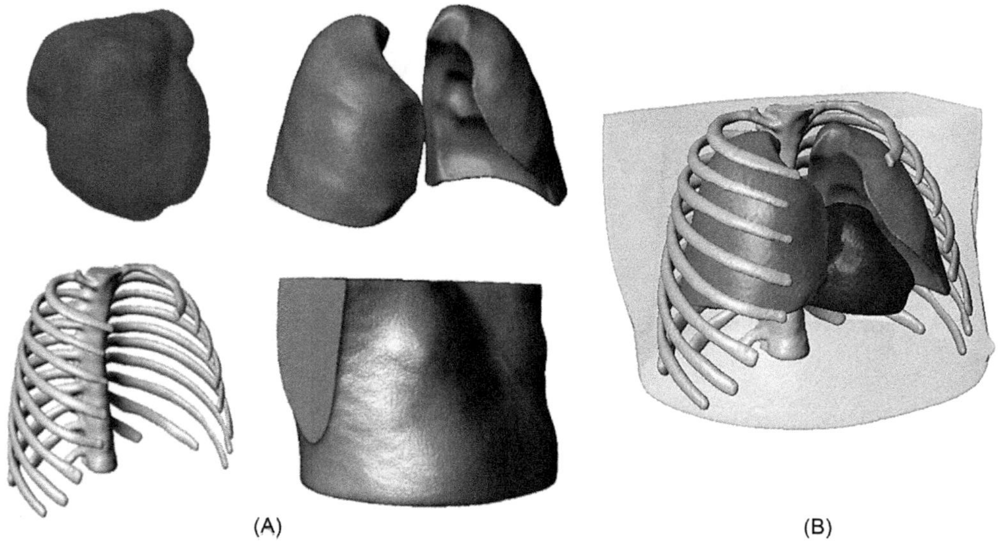

Figure 4.7 The morphologically accurate 3D computational domain for the action potential propagation and the electric field diffusion in the thorax (Section 4.4): (A) the constituent anatomic regions, and (B) the computational domain.

construction techniques for numerically solving the action potential propagation on the epicardium and the associated bioelectric field diffusion in the thorax (the D-ECG problem; Gulrajani, 1998)—Chapter 1, Physical, Mathematical and Numerical Modeling. This improvement is a step forward in the patient-specific numerical analysis, minimizing the numerical errors given by a drastic geometry simplification.

The computational domain may be generated using a high resolution CT image set. First, the myocardium domain is segmented out using a threshold filter, followed by several Boolean operations, morphological filtering and final adjustments with smoothing operations. The process is repeated for the spinal cord and ribs, the lungs, and the embedding thorax (Fig. 4.7A). The final model is made of four subdomains (Fig. 4.7B).

4.3 The electrical activity of the cardiac strand

The resting (electric) potential of an excitable cardiac cells characterizes its resting, equilibrium state, which may be disturbed when electrical currents intense enough to depolarize cross the membrane. This change of state triggers the initiation of an abrupt increase in the transmembrane voltage followed by a return, during the refractory period, to the resting state—the AP, or propagated impulse. Further on electrotonic mechanisms trigger energy interactions between adjacent cells that transmit the AP through atria and ventricles (Berne et al., 1998). In normal sinusal rhythm, the AP waves are sustained by the energy provided by the cell metabolism and triggered periodically by the pacemaker cells of the SA node.

The normal cardiac electrical activity manifests through electric field waves that propagate throughout atria and ventricles as synchronized APs. Arrhythmias are deviations from this self-sustained pattern of the AP wave initiation or propagation. The *alternans* is a long–short–long cycle of the AP duration, APD, during rapid pacing (Chialvo et al., 1990), and it is perceived as a forerunner of severe ventricular arrhythmias, such as ventricular tachycardia and ventricular fibrillation (Pastore et al., 1999). The origins of alternans, either *concordant* (in phase everywhere in the tissue) or *discordant* (in opposition in distinct spatial regions) APD (Watanabe et al., 2001), and accompanying reentries may be touched using one- and two-dimensional cardiac propagation models (Mocanu et al., 2007). The diastolic interval, DI, shortens the cell APD and, below a critical DI value, the cell no longer responds with an AP (Qu et al., 1999).

Alternans, called a "{2:2} rhythm", occur when two stimuli elicit two APs of different duration and shape. Increasing the stimulation frequency leads to conduction block and {2:1} synchronization, in which the cardiac tissue responds to every other stimulus (Mocanu et al., 2007). They are produced during high heart rates (pacing frequencies) when the inclination of the restitution curve is above one (Guevara et al., 1984; Strumillo and Ruta, 2002). Calcium channel blockers may contribute to flatten it and thus suppress the alternans (Garfinkel et al., 2000).

One-dimension action potential propagation

The AP transmitted by a one-dimensional filament of ventricular muscle has been modeled using an adaption of the cylindrical cable theory[1] (Mocanu et al., 2007; Schierwagen and Ohme, 2020), which yields the transmembrane voltage that may be calculated using

$$C_m \frac{\partial V_m}{\partial t} + I_{ion} = \frac{r}{2\rho} \frac{\partial^2 V_m}{\partial x^2}, \tag{4.10}$$

with the boundary conditions

$$\left. \frac{1}{\rho} \frac{\partial V_m}{\partial x} \right|_{x=0} = -I_{stim}, \quad \left. \frac{1}{\rho} \frac{\partial V_m}{\partial x} \right|_{x=L} = 0. \tag{4.11}$$

For the example shown here, $C_m = 1\ \mu\text{F/cm}^2$ is the membrane capacitance per unit area, $\rho = 0.25\ \text{k}\Omega \cdot \text{cm}$ is the intracellular axial resistivity per unit length, $r = 5\ \mu\text{m}$ is the cell radius, and $L = 7$ cm is the length of the cable. The ionic current density through the cell membrane, I_{ion} [$\mu\text{A/cm}^2$], sums up the contributions of all

[1] The telegraphist's (cable) equation was introduced in 1855 by Lord Kelvin, for the transatlantic telegraph cable.

ionic species (sodium, I_{Na}, calcium, I_{Ca}, and potassium, I_{K1} and I_{x_1}, and other nonspecific ionic components, I_s) (Beeler and Reuter, 1977)

$$I_{ion} = I_{Na} + I_{K_1} + I_{x_1} + I_s, \tag{4.12}$$

where $I_{K_1} = 1.4\frac{\exp(0.04(V_m+85))-1}{\exp(0.08(V_m+53))+\exp(0.04(V_m+53))}$, $I_{x_1} = 0.8x_1\frac{\exp(0.04(V_m+77))-1}{\exp(0.04(V_m+35))}$, $I_{Na} = G_{Na}m^3h(V_m - E_{Na})$, $I_s = G_s fd(V_m - E_s)$, x_1 is a gating variable, $G_{Na} = 15\ \text{mS/cm}^2$ is the maximal per unit area conductance of the Na^+ channels, $E_{Na} = 40$ mV is the equilibrium Nernst potential for the Na^+ ions; $G_s = 0.09\ \text{mS/cm}^2$ is the maximal membrane per unit area conductance of the Ca^{2+} channels, $E_s\,[\text{mV}] = -82.13 - 13.0287\ \ln c_{Ca,i}$ the font in the equation seems to be of smaller size is the equilibrium Nernst potential for the Ca^{2+} ions. The dynamics of the Ca^{2+} intracellular concentration c_{Ca}, the gating variables m and h (for Na^+ channels), f and d (for Ca^{2+} channels), and x_1 (for potassium channel) are ruled by

$$\frac{dc_{Ca,i}}{dt} = -10^{-7}I_s + 0.07\left(10^{-7} - c_{Ca,i}\right), \quad \frac{da}{dt} = (a_\infty(V_m) - a)/\tau_a(V_m); \quad a = \{m, h, d, f, x_1\}, \tag{4.13}$$

where a_∞ and τ_a for the different quantities are found in Beeler and Reuter (1977).

The initial conditions are $V_m(0) = -85$ mV (the resting potential of the cardiac cell), $c_{Ca,i}(0) = 3 \times 10^{-7}$ M, $m(0) = 0.01126$, $h(0) = 0.9871$, $d(0) = 0.003$, $f(0) = 1$, $x_1(0) = 0$. The ODEs Eq. (4.13) are solved numerically (Mocanu et al., 2007). The amplitude of the stimulus was twice the diastolic threshold, $I_{stim} = 70\ \mu\text{A/cm}^2$, with a duration of 2 ms (typical value used in experiments), for a 7-cm long cable.

For an S1−S2 stimulus protocol the APD decreases as the DI shortens (Fig. 4.8), and the APD versus DI slope eventually decreases below the critical value (1). Below

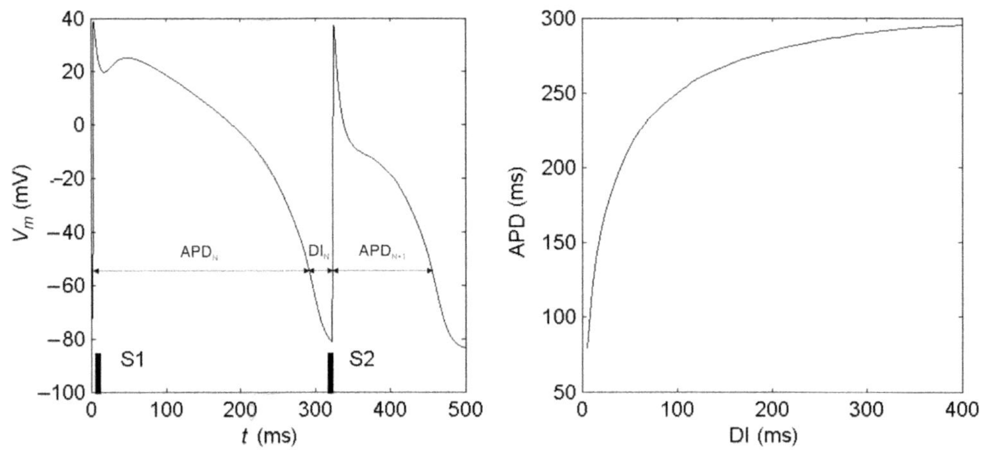

Figure 4.8 The AP restitution duration of cardiac cells. (Left) APs elicited by an S1−S2 stimulus protocol corresponding to a diastolic interval DI = 5 ms (S1 and S2 stimuli are shown with bars). (Right) Restitution curve $APD_{N+1} = f(DI_N)$. APD_{N+1} elicited by the S2 stimulus flows the decrease of the previous diastolic interval DI_N (Mocanu et al., 2007).

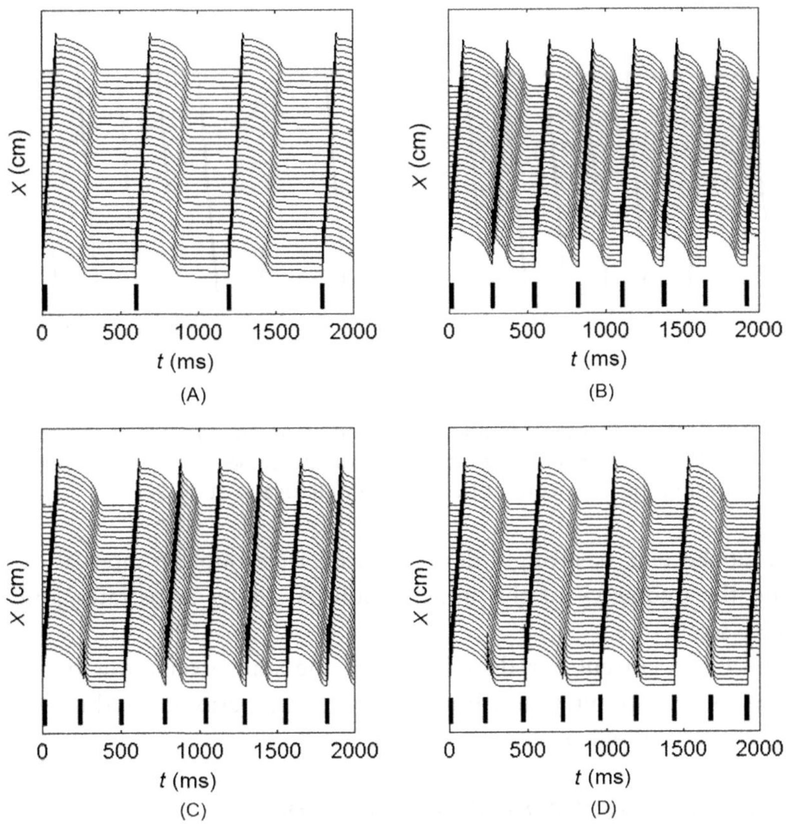

Figure 4.9 Space-time diagrams of AP propagation in a 7 cm cable paced at the bottom-end with a train of short current pulses at different frequencies. The stimulus train is shown with bars (Mocanu et al., 2007).

this margin the S2 stimulus occurs within the refractory period of the AP elicited by S1 and a new AP cannot be produced.

Fig. 4.9 renders the AP at 40 locations along the cable for cyclically pacing the fiber at $x = 0$ with a square-wave stimulus (shown with bars). At 1.6 Hz, the strand responds to every stimulus with an AP, resulting in a {1:1} normal sinus beat synchronization (Fig. 4.9A). This rhythm persists until 3.6 Hz (Fig. 4.9B), when concordant alternans emerge. At 3.8 Hz, synchronization shifts from the initial {2:1} to the {2:2} rythm (Fig. 4.9C). At 4.3 Hz an AP is triggered at every other stimulus, following a {2:1} rhythm (Fig. 4.9D).

The onset of discordant alternans is originated by an ectopic stimulus administered at the cable exit (at 500 ms in Fig. 4.10, left) shortly after a sinus wave concluded its travel, then followed by a 240 ms paced sinus rhythm stimuli administered at the beginning of the cable (Watanabe et al., 2001). The development of *discordant* alternans, with short–long–short AP duration at the beginning and long–short–long

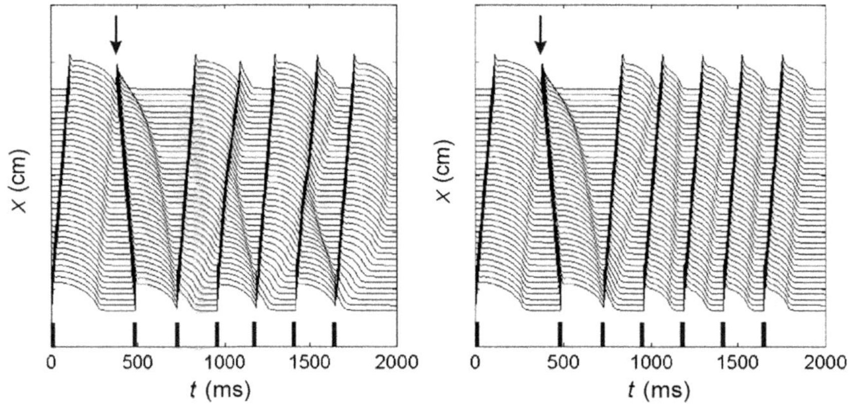

Figure 4.10 Discordant AP alternans duration produced by an ectopic stimulus (*arrow*) at the cable end followed by 240 ms pacing at the bottom end of the cable (left). AP alternans duration are suppressed by reducing the Ca^{2+} current by 25% (right) (Mocanu et al., 2007).

AP duration at the end, is strongly dependent on the coupling interval of the abnormal beat and the frequency of the sinus stimuli. At lower sinus rates, only *concordant* alternans were seen.

When the Ca^{2+} current is reduced by 25% the sinus excitations does not lead to discordant or concordant alternans, but rather a {1:1} periodic rhythm with a shorter APD occurs (Fig. 4.10, right).

Two-dimensional action potential propagation

A simplified model of the excitable cardiac tissue, which involves a fast variable u and a slow variable w, is introduced trough the continuous media, coupled PDEs (Aliev and Panfilov, 1996)

$$\frac{\partial u}{\partial t} = \frac{\partial^2 u}{\partial x^2} + \frac{\partial^2 u}{\partial y^2} \underbrace{- ku(u-a)(u-1)}_{\text{fast process } (AP \text{ upstroke})} - uw + \underbrace{i_{stim}}_{\text{applied stimulus}}, \tag{4.14}$$

$$\frac{\partial w}{\partial t} = \left[\underbrace{\varepsilon_0 + \mu_1 w / (\mu_2 + u)}_{\text{recovery phase}} \right] [-w - ku\ (u - a - 1)]. \tag{4.15}$$

The four parameters have no specific physiological meaning, and they are adjusted to render the dynamics of a square $L \times L$ excitable cell—here, $k=8$, $a=\ 0.15$, $\varepsilon_0\ =\ 0.002$, $\mu_1\ =\ 0.2$, $\mu_2\ =\ 0.3$. The time and the transmembrane potential are related to the model

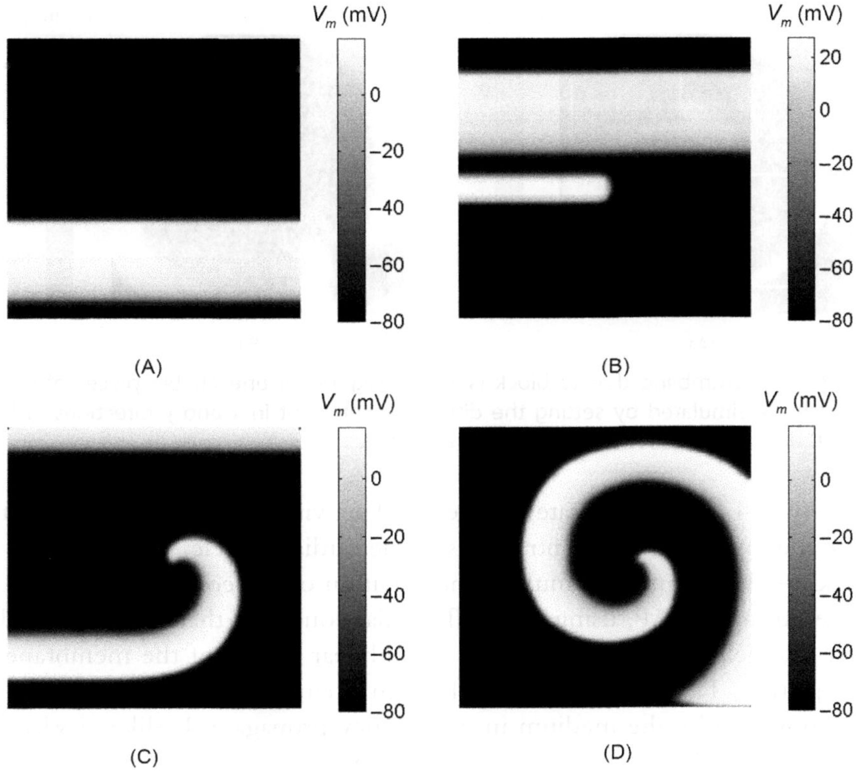

Figure 4.11 Reentry in a two-dimensional piece of cardiac tissue (Mocanu et al., 2007).

variables through $V_m[\text{mV}] = 100u - 80$ and $t[\text{ms}] = 12.9t$. The boundary conditions $\frac{\partial u}{\partial x}|_{x=0} = \frac{\partial u}{\partial x}|_{x=L} = \frac{\partial u}{\partial y}|_{y=L} = 0$ and $\frac{\partial u}{\partial y}|_{y=0} = i_{stim}$, where $i_{stim} = \begin{cases} 1, & \text{if } t \leq 5 \\ 0, & \text{else} \end{cases}$, and the initial conditions $u(0) = 0$ and $w(0) = 0$ close the mathematical model.

Fig. 4.11 presents the dynamics of an excitation wave perturbed by an early stimulation that is curled into a vortex (Aliev and Panfilov, 1996), which is seen as a potential cause of cardiac arrhythmias (Aguilar and Nattle, 2016).

A normal, AP wave travels upward (Fig. 4.11A) when, at $t = 930$ ms an early stimulus is delivered in the wake of the passing wave, in the section that just regained excitability (Fig. 4.11B). The new wave can travel in all directions except upward (down the cable), in the still refractory region. A critical point may eventually emerge and turn into the tip of a stable spiral wave (Fig. 4.11C and D), which is known to be the substrate of ventricular tachycardia (Aliev and Panfilov, 1996). This electric signal is a precursor of ventricular fibrillation, a deadly arrhythmia if not treated promptly (Aguilar and Nattle, 2016).

Moreover, the spiral wave interacts with a conduction block region (e.g., infarct, delineated with white rectangles) that paves the path to the transition to ventricular

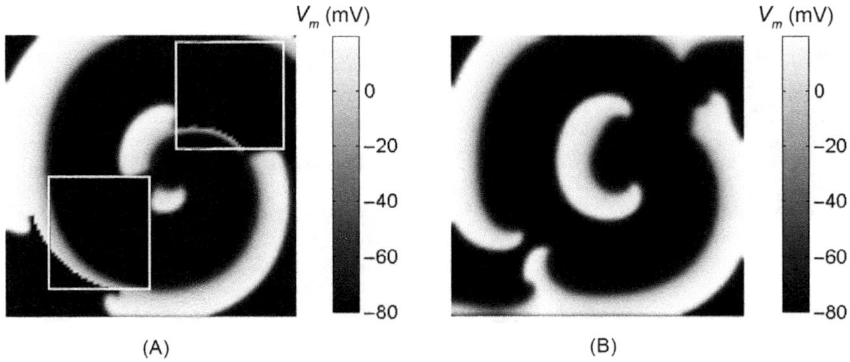

Figure 4.12 Spiral crumbling due to block (squares) regions of unexcitable pieces of tissue (e.g., infarct), which are simulated by setting the diffusion coefficient in x and y directions in Eq. (4.14) — before (A) and (B) after breakup (Mocanu et al., 2007).

fibrillation (Fig. 4.12). In this state the electrical activity turns chaotic, leading to high frequency and low amplitude contractions of the cardiac muscle.

Hodgkin—Huxley model simulates the variation of the conductance of the neural membrane during the AP, using four differential equations that describe the dynamics of ion channels. The model includes a nonlinear model of the membrane in the "cable" equation. The waves travelling through excitable media (autowaves) are self-sustained, powered by the medium in which they propagate. Unlike classical waves, which attenuate and distort after a certain distance, they do not dissipate, have no reflection and interference properties (when meeting they do not annihilate). Autowaves are characteristic not only of neural axons or cardiac tissues, but also occur in other nonlinear systems, chemical or physical. A famous example is the Belousov—Zabotinski chemical reaction (Bergé and Pomeau, 1984), which is an oscillating chemical system with spatial and temporal self-organizing properties that seem to question the second law of thermodynamics: an isolated system evolves irreversibly from order to disorder, and finally to thermal equilibrium (thermodynamic death)—a consequence of the fundamental law of thermodynamics. However, for biological systems, perpetuity is supported by other energy consuming physiological phenomena that develop in parallel, which are intended to regenerate the initial conditions.

Different abstractions are known to render the nonlinear dynamic of the electrophysiological model for the heart under various conditions using the two variables (excitation and inhibition) whose time evolution is described through two to four coupled PDEs for continuous media (Wiener and Rosenblueth, 1946; Fitzhugh, 1961; Zaikin and Zhabotinsky, 1970; Bergé and Pomeau, 1984; Keener, 1988; Bär et al., 1994; Boulakia et al., 2015; Chen et al., 2018), solved using numerical methods, (Bürger et al., 2010; Filippi and Cherubini, 2006, 2009;

Dowle et al., 1997; Scrale, 2009, Dobre et al., 2011a), that go back to Turing (1952) and Shiferaw and Karma (2006), and generic reaction-diffusion nonlinear chemical excitable systems of diffusing species (Alonso et al., 2013)

$$\frac{\partial V}{\partial t} = kI(V, w, p_i) + \nabla \cdot (D_V \nabla V), \tag{4.16}$$

$$\frac{\partial w}{\partial t} = R(V, w, p_i) + \nabla \cdot (D_w \nabla V), \tag{4.17}$$

where $I(V,w,p_i)$ and $R(V,w,p_i)$ are nonlinear functions and p_i are tuning parameters, which are typically solved in square (cubic) computational domains with homogeneous Neumann boundary conditions.

The tuning parameters are a limiting, arbitrary factor and their trial-and-error "tuning" may question the physical grounds of such endeavor. It is the price for using homogenization techniques needed to introduce continuous media. Nevertheless, numerical modeling on detailed "twins" of the actual electrophysiologic processes may have the ability to accurately render the complex responses of cardiac cells under normal and abnormal conditions.

4.4 Coupling the action potential with the electric field diffusion in the thorax

One of the applications of the direct problem of electrocardiography is to simulate the propagation of cardiac depolarization fronts with the aim to investigate arrhythmias, in the attempt to investigate common heart disorders, as seen before. Another important application is the modeling of electrical defibrillation, as emergency interventional therapy against extreme heart condition. *Ventricular fibrillation* is a cardiac arrhythmia that can be fatal if left untreated within minutes. The most effective therapy for ending ventricular fibrillation is the electric shock. The internal defibrillator is an electronic device that is implanted in cardiac patients prone to such arrhythmias with the aim to deliver on demand electric shocks. This device detects and terminates cardiac arrhythmias using electrodes inserted into the SVC and the right ventricle. A good model of the electrical activity of the heart is based on the adequate mathematical equations that best fit the studied phenomena. Choosing the right boundary and initial conditions are crucial too for tuning the numerical computation results (Mocanu et al., 2002).

The numerical study of the ECG problem starts with the AP propagation at the heart level, which is then coupled to the electric field diffusion within the thorax that is monitored using electrical signals numerically evaluated on the chest's surface.

The AP is provided, here, through the FitzHugh–Nagumo model that renders the normal electrical activity of the heart (Kogan et al., 1991)

$$\frac{\partial u}{\partial t} = \Delta u + (\alpha - u)(u - 1)u - w, \tag{4.18}$$

$$\frac{\partial w}{\partial t} = \varepsilon(\beta u - \gamma w - \delta), \tag{4.19}$$

where u is the "fast" variable which models the AP propagation, w is the "slow" variable which models the inhibitor that reflects the probability of a transmembrane ionic channel to conduct an ionic current, influencing the rest state comeback (Malmivuo and Plonsey, 1995). The $(\alpha-u)(u-1)u$ term is specific to the membrane depolarization of the myocardial nodal cells, α is the excitation threshold, ε is an empirical parameter which triggers the cells rest state comeback while β, γ, δ are simulating the electrical activity dynamics of the heart, when properly adjusted.

The AP propagation is happening very fast due to the short time intervals needed for the sodium ionic channel opening and conducting a transmembrane depolarization ionic current. On the other side, there are models for simulating the pathological regimes for the heart's electrical activity (e.g., the arrhythmias), described by self-sustained reentrant spiral waves. Landau–Ginzburg is one of the many oscillator models that describes these abnormal states (Aranson and Kramer, 2002)

$$\frac{\partial u}{\partial t} = \Delta\left(u - \chi_1 w\right) + u - \left(u - \chi_2 w\right)\left(u^2 + w^2\right), \tag{4.20}$$

$$\frac{\partial w}{\partial t} = \Delta\left(\chi_1 u + w\right) + w - \left(\chi_2 u + w\right)\left(u^2 + w^2\right), \tag{4.21}$$

where χ_1 and χ_2 are material property values that influence the solution existence and stability.

In general, Landau–Ginzburg equations are used to model chaotic phenomena and were introduced to describe the superconductivity theory and chaotic phenomena (Du et al., 1992). In other studies, the same equations model the LASER pulse generators functioning (Akhmediev et al., 2001) or the dynamics of Belousov–Zhabotinsky chemical reactions (Petrov et al., 1993).

In the numerical simulations of the AP propagation presented next homogeneous flux (Neumann) boundary conditions are set on the endocardium and epicardium (Fig. 4.13) assuming that no ionic current flows between the myocardium and the inside and outside neighboring domains of the heart. The computational domain is generated using the 3D model of the myocardium after applying voxel and volumetric marching cube (VoMaC)-based meshing algorithms (Keyak et al., 1990; Mueller and Ruegsegger, 1994; Simpleware, 2010).

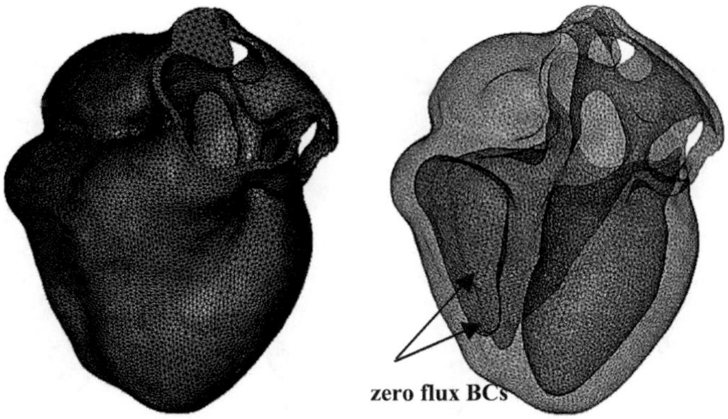

Figure 4.13 The computational domain for the action potential propagation on the epicardium, made of approximately 130 k tetrahedral elements. Opaque (left) and translucent (right) representations.

The AP propagation is solved for first using only the myocardium volume. The electric field diffusion in the thorax, given by the AP on the epicardium "echo," is computed by coupling the two physics, through boundary/interface conditions thus giving a fair approach to the numerical analysis of the ECG problem.

The electric field within the thorax (Fig. 4.14) is governed by Laplace equation, $\Delta V = 0$ (Chapter 1: Physical, Mathematical and Numerical Modeling). The AP propagation on the epicardium, described by either FitzHugh–Nagumo (4.18), (4.19) or Landau–Ginzburg models (4.20), (4.21), and the electric field diffusion inside the thorax are coupled by using the u variable as the only source of the electric field within the thorax. In the second problem, the electric field inside the thorax, the epicardium assumes Dirichlet BC, which specifies the AP distribution evaluated in the first problem (replacing

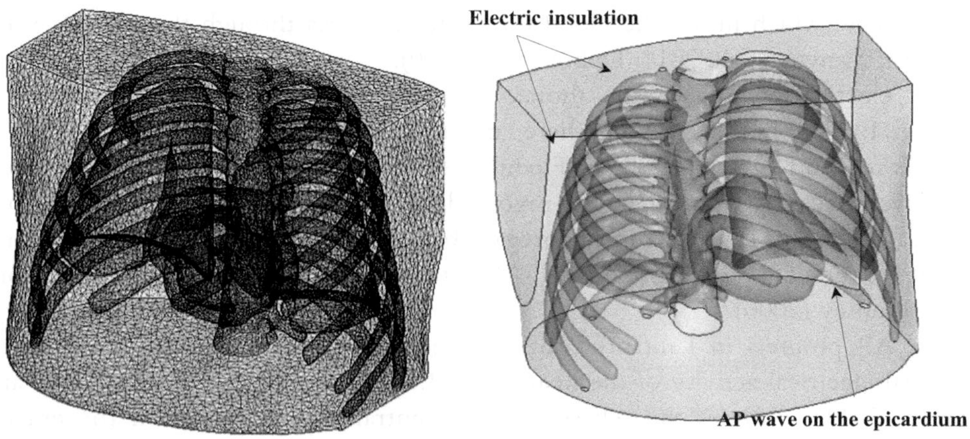

Figure 4.14 The computational domain and FEM mesh for the electric field diffusion in the thorax (left) and the associated BCs (right).

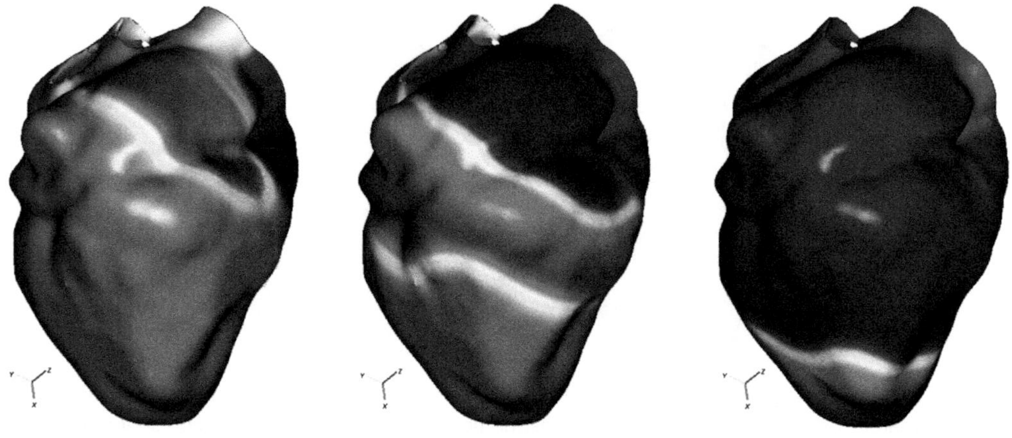

Figure 4.15 The AP on the epicardium for FitzHugh–Nagumo model at three moments: 50 ms (left), 100 ms (middle), and 150 ms (right).

the Neumann BC). This connection between the two problems defines a one-way coupling. The surface of the thorax is electrically insulated ($J_n = 0$) and the inner boundaries are set as continuity (interfaces between two different types of tissue).

The computational domain for the electric field diffusion in the thorax, here a volume conductor (Schwan and Kay, 1957; Plonsey, 2000), was generated similarly to the 3D model of the heart (Fig. 4.13), which is used for the AP propagation models. One extra challenge is to accurately define the SA node position inside the myocardium. Under normal (physiological) conditions, the depolarization wave propagates starting from the SA node to the apex of the heart, first triggering the contraction of the atria, which pumps the blood into the ventricles, followed by the contraction of the ventricles, which makes the blood flow into the aorta through the sigmoid valve (Fig. 4.15; Morega et al., 2011; Dobre et al., 2010).

The electric field inside the thorax is solved for after setting the AP, u, as Dirichlet BC on the epicardium (here, inner boundary). The electric field in the thorax is related to the AP, which (Fitzhugh–Nagumo model) is shaped as progressive waves that travel through the myocardium, from the SA node to the apex (Morega et al., 2011; Dobre and Morega, 2011) [behavior acknowledged in literature by Fenton et al. (2002)] (Fig. 4.16). The thorax is treated here as a volume conductor Plonsey (2000).

The AP progress in Landau–Ginzburg model (4.20), (4.21), associated to an abnormal electrical activity of the heart (Fig. 4.17; Morega et al., 2011; Dobre and Morega, 2011), outlines the self-sustained, reentrant spiral waves on the epicardium—very different dynamics pattern when compared to the FitzHugh–Nagumo model results.

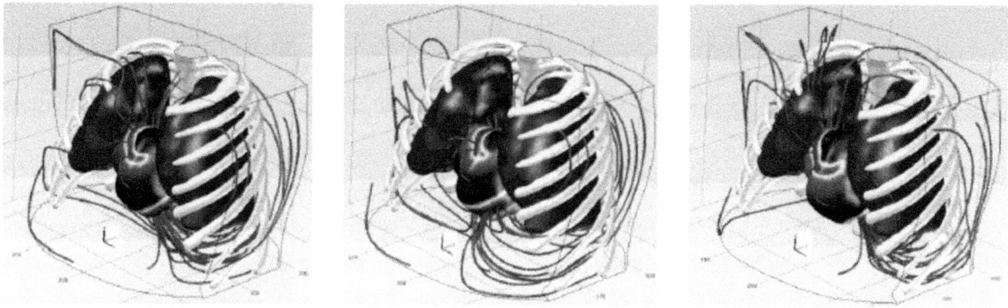

Figure 4.16 The electric field diffusion in the thorax given by the FitzHugh–Nagumo model at three moments: 450 ms (left), 475 ms (middle), and 500 ms (right).

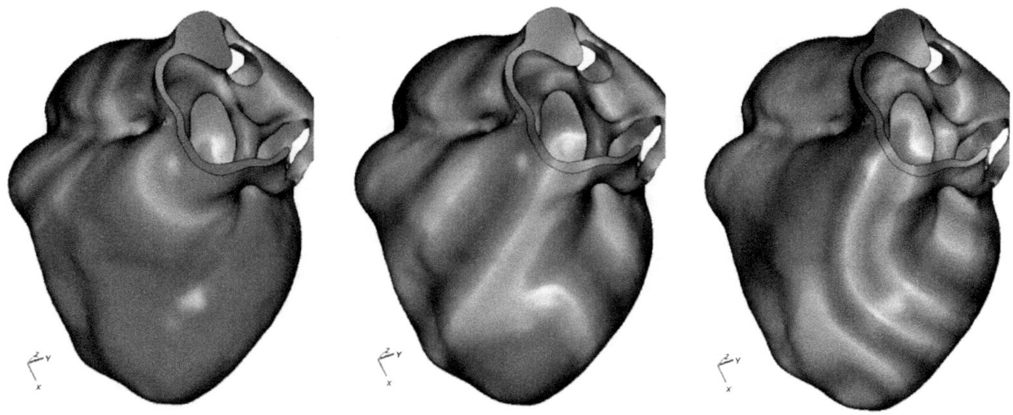

Figure 4.17 The AP for Landau–Ginzburg model at three moments: 50 ms (left), 100 ms (middle), and 150 ms (right).

This behavior is related to the presence of arrhythmias, fibrillations or tachycardia, which elicits the Landau–Ginzburg model a good candidate for noninvasive, reliable and cost-efficient optimization studies in the cardiac pacing or defibrillation procedures.

The electric field diffusion inside the thorax when coupled with the Landau–Ginzburg epicardial AP is presented in Fig. 4.18 (Morega et al., 2011; Dobre and Morega, 2011), which shows off the spiral waves on the epicardium.

The electric field calculated inside the thorax may be used to assess the ECG signal as measured regularly, using the Eindhoven triangle (Morega et al., 2011; Dobre and Morega, 2011).

The difference between the actual ECG leads and those obtained from the numerical simulations is mainly influenced by the simplified geometry, the approximate location of the SA node, and other simplifying hypotheses. The contact resistance between the electrodes and the skin is neglected here. Better agreement with experimental data is expected to occur should these limitations be alleviated.

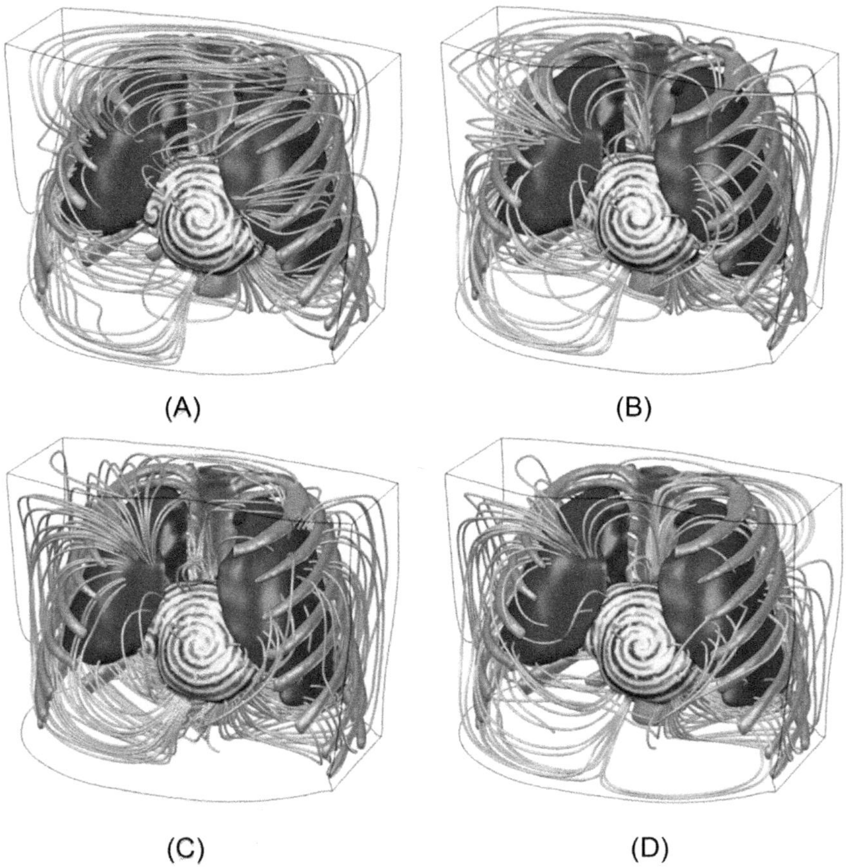

Figure 4.18 The electric field diffusion in the thorax given by the Landau–Ginzburg action potential propagation model for three different moments in time: 1250 ms (A), 1500 ms (B), 1750 ms (C), and 2000 ms (D).

4.5 Blood pressure pulse wave reflections

The cardiovascular system operation is strongly related to the electrical activity of the myocardium. The mechanical work (myocardium contraction and relaxation) required for the blood flow is sparked by the electrical signal of the heart (Kannel et al., 1971; Strandberg and Kaisu, 2003). The overall electromechanical activity of the heart is crucial for its function, and the hemodynamic parameters are a reflection of the leading electric activity heart that generated them.

The blood pressure wave

Among the numerous hemodynamic parameters used to assess the patient's health status, the most comprehensive and easy to acquire are the systolic (SP) and diastolic

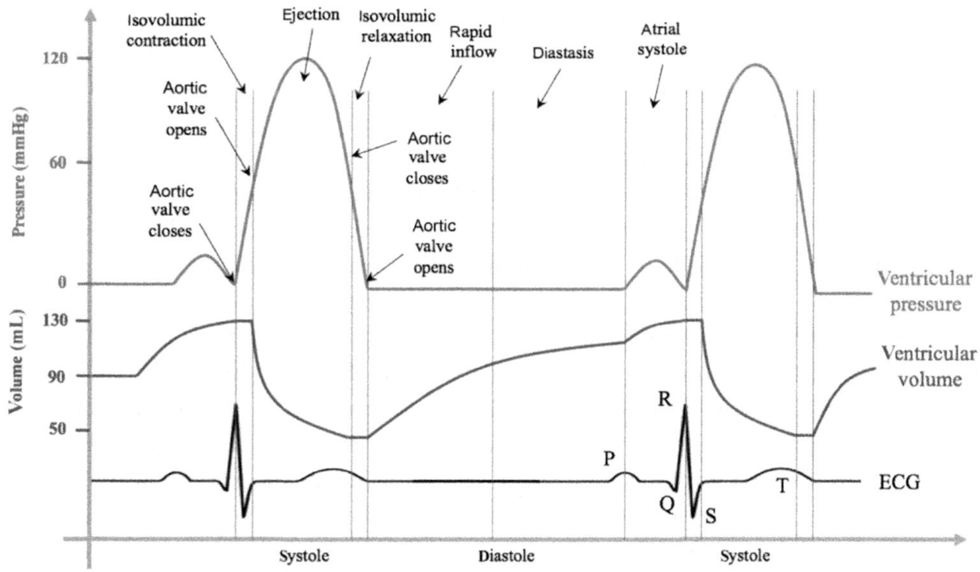

Figure 4.19 Vital signs during the cardiac cycle.

(DP) pressure values, Fig. 4.19. SP is associated with the peak value of the blood pressure during the systole, while DP indicates the lowest pressure value during the diastole, not the mean values, as sometimes confused.

Approximately 70% of the blood that fills the ventricles is passively received during the diastole, while the rest of the blood volume is given by the contraction of the atria. When the mitral and tricuspid valves close, the isovolumetric ventricular contraction takes place for about 50 ms as the myocardium contracts and increases the pressure until it exceeds 80 mm Hg in the left ventricle and 10 mm Hg in the right ventricle (Ganong, 2005). These are the threshold values when the ventricular ejection of blood starts. At this time, the pressure in the ventricles increases rampant until it reaches the peak values of ~120 mm Hg for the left ventricle and ~25 mm Hg for the right ventricle. Ahead of the isovolumetric ventricular relaxation stage, there is a protodiastole period (~40 ms), between the moment of the fully contracted ventricles and the aortic and pulmonary valves closing. The blood pressure keeps decreasing until it reaches another threshold value, below the atria pressure, when the AV valves open and the ventricles are filled with blood, ready for another cardiac contraction.

The cyclic cardiac muscle displacement is sustained through the fluid in the pericardial sac that has a lubricating effect. This enables the easy myocardium contraction, with minimal friction between the heart and surrounding thoracic viscera (Ganong, 2005).

The left ventricle contraction ejects the blood in the aorta and generates a pressure wave that propagates throughout the arterial network, expanding the blood vessel walls. This creates the arterial pulse, which can be sensed in different body regions

(Cox, 1971; Arts et al., 1979; Paquerot and Remoissenet, 1994). For example, at the wrist, the pulse is felt when touching the radial artery with a 100 ms delay from the moment of blood ejection in the aorta, due to different pressure wave velocity values along the arterial network (lower speeds in large arteries which increase as the blood vessel diameter shortens). Ageing hardens the smaller arterial walls and the pulse waves may travel faster for older subjects. Meaningful information on the cardiovascular condition may be obtained using pressure measurements (Benetos et al., 2012) performed at the ascending aorta level, or the central aortic pressure (CAP) and cardiovascular composite events and surrogate markers of cardiovascular disorders are linked significantly to CAP (Agabiti-Rosei et al., 2007; Ghiadoni et al., 2009; Yao et al., 2018), while peripheral vasculature and target organs [brain, heart, kidney (Mitchell, 2008)] are directly exposed to it. Unfortunately, the "gold standard" CAP measurements are invasive, only applicable during catheterization, and thus cannot be performed routinely therefore there is growing concern in the noninvasive measurement of CAP.

On the other hand, peripheral blood pressure (PBP) information can be obtained noninvasively, and many noninvasive methods for reliable measuring BP waves based on *oscillometry*, *auscultation* and *tonometry* are currently accessible (Bronzino, 2004; Lee and Nam, 2009). The auscultation method uses the stethoscope to analyze the noises that occur during the slow decompression of the cuff sited on the elbow. Oscillometry detects the maximum pulsations matching the phenomena in the systolic–diastolic sequence, sent out by the arteries compressed by a pneumatic sleeve (Stergiou et al., 2006; Sorvoja, 2006). Systolic blood pressure (SBP) is measured during the period of ventricular contraction. Diastolic blood pressure (DBP) is measured between periods of ventricular contraction.

However, PBP information (e.g., brachial blood pressure) per se, conveys less accurately CAP information, and shows off different response to certain drugs (O'Rourke, 2006; O'Rourke et al., 2001). To circumvent these difficulties and shortcomings and use PBP, recently, there is growing concern in the development of methods and devices for the noninvasive extract CAP out of PBP (Yao, et al., 2018), and the pulse wave reflection is the underlying signal that may be used to this aim.

The contraction of the heart increases the pressure in the aorta eliciting a direct, "ejected" pressure wave (EPW) that drives the blood flow throughout the body. The EPW may then be partially reflected from an artery bifurcation or from a peripheral artery: the structure and properties of the downstream vascular network, the peripheral vascular resistance, the elasticity and compliance of blood vessels may cause a reverse, reflected pressure wave (RPW), which interacts with the EPW and patterns the morphology of the pulse wave (Wang and Parker, 2004; Wang et al., 2004, 2006; van der Vosse and Stergiopulos, 2011).

The radial, ulnar, and brachial arteries stiffness does not change significantly with aging, hypertension, and exercise among subjects with similar physiological and pathological characteristics, which sets the theoretical grounds for the generalized transfer function

(GTF) to estimate CAP out of brachial or radial pressure waveform (Yao et al., 2018). Moreover, the need for a rigid structure in the vicinity of the artery, which may promote uniform compression and occlusion of the vessel, recommends the selection of an artery close to the body surface, for example, the radial or the brachial arteries (Stergiou et al., 2006; Lee and Nam, 2009; OMRON, 2020). The GTF implements a low-pass filter that cuts off the high harmonics of the pressure waveform traveling from central aorta to the periphery, and it can provide not only quantitative CAP but also CAP waveform (Chen et al., 1997; Lee and Nam, 2009). GTF is implemented in the first device accepted by US Food and Drug Administration for the estimation of CAP (Pauca et al., 2001), and it is the most widely used method so far (Yao et al., 2018). On the other hand, GTF was questioned in chronic kidney disease or arterial stiffness, and not all algorithms that implement GTFs have the same accuracy (Hope et al., 2003; Yao et al., 2018).

For adults from midlife onward, systolic CAP can be calculated via a regression equation using the second systolic peak as an independent variable because the RPW peak in the periphery approximates the central SBP—the pressure gradients in the arterial system are relatively small during late systole, and the late systolic shoulder represents the major peak for them (Pauca et al., 2001; OMRON, 2020). Other GTF specialized methods for CAP estimation are available (Yao et al., 2018): N-point moving average, NPMA (a first-order low-pass filter that removes the pressure wave reflections, providing only the central aortic SBP); adaptive transfer function (for tuning the GTF), individualized transfer function (ITF; uses an individualized physical transmission line for the aorta-brachial and aorta-radial model), blind system identification, BSI (reconstructs the input out of two or more outputs). Whatever method is used, the tonometry waveforms in carotid artery are calibrated to brachial SBP and DBP.

Applanation tonometry (AT), an oscillometric method, was introduced in ophthalmology to assess the pressure exerted by intraocular fluids on the cornea (Applanation Tonometry, 2020). Eventually the tonometric estimation was used to measure the pulse wave of a superficial artery noninvasively too (Kelly et al., 1989) and it evolved into the arterial applanation tonometry (AAT) (Pressman and Newgard, 1963). While the sphygmomanometer measures only AP and DP, the AAT provides continuous pulse waveform with pressure sensor placed over a superficial artery. AAT is use to diagnose atherosclerosis and the factors that can cause myocardial infarction, and it is aimed to estimate BCP (Kips et al., 2011; Cheng et al., 2013; Zayat et al., 2017) and solve for the disagreement between CAP and PBP that was evidenced to augment with the posology of vasoactive agents (Mackenzie et al., 2009).

The augmentation index

The AAT pressure readout is not identical to the invasively measured one, and the pressure applied to flatten the arterial wall and compress overlying tissues must be

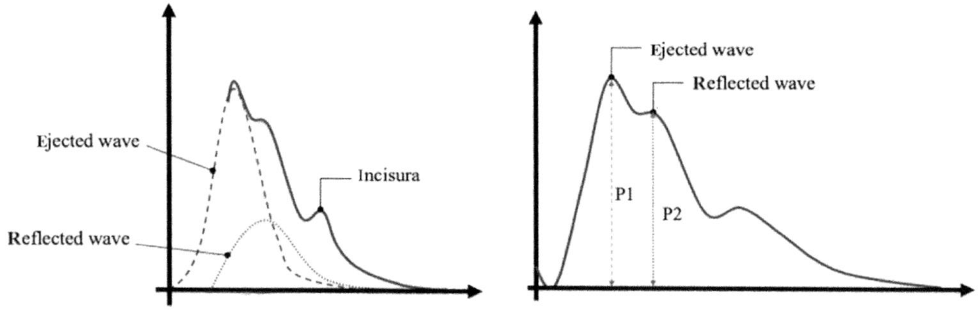

Figure 4.20 The pulse wave—a superposition of the ejected and reflected pulse waves (left)—and the augmentation index (right).

accounted for. The AAT master pressure wavy profile (Fig. 4.20) shows off an incisura (a deep indentation) associated with the RPW. The weights of the DPW and RPW define the augmentation index (AI), which is a measure of arterial stiffness: the larger the RPW the higher the AI, hence the stiffer the arteries are.

In fact, arterial stiffness progresses with aging and due to disorders such as hypertension, hypercholesterolemia, and diabetes. The left ventricular ejection pressure wave propagates faster through stiffer arteries, which leads to faster return of the RPW to the left ventricle. The RPW arriving during systole augments the late SBP (afterload) on the left ventricle. Moreover, the peak of the RPW approaches that of the EPW. Consequently, the heart has to enhance the myocardial contractility to increase the blood pressure, which poses a higher "load on the heart." If this action lasts longer the heart eventually gets strained. The reduction of coronary artery perfusion pressure leads to greater risk of angina, heart attack, stroke, and heart failure.

The pulse waveform obtained using AAT and GTF may satisfactorily estimate the arterial *compliance*. However, some concern is noted regarding AI recovery because the postprocessing relies on rendering the wave profile with higher fidelity. AI is calculated as the difference between the second and first systolic peak pressure (P2−P1) divided through the pulse pressure, expressed as percentage of the BCP (Sievi et al., 2015). Vital hemodynamic parameters may be thus reliably obtained through tonometry, which may be designed specifically to measure the cardiac output, the stroke volume (Zayat et al., 2017), the arterial blood pressure (Kemmotsu et al., 1991a,b), and others.

The generalized transfer function

The aortic pressure waveform for AI calculation can be estimated either from the radial artery waveform, using a transfer function, or from the common carotid waveform. The AT method (including the tonometer) may be seen as a metrological device that convolves some input (arterial pressure here) to be presented as a readout signal. Its functioning is actually a transfer function (TF) that maps the output signal (tonometer readout) onto the input signal (aortic pressure). Then an inverse of the TF may be used to deconvolve the readout and recover to input signal. Of course, noise filtering is provided in this process. The recommended linear ITF of "order 10" (10 previous successive readouts and 10 previous successive inputs), at 75 bpm (labeled AI@75) is defined as (Chen et al., 1997):

$$T(t) = -a_1T(t-1) - a_1T(t-2) - \cdots - a_1T(t-m) + b_1T(t-1) + b_1T(t-2) + \cdots + b_1T(t-n), \tag{4.22}$$

where $T(t)$ is the present readout, and $T(t-i)$, and $P(t-i)$, $i=\{1,2,\ldots,m\}$ are previous, known, outputs (tonometer readouts) and inputs (aortic pressure), "a" and "b" are the parameters, and m and n are the model order (here, 10). ITF is then convolved with a low-pass filter with a cut-off frequency such that the ITF gain function decreases below 1. Its inverse yields the aortic pressure

$$P(t) = -b_2/b_1P(t-1) - \cdots - b_n/b_1P(t-n) + a_1/b_1T(t-1) + \cdots + a_m/b_1T(t-m). \tag{4.23}$$

A GFT may be obtained by averaging the ITF from a population of participating patients (Chen et al., 1997). It has been asserted (Chen et al., 1997) that GTF is statistically more stable than other methods and yields dependable spectral estimates from limited data compared with nonparametric (Fourier transform) approaches (Karamanoglu et al., 1993; Sharman et al., 2006). The variance of the AI- and Fourier-derived spectra are similar only when larger data sets are used.

AI is an important indicator as it was associated with essential physiological parameters, either through univariate expressions, for example, SBP (nonlinear positive association), DPB, age (nonlinear positive association), pulse pressure (PP), central systolic blood pressure (cSBP), but it was negatively related with others such as sex, body mass index, BMI, and physical activity level, PAL (Sievi et al., 2015).

Using small size data collections to process the arterial flow evaluation

Small size samples (~20 subjects), targeting population of healthy subject without diagnosed cardiovascular diseases and nonprobability sampling technique, such as the *convenience sampling* that is prone to sampling bias, may though unveil the relations that exist between vital indicators recordable through AAT and related physiological signals: SBP (SYS), DBP (DIA), PP, central systolic blood pressure (cSBP), AI, and the pressure pulse values (PULSE) (Baran and Savastru, 2017), obtainable using devices such as OMRON 9000 AI (Fig. 4.21).

The characteristics of the population that are of interest here are three common anthropometric variables: the age (AGE), the height (H) and the weight (G) of the units, and normality is checked using graphic-analytical methods, based on the

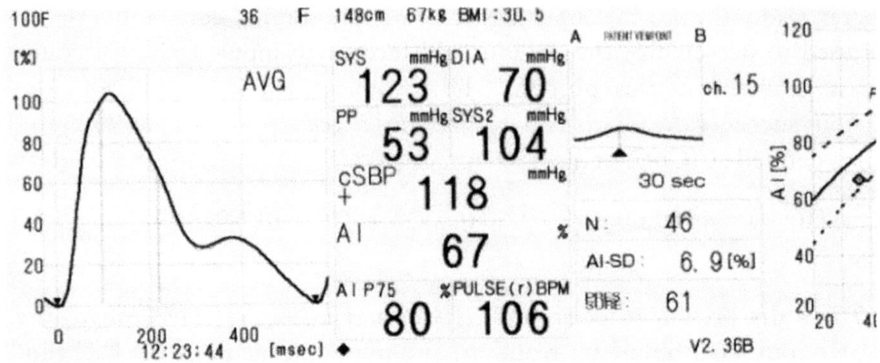

Figure 4.21 Measurement results generated by HEM OMRON AI 9000.

empirical cumulative distribution function and the use of goodness of fit tests. Frequently used in other areas, empirical cumulative function techniques are less common in biostatistics despite their advantages.

Quantile–Quantile plots (QQ plots) may be used for small size samples to fit univariate distributions and discern without bias if data show off a normal distribution (Chantarangsi et al., 2015). QQ plots become probability plots (PP plots) if the sample quantiles are function of theoretical quantiles computed for some distribution (e.g., normal distribution). And working with PP plots presents several advantages (Baran and Savastru, 2017): (1) fast concurrence check of the anticipated model with the sample distribution, (2) outliers and extreme values fast finding, (3) deviation type in case of theoretical model misfit (skewness, shorter, or longer than expected tails). PP-plots may single out extreme values and outliers, if any, and evidence the occurrence and sources of deviations from the normal repartition. They show off whether the variables present no outliers or extreme values and the null hypothesis that data collections has a normal repartition is acceptable, even for the cSBP and SBP (SYS) lacking an unbiased assessment of the deviation from normality.

This univariate analysis (a necessary step in any multivariate study), which relies on the empirical cumulative distribution function rather than on the empirical density function, validates, with a good confidence level, the normality of small correlated data samples, which is an objective difficultly attained otherwise. The normality of the marginal univariate distributions may be checked against the more frequently used Shapiro–Wilk test and the probability plots, combined with the refined Filliben test (based on the probability plot correlation coefficient) (Shapiro and Wilk, 1965; Filliben, 1975). This method is recommended for small size samples, and can distinguish between the normal model and other alternative asymmetric models.

4.6 Arterial function evaluation

The arterial hemodynamic

Unlike for ocular tonometry, in AAT a stern structure, like bone, is required to support the arterial vessel (Fig. 4.22).

The pressure field inside the artery acts upon the vessel wall and the produced stress propagates throughout the structure eventually reaching the tonometer transducer. An appropriate counter pressure is applied so as to partially flatten the artery and the sensor receives only internal arterial pressure.

The outlining physics that concur in modeling the *arterial function evaluation* from its source, the arterial flow, to its equivalent electrical signal has to account for the hemodynamic flow, the accompanying structural interactions produced by it that propagates throughout the anatomic structure to the sensing device, and the mechanoelectrical conversion occurring inside the pressure sensor. When numerical simulation is used, the computational domains may be constructed using imaging techniques, CAD, or both merged using fusion techniques (Chapter 3: Computational Domains). Image-based constructed domains are more realistic, they may be patient related, but their complexity increases the numerical calculation effort. Therefore to just touch the outlining physical insights in the arterial function pressure monitoring potential, CAD solutions oriented on graphic primitives that render the anatomical details (brachial—radial—ulnar system, along with the humerus—radial—ulnar bone system and muscle tissue), may be more efficient (Morega et al., 2015).

The brachial—radial—ulnar artery system, BUR, is an adequate area for AT measuring the pressure because the stiff formation (the bones) nearby the artery eases the uniform compression and vascular occlusion (Savastru et al., 2014). The CAD built

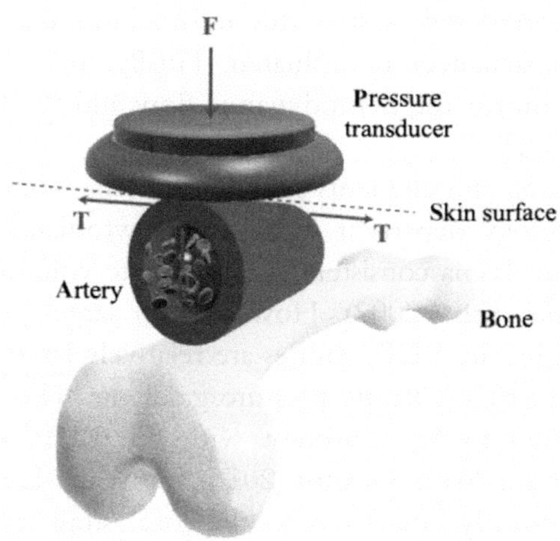

Figure 4.22 Applanation tonometry used in the evaluation of arterial flow dynamics—how it works.

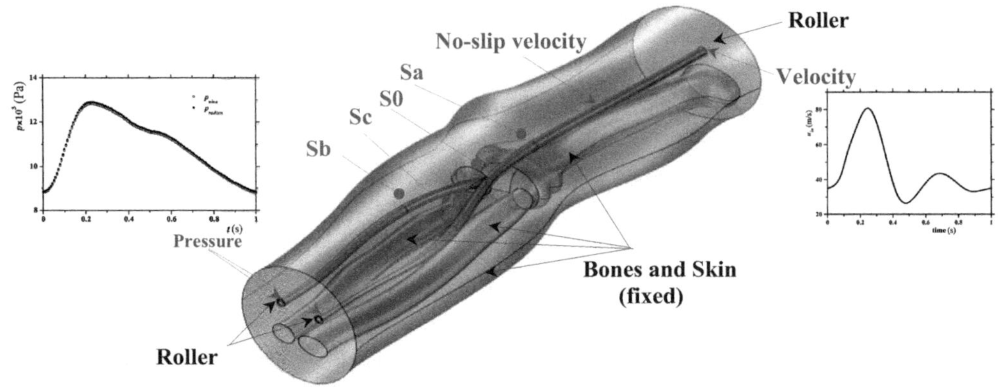

Figure 4.23 CAD-constructed computational domain and the boundary conditions in the arterial function evaluation. Sa, Sb, Sc, and S0 are transducers positions.

computational domain here is representative for an adult's forearm and arm segment: humerus ~36.34 cm, ulna bone ~28.2 cm, and radial bone ~26.4 cm. The BUR arteries sizes are in the range 0.35−0.53 cm (Lee and Nam, 2009; Savastru, 2016). The computational domain for the AT function evaluation is shown in Fig. 4.23.

The arterial function modeling implies the solution of the blood flow, the transmission of the stress field to the transducers, and the mechanoelectrics of the sensors. Moreover, the electric output of the array of sensors may be used to monitor the BUR bifurcation. The complex couplings between these "physics" may be simplified considering the time scales and the properties of the media. Thus a hemodynamic quasistationary problem is solved first. The transmission of the stress produced by the pressure field is then analyzed. Next, the mechanical deformation of the sensor (either PZT or capacitive) is considered. Using this information the electric signal (PZT voltage or change in capacitance) is evaluated. Finally, the electric output of the sensor array is shown to mirror the hemodynamic flow and the brachial−ulna−radius bifurcation (node).

Blood is made of plasma and elements such as platelets, white and red blood cells, and its macroscopic rheology departs it from the Newtonian behavior: its viscosity depends on the flow rate, plasma consistency, erythrocyte volume, platelet leukocytes, and erythrocyte distortion (Kim, 2002). However, for larger vessels the Newtonian model of fluid is acceptable, and BUR arteries are relatively large vessels with pulsating hemodynamic flow driven by oscillating pressure gradients. Three recommended rheological models considered to be consistent with the BUR region (Shibeshi and Collins, 2005; Morega et al., 2013; Savastru, 2016) are used: Newton, for fluids where the viscous stresses are linearly correlated to the local strain rate; Carreau−Yasuda,

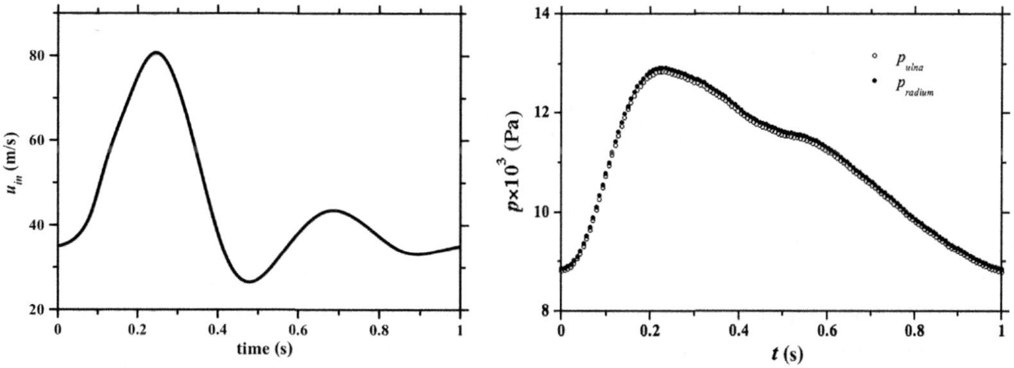

(A) (B)

Figure 4.24 (A) Inlet brachial velocity profile. (B) Outlet pressure profiles for ulnar and radial arteries. Inlet and outlet boundary conditions for the hemodynamic flow: (A) inlet brachial velocity profile, and (B) outlet pressure profiles for ulnar and radial arteries.

which implements a generalized Newton fluid, without elasticity, with a high shear stress, and Ostwald-de-Waele (or power law) that models fluids with higher shear rate and low viscosity. Carreau model exhibits Newtonian fluid behavior for a low shear rate and otherwise the model tends to Ostwald-de-Waele fluid (Chapter 1: Physical, Mathematical and Numerical Modeling). The arterial flow is presented through the momentum (Navier–Stokes) and mass conservation laws

$$\rho\left[\frac{\partial \mathbf{u}}{\partial t}+(\mathbf{u}\cdot\nabla)\mathbf{u}\right] = -\nabla p+\nabla\cdot\vec{\tau}, \quad \nabla\cdot\mathbf{u}=0, \tag{4.24}$$

where $\mathbf{u}$ is the velocity, p the pressure, $\vec{\tau}$ the shear stress tensor, ρ the mass density, and η the dynamic viscosity. The boundary conditions that close the flow model are no-slip (zero velocity) at the walls, inlet velocity (mass flow rate) uniform, time dependent profile for the brachial artery, and outlet pressure uniform, pressure dependent profiles for radius and ulna (Figs. 4.23 and 4.24).

Using these velocity and pressure profiles as boundary conditions introduces the direct and the reverse pressure waves. This approach may alleviate the difficulty of including a downstream lumped hydrodynamic circuit to model the pressure reflections (Olufsen, 1999).

Structural analysis

The bones are a rigid, nondeformable structure therefore the pressure wave leads to structural changes in the vessel wall and the surrounding muscular tissue only, and these tissues are assumed isotropic, almost incompressible. Their nonlinear deformations are modeled here as hyperelastic, neo-Hookean (Dobre et al., 2011b; Morega et al. 2010), which yields the strain energy density

$$W = \frac{1}{2} J^{-\frac{2}{3}}\left(I - \frac{1}{3}\bar{I}_1\, C^{-1}\right) + \frac{1}{2}\kappa J(J-1)C^{-1}, \tag{4.25}$$

where $J = \det(\mathbf{F})$ is the relative variation of the volume, $\mathbf{F}$ is the deformation gradient, $C = \mathbf{F}^T\mathbf{F}$ is the right Cauchy–Green tensor, and $I_1 = \text{trace}(C)$. For the muscle, the initial shear modulus is $\mu = 719{,}676$ Pa, $\bar{I}_1 = I_1\bar{J}^{2/3}$, and the initial bulk modulus is $\kappa = 14{,}393{,}520$ Pa that corresponds to the Poisson ratio $\nu = 0.45$ (Bangash et al., 2007). The stress, S, is then the derivative of the strain energy density function, W, with respect to Green-strains, E, that is, $S = \partial W/\partial E$.

The boundary conditions in this structural analysis are shown in Fig. 4.23. The outer surface of the arm (the skin) is set "free," and "roller" type BCs are set for the cross-sections. The total stress solved for in the first step is used as BC (load) in the structural analysis of the arterial walls deformations produced by the pulsatile flow.

Pressure transducers and their positioning

Piezoelectric or capacitive pressure transducers may be used to determine pressure variation by detecting oscillations in the skin and convert them into measurable electrical quantities (Figliola and Beasley, 1994). The force transmission structure of the transducer (sometimes called artery rider) is smaller than the flattened area of the artery, and centered over the flattened area (Eckerle, 2006; Lee and Nam, 2009). The capacitive sensor consists of a pair of armatures, one fastened and one mobile, compliant to the skin displacement. The distance between the armatures changes yielding a pressure induced change in the electric capacity. The advantages of capacitive sensors over piezoelectric sensors are increased sensitivity, simplicity, at lower costs (Webster, 2006). Metrological properties such as sensitivity, stability, and linearity recommend these sensors (Kumar, 2000).

The precision of the transducers readouts is highly influenced by the sensors proper positioning with respect to the ROI. Their precise centered positioning with respect to the artery is needed for reliable recordings (Kelly et al., 1989), and arrays of fixed sensors might be used to best measure the radial pulse wave that indents the radial artery (Terry et al., 1990; Kemmotsu et al., 1991a,b; Webster, 2006). One or more sensors are placed over the artery and recognized by comparing the measured pressure and the pressure distribution of each sensor at a diastolic interval. Several accompanying, inherent factors (physiological and psychological variations, movements) may affect the measurement accuracy and care should be devoted to reduce or eliminate them (Lee and Nam, 2009). Standardized measurement settings are suggested to shed the subject's variation (Van Bortel et al., 2002).

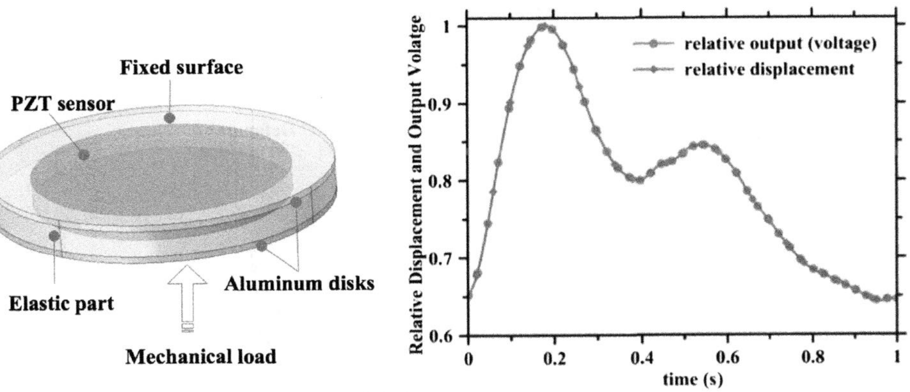

Figure 4.25 The piezoelectric transducer (left) and its output, nondimensional form (right) (Morega et al., 2014).

The PZT transducers (cylindrical, 8-mm diameter and 2-mm height) that record the arterial function are located above the arteries (Fig. 4.23). The central part is the piezoelectric core, embedded in an elastic material, and capped by two aluminum disks (armatures) (Fig. 4.25). One of the disks is fixed (e.g., with a cuff) and the other one contacts the skin. The pressure pulse wave that drives the acceleration phase of the blood flow is transmitted through the vessel wall and surrounding tissue to the PZT cell, causing its compression and thus producing a voltage drop at the sensor terminals that may be conditioned and used for monitoring purposes (Pro-Wave Electronics Corp, 1998; Sur and Ghatak, 2020; Mohammadi et al., 2013).

The direct PZT coupling of electric and mechanical stress fields that, in the quasistatic, linear approximation is described through the first-order Onsager relations (Chapter 1: Physical, Mathematical and Numerical Modeling) in "stress-charge" form

$$\sigma = c_E \varepsilon_{strain} - e^T \mathbf{E}, \quad \mathbf{D} = e\varepsilon_{strain} + \varepsilon_0 \varepsilon_r \mathbf{E}, \tag{4.26}$$

where σ is the normal stress, e is a coupling matrix, c_E the elasticity matrix, ε_{strain} the relative strain, $\mathbf{D}$ the electric flux density, $\mathbf{E}$ the electric field strength, ε_0 the permittivity of free space, and ε_{rS} is the relative permittivity of the PZT material (here linear, homogeneous, isotropic).

Because the sensor works at the pulse rate of the circulatory system and the PZT uses an elastic linear material with small deformations, the electrical part of problem may be assumed static, decoupled

$$-\nabla(\varepsilon_0 \varepsilon_r \nabla V) = \rho_v, \tag{4.27}$$

where V is the electric potential and ρ_V is the electric charge density. The boundary conditions that close the model are mechanical stress for the armature that contacts the

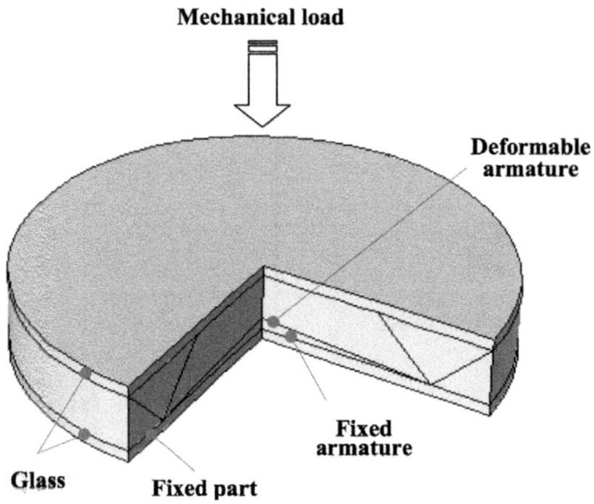

Figure 4.26 Capacitive mechanoelectrical transducer (Savastru et al. 2014).

skin (the skin is flattened); the opposite face is fixed, prestressed (the cuff), and the lateral side is unconstrained (Fig. 4.23). The electrical state assumes floating potential for the armature in contact with the skin; ground for the opposite face, and electrical insulation (zero charge density) for the lateral side.

An important issue with sensors is the sensitivity and linearity of their response—here, the electrical signal (voltage drop) versus the mechanical stress—and Fig. 4.25 (right) shows off the voltage sourced by the PZT when subject to a mechanical load that models a pulsating hemodynamic flow (Morega et al., 2014).

Precision capacitive pressure sensors may be used to convert mechanical quantities (displacements, stress). Capacitive sensors are passive devices. Fig. 4.26 shows a capacitive sensor that proves a concept: two planar armatures sandwich a deformable dielectric [e.g., Kapton P-HN polyimide (Dupont, 2020)]. The device here is cylindrical, 1.5-mm high (the flexible part is made of 20-μm thick either polyimide P-HN or silicon), its radius is 5 mm, with an initial 1.5 mm distance between armatures.

The material properties are listed in Table 4.1 (Savastru, 2016; Dupont, 2020).

The boundary conditions for the capacitive sensor in the mechanical problem are normal load for the armature that contacts the skin, the opposite side of the device is

Table 4.1 Mechanical properties for the capacitive sensor parts.

Property	Silicon	Kapton HN	Glass
Poisson ratio, ν	0.27	0.34	0.244
Young modulus, E (GPa)	0.131	2.5	86.667
Mass density, ρ (kg/m^3)	2330	1.42	2600
Thermal expansion coefficient, α (K^{-1})	4.51×10^{-6}	20×10^{-6}	3.41×10^{-6}

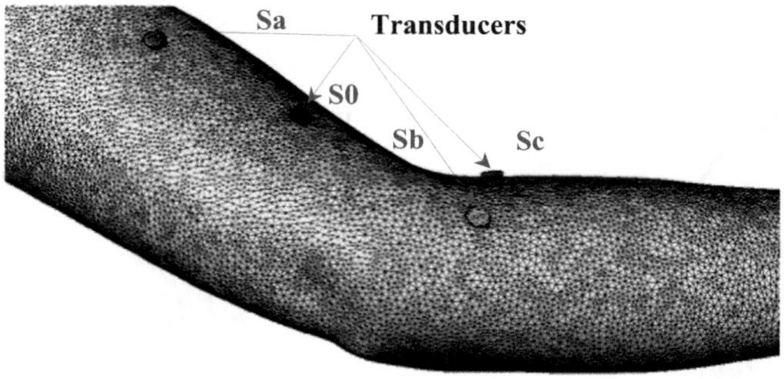

Figure 4.27 Sensors positions in the BUR region.

fixed and the side area is free. For the electrical problem, the armature that contacts the skin has ground condition, upper armature is set at 1 V, and the sides are electrically insulated (Fig. 4.26).

The hemodynamic pressure pulse wave is eventually sensed as a significant mechanical signal (load) by the capacitive transducer. The relatively slow change in the capacitance (at the pulse rate) recommends quasistatic working conditions (4.26) to calculate it. The Arbitrary Lagrangian–Eulerian technique (Gadala and Wang, 1998) was used to solve the deformable component of the capacitive sensor Comsol. Thus the FEM mesh can be deformed inside the computational domain whereas on the boundary of the domain (the deformable component) the mesh accurately follows the limits of the computational domain (Savastru et al., 2014). Numerical simulations have evidenced the linearly proportional changes in the capacitance of this sensor, which qualifies then for a precise arterial functional monitoring (Savastru, 2016).

The sensors, PZT or capacitive, are conveniently positioned (Bronzino, 2004) to monitor the hemodynamic flow throughout BUR junction (Fig. 4.27).

The size of the sensor-to-skin contact surface is correlated with the arteries sizes, and is an important factor in ensuring the measurement accuracy. Having in view the vessels sizes here (Chami et al., 2009; Ashraf et al., 2010), and assuming a circular contact, its diameter is in the range 3.5–5.3 mm, which facilitates a quasiuniform mechanical loading.

Arterial flow evaluation

The arterial flow is rendered in Fig. 4.28 for the Newtonian model, at to remarkable moments—maximum and minimum flow rates. Fluctuating recirculation cells are observed in the bifurcation region, in the entrance regions of ulnar and radial arteries, which are prone to atherosclerosis (Morega et al., 2013).

The cross-sectional average pressure was recorded at stations Sa, Sb, Sc (Fig. 4.23) along the brachial flow, for each rheological model. Fig. 4.29 renders the results at Sa.

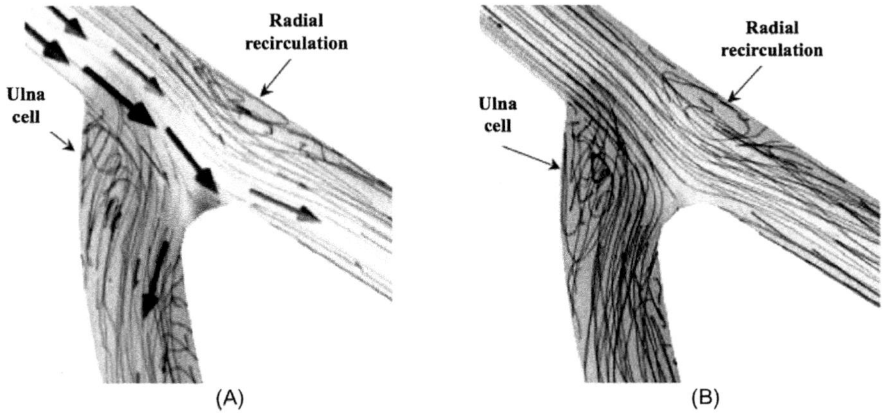

Figure 4.28 (A) Maximum flow rate, at $t = 0.25$ s. (B) Minimum flow rate, at $t = 0.5$ s. The hemodynamic of the BUR system at—oscillating recirculation cells are observed at the entrance in the radial and ulnar arteries (Morega et al., 2013).

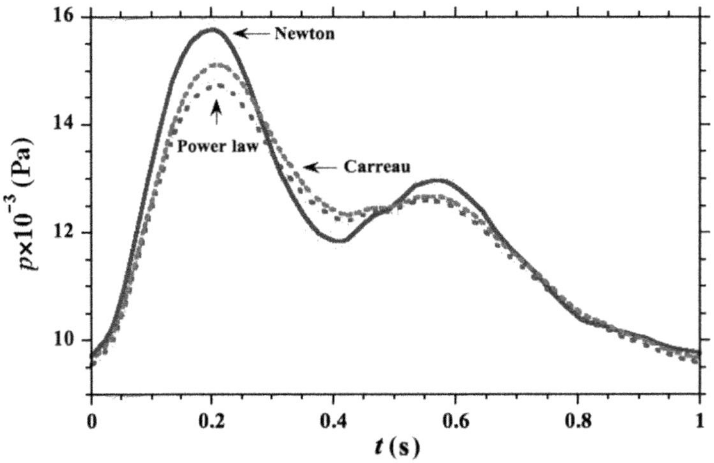

Figure 4.29 Pressure recorded at station Sa (Fig. 4.23) for different rheological models—numerical simulation results (Morega et al., 2013).

Although the Ostwald-de-Waele fluid provides, in the present circumstances, for the lowest flow resistance (lowest pressure drop), the pressures recorded (numerical simulations) by the sensors at all stations do not indicate major discrepancies between the three rheological models. They lead to comparable results, however, some existing variability suggests differences in revealing factors such as the shear stress, velocity profile, wall stress, recirculation cells, residence time, and others, which may influence the development of cardiovascular disease (Chien et al., 1998).

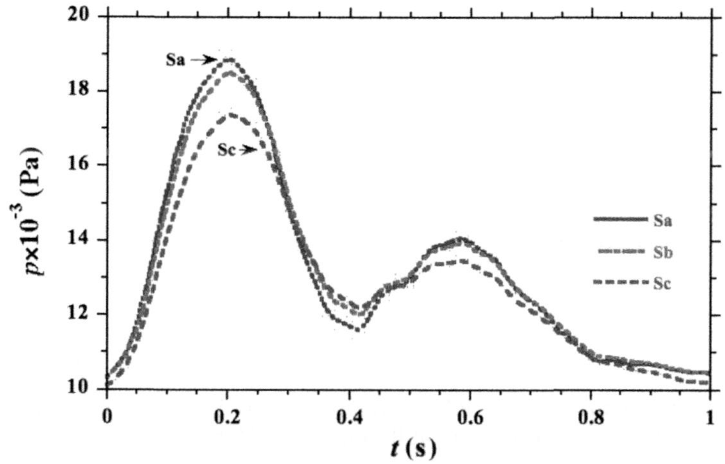

Figure 4.30 The average pressure when the radial artery is occluded; values are referred with respect to the absolute outlet pressure (Fig. 4.24) (Morega et al., 2013).

Fig. 4.30 shows the pressure sensed at stations Sa, Sb and Sc, when the radial artery is completely occluded—arterial tonometry requires its applanation (partial occlusion) only—for the power law fluid.

Two recirculation cells are seen by and upstream the ulna and radius arteries, and they intensify during the slow flow that sets in 0.02 and 0.4 s. In a real-life scenario, depending on downstream conditions, local vascular morphology and walls stiffness, and blood specific rheology, the bifurcation may promote or intensify an RPW, as detected and measured through the AAT and evidenced by the AI. The average value of blood pressure at the three stations is seen in Fig. 4.31, and the voltages produced by the PZT sensors and the

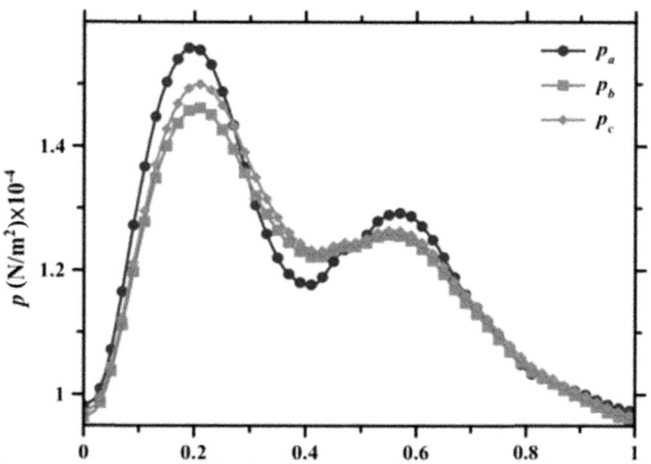

Figure 4.31 Average pressure calculated at Sa, Sb, Sc, for Ostwald-de-Waele fluid (Morega et al., 2013).

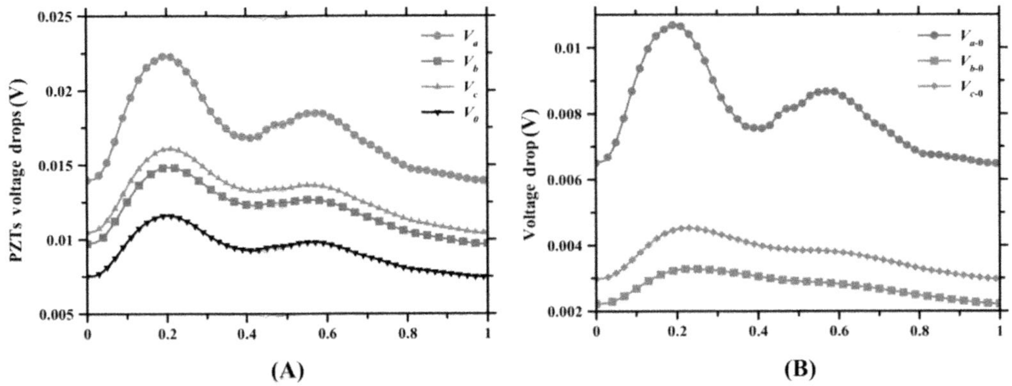

Figure 4.32 Voltages produced by the PZTs at stations Sa, Sb, Sc (A), and the voltage drops at the star circuit terminals that connect them (B) at BUR junction (Morega et al., 2013).

voltage drops at the star electric circuit terminals made by them, when their ground armatures are connected, are plotted in Fig. 4.32.

The largest voltage drop is across the brachial branch, which may be used to model an electric circuit analogue to the arterial ramification and calculate its hydrodynamic impedances.

Same quality results are obtained when capacitive sensors are used (Savastru et al., 2014). For the sensor introduced earlier, the change in capacity is ~31 pF for 11.2 mN compression force. The larger the pressure variations are the smaller the distance between the stations should be, and the larger the change in capacity is. A glimpse in the mechanical interaction between the transducer and the arm is shown by the structural deformation at $t = 0.55$ s (Fig. 4.33).

Displacements larger than 3.2 times were calculated for the K-HN sensor as compared to the Si one. However, in both cases, the transfer function of the capacitive sensor (capacitance–mechanical load) is linear (Fig. 4.34).

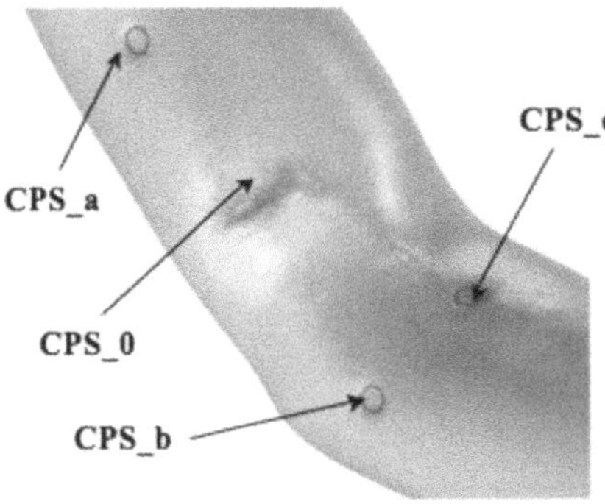

Figure 4.33 Firm transducers positions against arm deformation at $t = 0.55$ s (Fig. 4.24)—amplified 30 times.

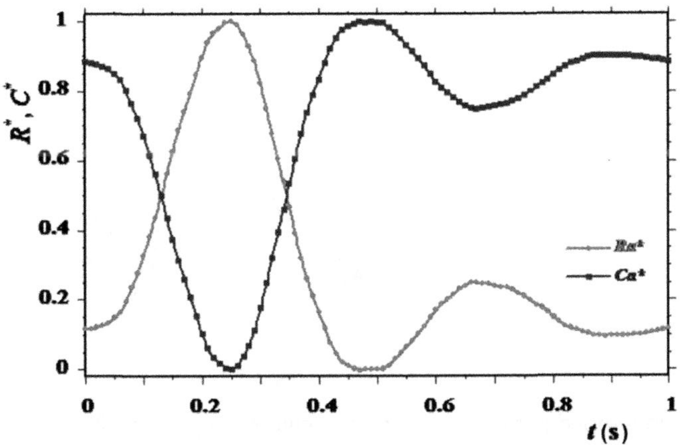

Figure 4.34 Si and K-HN capacitive sensor linear response, nondimensional values—mechanical load (R^*) and capacitance (C^*) (Savastru et al., 2014).

The changes in pressure following the cardiac cycle are detected by the capacitive sensor, which transforms them into electrical signals. The mechanical load (R^*) and the capacitance (C^*) may be scaled using

$$R^* = \frac{R_{\max} - R}{R_{\max} - R_{\min}}, \quad C^* = \frac{C_{\max} - C}{C_{\max} - C_{\min}}, \tag{4.28}$$

where $(\cdot)_{\max}$ and $(\cdot)_{\min}$ are quantities specific to the sensor. For scaling consistency, here, the Si capacitive sensor that is situated above the brachial artery is selected as reference, $C_{\max} = 30.9077$ pF ($t = 0.255$ s), $C_{\min} = 30.9021$ pF ($t = 0.285$ s), $R_{\max} = 139.349$ μm, and $R_{\min} = 139.344$ μm.

Moreover, the mass flow rate (calculated here) may be measured experimentally.

A equivalent lumped parameters electric circuit

In general, lumped parameters circuits are built using ideal circuit elements (Chapter 1: Physical, Mathematical and Numerical Modeling and Chapter 6: Magnetic Drug Targeting). The analogy that exists in the physics of the flux (electrical current)—gradient (voltage drop) electromagnetic field, on one hand, and the flux (mass flow rate)—gradient (pressure drop) fluid mechanics, on the other hand, may be used to introduce and use hydrodynamic circuits made of impedances (vascular system) and power sources (the heart) can be used for the calculation of the direct and inverse pressure waves (Olufsen, 1999).

In view of this analogy, the BUR junction may be modeled as a star fluidic circuit, which admits an equivalent electric circuit. Pressure readouts at the four stations are available as the mass flow rates do (either calculated, in a numerical experiment, or

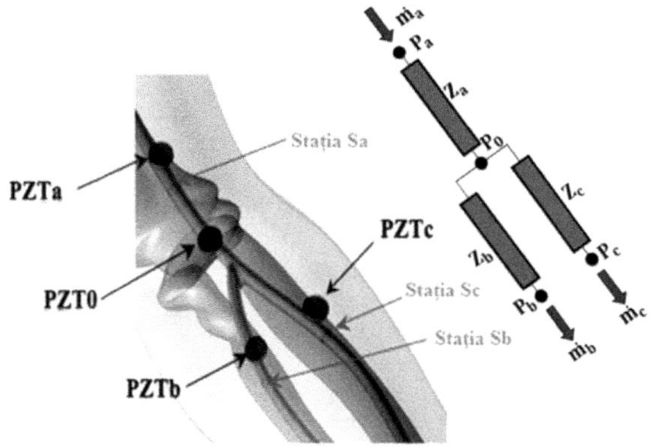

Figure 4.35 The twin-lumped parameters electric circuit (Morega et al., 2014).

measured, in a physical experiment). The impedances of the arterial bifurcation branches may be defined using pressure drop to mass flow rate ratios. A system of four voltages may be obtained when PZTs are used, and a twin lumped parameters passive electrical circuit may be then constructed Fig. 4.35—an inverse, synthesis problem has to be solved. This passive electric circuit can then be used to study direct and inverse voltage (pressure) wave's propagation.

Nondimensional quantities and parameters may be used. Thus $\tilde{V} = \tilde{p}$ and $\tilde{Z} = \tilde{p}/\tilde{\dot{m}}$, where $\tilde{V} = V/V_{max}$ is the nondimensional voltage; V is the PZT voltage output, V_{max} is the maximum voltage produced by the PZTs; $\tilde{p} = p/p_{max}$ is the nondimensional pressure; p is the average local pressure, p_{max} is the maximum pressure; and $\tilde{\dot{m}} = \dot{m}/\dot{m}_{max}$ is the nondimensional mass flow rate, $\dot{m}_{max}$ is the maximum mass flow rate.

The impedance dynamics for the brachial, ulna, and radial arterial segments (left) and their derivatives (right) are seen in Fig. 4.36. The derivatives highlight aspects

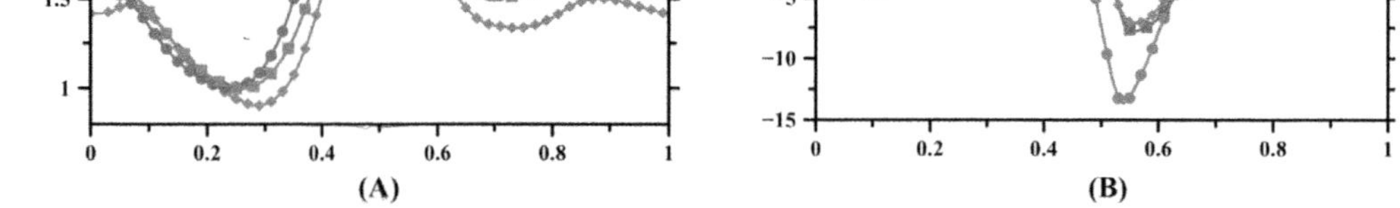

Figure 4.36 Hydrodynamic impedances, nondimensional values (Morega et al., 2014). (A) Brachial–ulnar–radial impedances, and (B) their derivatives.

specific to hemodynamic events associated with the heart cycle, for example, at times $t \sim 0.55$ s and $t \sim 0.85$ s.

The analysis of hemodynamic elements through the fluid circuit twin provides information about measuring blood pressure, or estimates the DPW and RPW by solving the companion ODE models.

The cardiac electrical activity is macroscopically manifested as action potentials that propagate in a synchronized manner. The tuning parameters introduced by the outlining models proposed to present them may raise concerns on their physical justification hence usage. It is the penalty that has to be accepted for using homogenization techniques needed for using continuous, homogenized media. Even so, numerical modeling on detailed "avatar physics" of the actual electrophysiologic processes may have the merit to render the cardiac cells behavior under normal and abnormal conditions. This behavioral prototype may be used to provide the additional information needed to optimize the inversion techniques that aim to predict the cardiac electric activity out of measurements performed on the body surface (the EEG inverse problem), to assist the medical staff in the analysis of complex cardiac problems and to optimize existing pacing or defibrillation methods.

The electric activity of the heart triggers the mechanical activity of the hemodynamic system, whose state is its reflection. Noninvasive methods aim to qualify and quantify the vascular system state, which may be used, in turn, to assert the cardiac state. This complex electromechanical interaction may suggest a "holistic" approach in modeling cardiovascular interactions and processes—from the triggering excitable cardiac tissue to the pulse pressure wave response and the associated feedback. Here and not the least, the hemodynamic system and flow may be well presented through analogue, twin electric circuits, which facilitate their mathematical modeling and synthesis.

References

Agabiti-Rosei, E., Mancia, G., O'Rourke, M.F., Roman, M.J., Safar, M.E., Smulyan, H., et al., 2007. Central blood pressure measurements and antihypertensive therapy. A consensus document. Hypertension 50, 154–160.

Aguilar, M., Nattle, S., 2016. The pioneering work of George Mines on cardiacarrhythmias: groundbreaking ideas that remain influentialin contemporary cardiac electrophysiology (symposium review). J. Physiol. 594 (9), 2377–2386.

Akhmediev, N., Soto-Crespo, J.M., Town, G., 2001. Pulsating solitons, chaotic solitons, period doubling, and pulse coexistence in mode-locked lasers: complex Ginzburg-Landau equation approach. Phys. Rev. E 63.

Aliev, R.R., Panfilov, A.V., 1996. A simple two-variable model of cardiac excitation. Chaos Soliton. Fract. 7 (3), 293–301.

Alonso, S., Bär, M., Panfilov, A.V., 2013. Negative tension of scroll wave filaments and turbulence in three-dimensional excitable media and application in cardiac dynamics. Bull. Math. Biol. 75, 1351–1376.

Applanation Tonometry, 2020. Mayo Clinic <https://www.mayoclinic.org/tests-procedures/eye-exam/multimedia/applanation-tonometry/img-20006176>.

Aranson, I.S., Kramer, L., 2002. The world of the complex Ginzburg-Landau equation. Rev. Mod. Phys. 74, 99–143.

Arts, T., Kruger, R.T., van Gerven, W., Lambregts, J.A., Reneman, R.S., 1979. Propagation velocity and reflection of pressure waves in the canine coronary Artery. Am. J. Physiol. Heart Circ. Physiol. 237 (4), 469–474.

Ashraf, T., Panhwar, Z., Habib, S., Memon, M.A., Shamsi, F., Arif, J., 2010. Size of radial and ulnar artery in local population. J. Pak. Med. Assoc. 60 (10), 817–819.

Bangash, M.Y.H., Bangash, F.N., Bangash, T., 2007. Trauma. An Engineering Analysis with Medical Case Studies Investigation. Springer, ISBN 3-540-36305-X.

Bär, M., Gottschalk, N., Eiswirth, M., Ertl, G., 1994. Spiral waves in a surface reaction: model calculations. J. Chem. Phys. 100 (2), 1202–1214.

Baran, I.M., Savastru, C., 2017. Procedure for assessing normality in small size samples of physiological data. U.P.B. Sci. Bull., Ser. C. 79 (1), 161–172.

Beeler, G.W., Reuter, H., 1977. Reconstruction of the action potential of ventricular myocardial fibres. J. Physiol. 268 (1), 177–210.

Benetos, A., Gautier, S., Labat, C., Salvi, P., Valbusa, F., Marino, F., et al., 2012. Mortality and cardiovascular events are best predicted by low central/peripheral pulse pressure amplification but not by high blood pressure levels in elderly nursing home subjects. J. Am. Coll. Cardiol 60 (16), 1503–1511.

Bergé, P., Pomeau, Y., 1984. Vidal Ch., Order out of Chaos. Towards a Deterministic Approach to Turbulence. John Wiley & Sons, New York, Hermann, Paris.

Berne, R.M., Levy, M.N., Koeppen, B.M., Stanton, B.A. (Eds.), 1998. Physiology. fourth ed. Mosby Inc.

Boulakia, M., Fernandez, M.A., Gerbeau, J.F., Zemzemi, N., 2008. Direct and inverse problems in electrocardiography. AIP Conf. Proc. 1048, 113. Available from: https://doi.org/10.1063/1.2990868.

Boulakia, M., Genadot, A., Thieullen, M., 2015. Simulation of SPDEs for excitable media using finite elements. J. Sci. Comput. 65, 171–195.

Bronzino, J.D., 2004. The Biomedical Engineering Handbook second ed., vol. 1. CRC Press LLC, pp. 1174–1213.

Burnes, J.E., Taccardi, B., Rudy, Y., 2000. A noninvasive imaging modality for cardiac arrhythmias. Circulation 102, 2152–2158.

Bürger, R., Ruiz-Baier, R., Schneider, K., 2010. Adaptive multiresolution methods for the simulation of waves in excitable media. J. Sci. Comput. 43, 261–290.

Capasso, J.M., Malhotra, A., Remily, R.M., Scheuer, J., Sonnenblick, E.H., 1983. Effects of age on mechanical and electrical performance of rat myocardium. Am. J. Physiol. Heart Circ. Physiol. 245 (1), H72–H81.

Carvalhaes, C., de Barros, J.A., 2014. The surface laplacian technique in EEG: theory and methods. I. J. Psychophysiol. Available from: https://arxivorg/abs/1406.0458.

Chami, H.A., Keyes, M.J., Vita, J.A., Mitchell, G.F., Larson, M.G., Fan, S., et al., 2009. Brachial artery diameter, blood flow and flow-mediated dilation in sleep-disordered breathing. Vasc. Med. 14, 351–360.

Chantarangsi, W., Liu, W., Bretz, F., Kiatsupaibul, S., Hayter, A., Wan, F., 2015. Normal probability plots with confidence. Biometrical J. 57, 52–63.

Chen, C.-H., Nevo, E., Fetics, B., Pak, P.H., Yin, F.C.P., Lowell Maughan, W., et al., 1997. Estimation of central aortic pressure waveform by mathematical transformation of radial tonometry pressure. Validation of generalized transfer function. Circulation 95 (7), 1827–1836.

Chen, J.-X., Xiao, J., Qiao, L.-Y., Xu, J.-R., 2018. Dynamics of scroll waves with time-delay propagation in excitable media. Commun. Nonlinear Sci. Numer. Simulat. 59, 331–337.

Cheng, H.-M., Lang, D., Tufanarau, C., Pearson, A., 2013. Measurement accuracy of non-invasively obtained central blood pressure by applanation tonometry: a systematic review and meta-analysis. Int. J. Cardiol. 167, 1867–1876.

Chialvo, D.R., Gilmour, R.F., Jalife, J., 1990. Low dimensional chaos in cardiac tissue. Nature 343, 653–657.

Chien, S., Li, S., Shyy, J.Y.-J., 1998. Effects of mechanical forces on signal transduction and gene expression in endothelial cells. Hypertension 31, 162–169.

Cox, R.H., 1971. Determination of the true phase velocity of arterial pressure waves *in vivo*. Circulation Res. 29 (4), 407–418.

Di Francesco, D., Noble, D., 1985. A model of cardiac electrical activity incorporating ionic pumps and concentration changes. Phil. Trans. R. Soc. B Biol. Sci. 307 (1133).

Dobre, A.A., Morega, A.M., 2011. Numerical simulation of the subclavian aneurysm blood flow. In: International Conference on E-Health and Bioengeering, EHB 2011, Iasi, Romania, 24–26 November 2011.

Dobre, A.A., Ipate, M.C., Morega, A.M., A numerical study of the electrical activity of the heart. In: Proceedings of the Biomaterials, Tissue Engineering and Medical Devices—BiomMed 2010 Conference, 23–25 September 2010, Sinaia, Romania.

Dobre, A.A., Morega, A.M., Morega, M., Ipate, C.M., 2011a. Numerical simulation in electrocardiography. Rev. Roum. Sci. Techn. Électrotechn. et. Énerg. 56 (2), 209–218.

Dobre, A.A., Morega, A.M., Morega, M., Ipate, M.C., 2011b. Thoracic mapping of the nonlinear electrical activity of the heart, In: Proceedings of the 7th International Symposium on Advanced Topics in Electrical Engineering—ATEE 2011, 12–14 May 2011, Bucharest, Romania.

Dowle, M., Mantel, R.M., Barkley, D., 1997. Fast simulations of waves in three-dimensional excitable media. Int. J. Bifurc. Chaos 7 (11), 2529–2545.

Du, Q., Gunzburger, M.D., Peterson, J.S., 1992. Analysis and approximation of the Ginzburg-Landau model of superconductivity. SIAM Rev. 34 (1), 54–81.

Dupont Kapton®, 2020. HN polyimide. <https://www.dupont.com/products/kapton-hn.html> (accessed May 2020).

Eckerle, J.S., 2006. Tonometry, arterial. In: Webster, J.G. (Ed.), Encycopedia of Medical Devices and Instrumentation, second ed. Wiley, pp. 402–409.

Einthoven, W., 1908. Weiteres über das Elektrokardiogram. Pflüger Arch. Ges. Physiol. 122, 17–48.

Ettinger, P.O., Regan, T.J., Oldewurtel, H.A., 1974. Hyperkalemia, cardiac conduction, and the electrocardiogram: a review. Am. Heart J. 88 (3), 360–371.

Fenton, F.H., Cherry, E.M., Hastings, H.M., Evans, S.J., 2002. Multiple mechanisms of spiral wave breakup in a model of cardiac electrical activity. Chaos 12 (3), 852–892.

Figliola, R.S., Beasley, D.E., 1994. Theory and Design for Mechanical Measurements, second ed. J. Wiley & Sons, Inc.

Filliben, J.J., 1975. The probability plot correlation coefficient test for normality. Technometrics 17 (1), 111–117.

Filippi, S., Cherubini, C., 2006. Multiphysics Models of Biological Systems. Milano.

Filippi, S., Cherubini, C., 2009. Electrical signals in a heart. Comsol Model Library, Equation-Based Models, Comsol Multiphysics, v.3.5a.

Fitzhugh, R., 1961. Impulses and physiological states in theoretical models of nerve membrane. Biophysical J. 1, 445–466.

Gadala, M.S., Wang, J., 1998. A practical procedure for mesh motion in arbitrary Lagrangian-Eulerian method. Eng. Comput. 14, 223–234.

Gao, P., Hao, D., An, Y., Wang, Y., Qiu, Q., Yang, L., et al., 2017. Comparison of electrohysterogram signal measured by surface electrodes with different designs: a computational study with dipole band and abdomen models. Sci. Rep. 7 (17282), 10 pp.

Garfinkel, A., Kim, Y.H., Voroshilovsky, O., Voroshilovsky, O., Qu, Z., Kil, J.R., et al., 2000. Preventing ventricular fibrillation by flattening cardiac restitution. Proc. Natl. Acad. Sci. USA 97, 6061–6066.

Ganong, W.F., 2005. Review of Medical Physiology, 22nd ed. Lange Medical Books/McGraw Hill, Medical Publishing Division, ISBN 0-07-144040-2.

Gesselowitz, D.B., 1967. On bioelectric potentials in an inhomogeneous volume conductor. Biophys. J. 7 (1), 1–11.

Ghiadoni, L., Bruno, R.M., Stea, F., Virdis, A., Taddei, S., 2009. Central blood pressure, arterial stiffness, and wave reflection: new targets of treatment in essential hypertension. Curr. Hypertens. Rep. 11, 190–196.

Gong, Y., Christini, D.J., 2004. Functional reentrant waves propagate outwardly in cardiac tissue. Phys. Lett. A 331 (3–4), 209–216.

Guevara, M.R., Ward, G., Shrier, A., Glass, L., 1984. Electrical alternans and period doubling bifurcations. IEEE Comp. Cardiol. 562, 167–170.

Gulrajani, R.M., 1998. The forward and inverse problems of electrocardiography. IEEE Eng. Med. Biol. 17 (5), 84–101.

He, B., Yu, X., Wu, D., Mehdi, N., 1997. Body surface laplacian mapping of bioelectrical activity. Methods Inf. Med. 36 (04/05), 326–328.

Hope, S.A., Tay, D.B., Meredith, I.T., Cameron, J.D., 2003. Use of arterial transfer functions for the derivation of aortic waveform characteristics. J. Hypertension 21 (7), 1299–1305.

Iakovidis, I., Martin, C.F., 1991. A model study of instability of the inverse problem in electrocardiography. Math. Biosci. 107, 127–148.

Johnson, C.R., MacLeod, R.S., 1996. Local regularization and adaptive methods for the inverse Laplace problem. In: Ghista, D.N. (Ed.), Biomedical and Life Physics. Vieweg + Teubner Verlag.

Johnston, P.R., Walker, S.J., Hyttinen, J.A.K., Kilpatrick, D., 1994. Inverse electrocardiographic transformations: dependence on the number of epicardial regions and body surface data points. Math. Biosci. 120, 165–187.

Kannel, W.B., Gordon, T., Schwartz, M.J., 1971. Systolic versus diastolic blood pressure and risk of coronary heart disease: the Framingham study. Am. J. Cardiol 27 (4), 335–346.

Karamanoglu, M., O'Rourket, M.F., Avolio, A.P., Kelly, R.P., 1993. An analysis of the relationship between central aortic and peripheral upper limb pressure waves in man. Eur. Heart J. 14, 160–167.

Katz, A.M., 2011. Physiology of the Heart, fifth ed. Wolters-Kluwer.

Keener, J.P., 1988. The dynamics of three-dimensional scroll waves in excitable media. Phys. D. 31, 269–276.

Keener, J.P., Sneyd, J., 2009. Mathematical Physiology II: Systems Physiology, second ed. Springer.

Kelly, R., Hayward, C., Avolio, A., O'Rourke, M., 1989. Noninvasive determination of age-related changes in the human arterial pulse. Circulation 80 (6), 1652–1659.

Kemmotsu, O., Ueda, M., Otsua, H., Yamamura, T., Winter, D.C., Eckerle, J.S., 1991a. Arterial tonomtery for noninvasive, continuous blood pressure monitoring during anesthesia. Anesthesiol, Lab. Rep. 75, 333–340.

Kemmotsu, O., Ueda, M., Otsuka, H., Yamamura, T., Okamura, A., Ishikawa, T., et al., 1991b. Blood pressure measurement by arterial tonometry in controlled hypotension. Anesth. Analg. 73–544.

Keyak, J.H., Meagher, J.M., Skinner, H.B., Mote Jr., C.D., 1990. Automated three-dimensional finite element modelling of bone: a new method. J. Biomed. Eng. 12, 389–397.

Kim, S., 2002. *A Study of Non-Newtonian Viscosity and Yield Stress of Blood in a Scanning Capillary-Tube Rheometer* (PhD Thesis). Drexel University, Philadelphia, PA.

Kips, J., Schutte, A., Vermeersch, S., Huisman, H., Van Rooyen, J., Glyn, M., et al., 2011. Comparison of central pressure estimates obtained from SphygmoCor, Omron HEM-9000AI and carotid applanation tonometry. J. Hypertens. 29 (6), 1115–1120.

Kogan, B.Y., Karplus, W.J., Billett, B.S., Pang, B.S., Karagueuzian, H.S., Khan, S.S., 1991. The simplified FitzHugh-Nagumo model with action potential duration restitution: effects on 2D wave propagation. Phys. D Nonlinear Phenom. 50 (3), 327–340.

Kumar, A., 2000. Biosensors based on piezoelectric crystal detectors: theory and application. JOM-E 52 (10).

Lee, J., Nam, K.C., 2009. Tonometric vascular function assessment. Biomed. Eng. 30, 549–566.

Li, G., Wang, Y., Lin, L., Jiang, W., Wang, L.L., Slu, C.-Y., et al., 2005. Active laplacian electrode for the data-acquisition system of EHG. J. Physics: Conf. Ser. 13, 330–335.

Luo, S., Johnston, P., 2010. A review of electrocardiogram filtering. J. Electrocardiol. 43 (6), 486–496. Available from: https://doi.org/10.1016/j.jelectrocard.2010.07.007.

Mackenzie, I.S., McEniery, C.M., Dhakam, Z., Brown, M.J., Cockcroft, J.R., Wilkinson, I.B., 2009. Comparison of the effects of antihypertensive agents on central blood pressure and arterial stiffness in isolated systolic hypertension. Hypertension 54 (2), 409–413.

Makeyev, O., 2018. Solving the general inter-ring distances optimization problem for concentric ring electrodes to improve laplacian estimation. BioMed. Eng. OnLine 17 (117), 21 pp.

Malmivuo, J., Plonsey, R., 1995. Bioelectromagnetism. Principles and Applications of Bioelectric and Biomagnetic Fields. Oxford University Press.

Mitchell, G.F., 2008. Effects of central arterial aging on the structure and function of the peripheral vasculature: implications for end-organ damage. J. Appl. Physiol. (1985) 105 (5), 1652–1660.

Mocanu, D., 2002. Electromegnetic Models in Biomedical Engineering. The Analysis of Inverse Problems of Electrocardiography and X Ray Computerized Tomography. Doctoral Thesis, Faculty of Electrical Engineering, University Politehnica of Bucharest, Bucharest, Romania.

Mocanu, D., Kettenbach, J., Eisenberg, S.R., Morega, A.M., 2002. A total variation approach to inverse electrocardiography, In: 2nd European Medical & Biological Engineering Conference, 4–8 December, Vienna, Paper no. 1339.

Mocanu, D., Kettenbach, J., Eisenberg, S., Morega, A.M., Morega, M., 2005. Nonsmooth regularization in electrocardiographic imaging. Rev. Roumaine Sci. Techn. Electrotech. et. Energ. 50 (2), 249–260.

Mocanu, D., Morega, A.M., Morega, M., 2007. Electrical rhythms in paced cardiac fibers, In: 1st International Conference on Advances of Medical and Health Care through Technology, MediTech, 27–29 September 2007, Cluj-Napoca, Romania, pp. 315–318.

Mohammadi, V., Mohammadi, S., Barghi, F., 2013. Piezoelectric pressure sensor based on enhanced thin-film PZT diaphragm containing nanocrystalline powders. INTECH. Available from: http://dx.doi.org/10.5772/54755.

Morega, A.M., Savastru, C., Morega, M., 2015. Measuring the hemodynamic flow in the brachial-ulnar-radial arterial system. An. St. Univ. Ovidius Constanţa 23 (3), 113–127.

Morega M., 1999. *Bioelectromagnetism*, MatrixRom, Bucharest (in Romanian).

Morega A.M., Mocanu, D., Morega, M., 2000. A 3D realistic ANSYS finite element reconstruction of the human thorax for the extraction of the transfer coefficients for electrocardiographic lead vectors. In: The 9th International ANSYS Conference and Exhibition, 28–30 August, Pittsburgh, PA.

Morega A.M., Dobre A.A., Morega M., 2010. Numerical simulation of magnetic drug targeting with flow—structural interaction in an arterial branching region of interest, In: COMSOL Conference, 17–19 November 2010, Versailles, France.

Morega, A.M., Dobre, A.A., Ipate, M.C., Morega, M., 2011. Electrical activity of the heart by numerical simulation. In: Ion, S., Popa, C. (Eds.), Topics in Mathematical Modeling of Life Science Problems. MatrixRom, Bucharest.

Morega, A.M., Savastru, C., Morega, M., 2013. Numerical simulation of flow dynamics in the brachial-ulnar-radial arterial system, In: The 4th IEEE International Conference on E-Health and Bioengineering—EHB 2013, G.T. Popa University of Medicine and Pharmacy, Iaşi, Romania, 21–23 November.

Morega, A.M., Savastru, C., Morega, M., 2014. A numerical simulation of the electrical monitoring of the brachial-ulnar-radial arterial hemodynamic. In: Vlad, S., Ciupa R., (Eds.) International Conference on Advancements of Medicine and Health Care through Technology, 5–7 June, Cluj-Napoca, Romania, IFMBE Proceeding, vol. 44, Springer, Cham.

Mueller, R., Ruegsegger, P., 1994. Three dimensional finite element modeling of non-invasively assessed trabecular bone structures. J. Med. Eng. Phys. 17, 126–133.

Mukherjee, S., Farr, J.B., Yao, W., 2016. A study of total-variation based noise-reduction algorithms for low-dose cone-beam computed tomography. Int. J. Image Process. (IJIP) 10 (4), 188–204.

Murillo, M., Cai, X.C., 2004. A fully implicit parallel algorithm for simulating the non-linear electrical activity of the heart. Numer. Linear Algebra Appl. 11 (2–3), 261–277.

Nagumo, J., Arimoto, S., Yoshizawa, S., An active pulse transmission line simulating nerve axon, *Proc. IRE.*, lf, pp. 2061–2070, 1962.

Olufsen, M.S., 1999. Structured tree outflow condition for blood flow in larger systemic arteries. Am. J. Physiol. Heart Circ. Physiol. 276, H257–H268.

OMRON, 2020. <https://medaval.ie/docs/manuals/Omron-HEM-9000AI-Manual.pdf> (accessed May 2020).

O'Rourke, M., 2006. Differential impact of blood pressure—lowering drugs on central aortic pressure and clinical outcomes. Principal results of the conduit artery function evaluation (CAFE) study, CAFE. Investigators. Circulation 1213—1225.

O'Rourke, M.F., Pauca, A., Jiang, X.-J., 2001. Pulse wave analysis. J. Clin. Pharmacol. 51, 507—522.

Oster, H.S., Taccardi, B., Lux, R.L., Ershler, P.R., Rudy, Y., 1998. Electrocardiographic imaging. Noninvasive characterisation of intramural myocardial activation from inverse-reconstructed epicardial potentials and electrograms. Circulation 97, 1496—1507.

Paquerot, J.F., Remoissenet, M., 1994. Dynamics of nonlinear blood pressure waves in large arteries. Phys. Lett. A 194 (1—2), 77—82. 600.

Pastore, J.M., Girouard, S.D., Laurita, K.R., Akar, F.G., Rosenbaum, D.S., 1999. Mechanism linking T-wave alternans to the genesis of cardiac fibrillation. Cir. Res. 99, 1385—1394.

Pauca, A.L., O'Rourke, M.F., Kon, N.D., 2001. Prospective evaluation of a method for estimating ascending aortic pressure from the radial artery pressure waveform. Hypertension 932—937.

Petrov, V., Gáspár, V., Masere, J., Showalter, K., 1993. Controlling chaos in the Belousov-Zhabotinsky reaction. Nature 361, 240—243.

Plonsey, R., 1963. Reciprocity applied to volume conductors and the *ECG*. IEEE Trans. BioMed. Electron. 10 (1), 9—12.

Plonsey, R., 2000 Volume conductor theory, The Biomedical Engineering Handbook, second ed., 1, (Chapter 9), CRC Press LLC.

Prakosa, A., et al., 2014. Methodology for image-based reconstruction of ventricular geometry for patient-specific modeling of cardiac electrophysiology. Prog. Biophys. Mol. Biol. 115 (2—3), 226—234.

Pressman, G., Newgard, P., 1963. A transducer for the continuous external measurement of arterial blood pressure. IEEE Trans. Bio-Med. Electron. BME-10, 73—80.

Pro-Wave Electronics Corp, 1998. Piezoelectric blood pressure sensors. <http://www.prowave.com.tw/english/products/pp/bps.htm> (accessed May 2020).

Pullan, A.J., Buist, M.L., Cheng, L.K., 2005. Mathematically Modelling the Electrical Activity of the Heart. From Cell to Body Surface and Back Again. World Scientific Publishing.

Qu, Z., Weiss, J.N., Garfinkel, A., 1999. Cardiac electrical restitution properties and stability of reentrant spiral waves: a simulation study. Am. J. Physiol. 276, pp. 268-263.

Rogozea, L.M., Diaconescu, D.E., Dinu, E.A., Badea, O., Popa, D., Andreescu, O., et al., 2015. Bioethical dilemmas in using animal in medical research. Challenges and opportunities. Romanian J. Morphol. Embryol.—Rev. Roumaine de. Morphologie et. Embryologie 56 (3), 1227—1231.

Royston Maloeuf, N.S., 1935. The myogenic automatism of the contraction of the heart of insects. Ann. Entomological Soc. Am. 28 (3), 332—337.

Rudin, L.I., Osher, S., Fatemi, E., 1992. Nonlinear total variation based noise removal algorithms. Phys. D. 60, 259—268.

Sato, S., Doi, S., Nomura, T., 1994. Bonhoeffer-van der Pol oscillator model of the sino-atrial node: A possible mechanism of heart rate regulation. Methods Inf. Med. 33 (1), 116—119.

Savastru, C., 2016. Electromagnetic Field, Heat Transfer and Structure Interactions in Biomedical Applications (Doctoral thesis), Doctoral School of Electrical Engineering, University Politehnica of Bucharest, Bucharest, Romania.

Savastru, C., Morega, A., Morega, M., 2014. Sensing the hemodynamic of the brachial-ulnar-radial arterial system. In: Proceedings of the 8th International Conference and Exposition on Electrical and Power Engineering, IEEE-EPE, Iaşi, Romania, October.

Schierwagen, A., Ohme, M., 2020. A model for the propagation of action potentials in nonuniform axons. Core available from: <https://core.ac.uk/> (accessed in May 2020).

Schwan, H.P., Kay, C.F., 1957. Capacitive properties of body tissues. Circulation Res. 6, 439—443.

Scrale, S., 2009. Implications of the Turing completeness of reaction-diffusion models, informed by GPGPU simulations on an XBox 360: Cardiac arrhythmias, re-entry and the Halting problem. Comp. Biol. Chem. 33, 253—260.

Shahidi, A.V., Savard, P., Nadeau, R., 1994. Forward and inverse problems of electrocardiography: modeling and recovery of epicardial potentials. IEEE Trans. Biomed. Eng. 41, 249—256.

Shapiro, S.S., Wilk, M.B., 1965. An analysis of variance test for normality (complete samples). Biometrika 52 (3/4), 591–611.

Sharman, J.E., Lim, R., Qasem, A.M., Coombes, J.S., Burgess, M.I., Franco, J., et al., 2006. Validation of a generalized transfer function to noninvasively derive central blood pressure during exercise. Hypertension 47, pp. 1203-1028.

Shiferaw, Y., Karma, A., 2006. Turing instability mediated by voltage and calcium diffusion in paced cardiac cells. Proc. Natl. Acad. Sci. U.S.A. 103 (15), 5670–5675.

Shibeshi, S.S., Collins, W.E., 2005. The rheology of blood flow in a branched arterial system. Appl. Rheol. 15 (6), 398–405.

Sievi, N.A., Franzen, D., Kohler, M., Clarenbach, C.F., 2015. Physical inactivity and arterial stiffness in COPD. Internatl. J. COPD 10, 1891–1897.

Simpleware 4.2, *Simpleware Ltd.*, U.K., 2010.

Sorvoja, H., 2006. Noninvasive Blood Pressure Pulse Detection and Blood Pressure Determination. PhD thesis). Oulun Yliopisto, Oulu, Finland.

Spielman, B.J., 2007. Bioethics in Law. Humana Press Inc, ISBN 9781597452953.

Srinidhi, K., 1999. Convex Optimization Algorithms for Inverse Electrocardiography with Physiologically Relevant Constraints (PhD thesis). Northeastern University, Boston, MA.

Stergiou, G.S., Yiannes, N.G., Rarra, V.C., 2006. Validation of the Omron 705 IT oscillometric device for home blood pressure measurement in children and adolescents: The Arsakion School Study, Blood Press. Monit., 11. pp. 229–234.

Strandberg, T.E., Kaisu, P., 2003. What is the most important component of blood pressure: systolic, diastolic or pulse pressure? Curr. Opin. Nephrol. Hypertens. 12 (3), 293–297.

Strumillo, P., Ruta, J., 2002. Poincaré mapping for detecting abnormal dynamics of cardiac repolarization. IEEE Eng. Med. Biol. 62–65.

Sur, S., Ghatak, S.K., 2020. An inexpensive arterial pressure wave sensor and its application in different physiological condition, <https://arxiv.org/abs/physics/0512071> (accessed May 2020).

Szilagyi, S.M., Benyo, Z., Szilagyi, L., 2002. Comparison of malfunction diagnosis sensibility for direct and inverse ECG signal processing methods. In: Proceedings of the Second Joint 24th Annual Conference and the Annual Fall Meeting of the Biomedical Engineering Society, Houston, TX, vol. 1, 2002, pp. 244–245.

Takembo, C.N., Mvogo, A., Fouda, H.P.E., Kofane, T.C., 2019. Localized modulated wave solution of diffusive FitzHugh–Nagumo cardiac networks under magnetic flow effect. Nonlinear Dyn. 95 (2), 1079–1098.

Terry, S., Eckerle, J.S., Kornbluh, R.D., Low, T., Ablow, C.M., 1990. Silicon pressure transducer arrays for blood-pressure measurement. Sens. Actuators A21–A23, 1070–1079.

Thomas, M.D., Fox, K.F., Coats, A.J.S., Sutton, G.C., 2004. The epidemiological enigma of heart failure with preserved systolic function. Eur. J. Heart Fail. 6, 125–136.

Tournoux, F., Donal, E., Leclercq, C., De Place, C., Crocq, C., Solnon, A., et al., 2007. Concordance between mechanical and electrical dyssynchrony in heart failure patients: a function of the underlying cardiomyopathy. J. Cardiovasc. Electrophysiol. 8 (10), 1022–1027.

Turing, A.M., 1952. The chemical basis of morphogenesis. Phil. Trans. R. Soc. London. Ser. Biol. Sci 237 (641), 37–72.

Vadakkumpadan, F., Arevalo, H., Prassl, A.J., Chen, J., Kickinger, F., Kohl, P., et al., 2010. Image-based models of cardiac structure in health and disease. Wire's Syst. Biol. Med. 2 (4), 489–506.

Valenzuela, F., Kabela, E., 1984. Automatism: an intrinsic property of cardiac tissues. Arch. Inst. Cardiol. Mex. 54 (6), 601–613.

Van Bortel, L.M., Duprez, D., Starmans-Kool, M.J., Safar, M.E., Giannattasio, C., Cockfroft, J., et al., 2002. Applications of arterial stiffness, Task Force III: recommendations for user procedures. Am. J. Hypertens. 15, 445–452.

Van der Pol, B., Van der Mark, J., 1928. LXXII. The heartbeat considered as a relaxation oscillation, and an electrical model of the heart. London, Edinburgh, Dublin Philos. Mag. J. Sci. Series 7 (6), 38.

Van de Vosse, F.N., Stergiopulos, N., 2011. Pulse wave propagation in the arterial tree. Annu. Rev. Fluid Mech. 43, 467–499.

Velipasaoglu, E.O., Sun, H., Zhang, F., Berrier, K.L., Khoury, D.S., 2000. Spatial regularization of the electrocardiographic inverse problem and its application to endocardial mapping. IEEE Trans. Biomed. Eng. 47 (3), 327–337.

Vigmond, E., Vadakkumpadan, F., Gurev, V., Arevalo, H., Deo, M., Plank, G., et al., 2009. Towards predictive modeling of the electrophysiology of the heart. Exp. Physiol. 94, 563–577.

Wang, J.J., Parker, K.H., 2004. Wave propagation in a model of the arterial circulation. J. Biomech. 37, 457–470.

Wang, J.-J., Liu, S.-H., Chern, C.-I., Hsieh, J.-H., 2004. Development of an arterial applanation tonometer for detecting arterial blood pressure and volume. Biomed. Eng. Appl. Basis Comm. 16 (6), 322–330.

Wang, J.-J., Liu, S.-H., Kao, T., Hu, W.-C., Liu, C.-P., 2006. Noninvasive determination of arterial pressure-dependent compliance in young subjects using an arterial tonometer. Biomed. Eng. Appl. Basis Comm. 18 (3), 111–118.

Watanabe, M.A., Fenton, F.H., Evans, S.J., Hastings, H.M., Karma, A., 2001. Mechanisms for discordant alternans. J. Cardiovasc. Electrophysiol. 12, 196–206.

Webster, J.G., 2006. Encyclopedia of Medical Devices and Instrumentation second ed., 6. John Wiley & Sons, Inc.

Wiener, N., Rosenblueth, A., 1946. The mathematical formulation of the problem of conduction of impulses in a network of connected excitable elements, specifically in cardiac muscle. Arch. Inst. Cardiol. Méx 16, 205–265.

World Health Organization, 2017. <https://www.who.int/news-room/fact-sheets/detail/cardiovascular-diseases-(cvds)>, (accessed 6.02.2020).

Woosley, J.K., Roth, B.J., Wikswo Jr., J.P., 1985. The magentic field of a single axon: A volume conductor model. Math. Biosci. 76 (1), 1–36.

Yancy, C.W., 2000. Heart failure in African Americans: A cardiovascular Enigma. J. Card. Fail. 6 (3), 183–186.

Yamashita Y., 1982. Theoretical studies on the inverse problem in electrocardiography and the uniqueness of the solution, IEEE Trans. Biom. Eng. 29(11), 719–725

Yao, Y., Wang, L., Hao, L., Xu, L., Zhou, S., Liu, W., 2018. The noninvasive measurement of central aortic blood pressure waveform. In: Artis A.S., (Ed.), *Blood Pressure—From Bench to Bed*, vol. 18. Istanbul Medeniyet University, Turkey (Chapter 1). <https://doi.org/10.5772/intechopen.76770>, (accessed May 2020).

Zaikin, A.N., Zhabotinsky, A.M., 1970. Concentration wave propagation in two-dimensional liquid-phase self-oscillating system. Nature 225 (5232), 535–5377.

Zayat, R., Goetzenich, A., Lee, J.-Y., Kang, H.J., Jansen-Park, S.-H., Schmitz-Rode, T., et al., 2017. Comparison between radial artery tonometry pulse analyzer and pulsed-Doppler echocardiography derived hemodynamic parameters in cardiac surgery patients: a pilot study. Peer J. 16.

Zduniak, B., Bodnar, M., Forys, U., 2014. A modified van der Pol equation with delay in a description of the heart Action. Int. J. Appl. Math. Comp. Sci. 24 (4), 853–863.

CHAPTER 5

Bioimpedance methods

5.1 Introduction

The increase in the prevalence of chronic diseases has driven the interest in using new technological advances (communication technologies, consumer electronics, wearable systems, etc.) to enhance the quality and affordability of the healthcare systems (Majumder et al., 2019; Punj and Kumar, 2019), either in a clinical environment or domestic conditions (Kyle et al., 2004). With respect to this, bioimpedance techniques have gained an important role because they may provide insights about the internal processes of the body and the living matter in a noninvasive manner (Grimnes and Martinsen, 2008; Naranjo-Hernández et al., 2019; Piuzzi et al., 2019; Rapin et al., 2019). The impedance of a biological medium (cell culture, tissue, body, and also inorganic media) depends on the frequency of the electric signal (in general, a low amplitude, alternating current) used to measure it, and this behavior may provide insights about the physiology and pathology of cells and tissues.

Bioimpedance techniques are used in the body composition analysis (BIA), to evaluate the hydration and nutritional status in many clinical areas, as reviewed by Kyle et al. (2004): obstetrics, critical care, postoperative monitoring, pregnancy, lactation, nutrition, gastroenterology, obesity, chronic inflammation, skin water content, blood volume, ablation monitoring, tissue ischemia, viability of transplanted organs monitoring, sleep apnea detection (Ahmad et al., 2013), chronic kidney diseases management (López-Gómez, 2011), or sports science (Di Vincenzo et al., 2019).

Clinical laboratory tools including lab-on-chip devices (Kassanos et al., 2014) utilize bioimpedance techniques (Kyle et al., 2004), such as cell culture monitoring systems and hematocrit meters. Electrical bioimpedance platforms are used to measure the cell cultures growth, motility, activity, and viability, for the detection of interactions with drugs (Alexander et al., 2013) and, more recently, they drive the research for different types of cancer (Hong et al., 2011; Huertas et al., 2015; Yu et al., 2016). This method is sensible, noninvasive, and it allows for online monitoring of the electrical and morphological parameters of cell monolayers. It uses low amplitude electrical currents, in the frequency range $1 - 10^7$ Hz to measure the impedance of the cells grown on electrodes. Based on the dual properties of the noble materials (conductors and plasmon structures), recent advances outline the advantages of using the couplings between the alternating electric fields and the surface plasmons, which seem to pave the way to a new analysis technique called *plasmonic electroimpedance* (P-EIS) (Roşu-Hamzescu, 2019), that may result in the amplification of the analytic capabilities to evidence the electric and morphologic properties of the cellular structures.

Computational Modeling in Biomedical Engineering and Medical Physics
DOI: https://doi.org/10.1016/B978-0-12-817897-3.00005-1

Bioimpedance techniques are prominent tools in the noninvasive monitoring of cardiovascular dynamics, which is of great clinical interest. For example, the impedance cardiography (ICG) is used in conjunction with the ECG (Einthoven's Triangle & Cardiac Monitoring, 2020) to characterize the mechanical function of the heart, such as stroke volume, left-ventricular ejection time, cardiac output, or systolic time ratio, of importance in the evaluation of patients with cardiovascular diseases (Bernstein, 2010; Kubicek et al., 1966; Sramek, 1986). The pulse wave velocity (PWV) is used to evaluate the arterial stiffness, which is a useful parameter in arrhythmia diagnosis, hypertension, and stroke events (Lee and Cho, 2015).

Another line of applications is the medical imaging of human body inside, enabled by the electrical impedance tomography (EIT) (Malmivuo and Plonsey, 1995). An electrode array placed on the torso provides for boundary voltage measurements that may be used to find the electrical conductivity distribution (the solution to an inverse problem), hence anatomic structure, which is of great interest, nevertheless, the low resolution still provided.

Numerous reviews, for example, work by Bera (2014), Cichoż-Lach and Michalak (2017), Ward (2019), Di Vincenzo et al. (2019), Kyle et al. (2004), Križaj (2018), prove that the bioimpedance concept, model, method, and implementation in systems, equipments, and devices are part commonly used in the biomedical practice. The advantages that distinguish the bioimpedance techniques (noninvasive, low-cost, portable, user-friendly) are key to their rapid acceptance. Their development though has still to overcome significant challenges, such as miniaturization, efficient algorithms, new parameters, novel sensing technologies, increased sensitivity, reduced power consumption, improved circuit designs, etc.

5.2 The electrical impedance

Lumped parameters may be defined for boundary value problems without internal sources when the interaction with the surroundings occurs through ports on the boundary, in a gradient-flux context, as the ratio between driving gradients and the conjugated fluxes. For example, for a system where the electromagnetic field interaction with the surroundings occurs at two terminals (ports) level only (the electromagnetic field is contained in the system, does not "cross" the boundary), an electric impedance is calculated as the ratio of voltage drop (gradient), $U(t)$, over the electric current (flux), $I(t)$, that is, $Z(t) = U/I$ [Ω]—although U and I are not with respect to the same terminals.

Fig. 5.1 suggests, graphically, the definition and meaning of the electrical impedance in the problem of the direct current flow through an electroconductive cylinder, between two current ports on the boundary—the rest of the boundary is assumed electrically insulated—powered by an electric source. A voltage drop, U, is sensed by a pair of measurement, equipotential electrodes. It is assumed that the electric current that may cross out through the voltage electrodes is negligibly small. The surfaces of the voltage electrodes (V_1 and V_2) are corresponding surfaces of a flux tube, and the electrical current tube in between them is approximately equal to the current at the terminals.

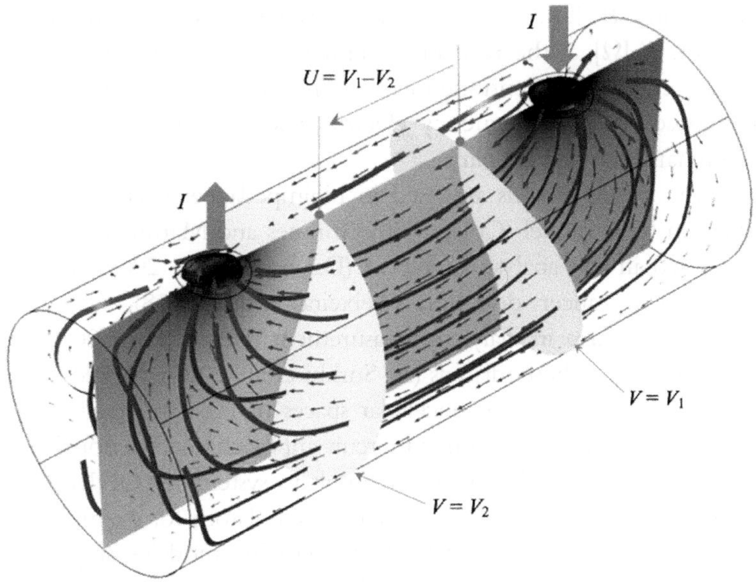

Figure 5.1 The impedance calculated as a voltage drop divided through the current tube—the standard four electrodes configuration. The voltage drop (at the measurement terminals) is with respect to two constant voltage surfaces and the electric current is conveyed through the excitation terminals.

Considering the gradient (voltage drop)—flux (electric current) relation at the terminals, the model in Fig. 5.1 exemplifies the introduction of a dipolar circuit element—a lumped-parameter model—with (electromagnetic) field effects. All physics inside the physical domain (electric field here) are concisely presented through $U = U(I)$. Using such circuit elements with field effects may reduce a boundary and initial value problem to an equivalent circuit model, where the interactions occur at a terminal level only. The lumped-parameter model reduces the description—configuration, geometry, field, and properties—of the physical systems to the topology of a distributed companion system, whose nodes represent the terminals (ports) and the branches are the lumped elements. Subsequently, the partial differential equation(s), PDE(s) models are reduced to ordinary differential equation(s) ODE(s) ones, usually easier to tackle with since Cauchy-type problems have to be solved. There are advantages (ODEs instead of PDEs) and disadvantages (for instance, no spatial information results, the lumped circuits are known only with respect to some ports) that come with this approach.

For time-harmonic excitation, when the electric sources operate at the same frequency, it is customary to represent and solve for the complex, simplified representation of the ODE. The dynamic impedance, $Z(t)$, is represented by its complex image (Chapter 1: Physical, Mathematical and Numerical Modeling)

$$\underline{Z} = R + jX, \tag{5.1}$$

where the real part, R [Ω], is the electric resistance (electro-thermal effect), and the imaginary part, X [Ω], is the reactance—either inductive (magnetic field) or capacitive (electric field) effects. As mentioned earlier, the body tissues behave like conductive media when exposed to low frequency electric field; the electrical impedance is thus represented dominantly by its real part.

Lumped electric circuit models are used in bioimpedance spectroscopy. For example, the single dispersion Cole model (Cole, 1940; Grimnes and Martinsen, 2008; Ivorra et al., 2004) has been used in the analysis of blood (Dai and Adler, 2009), body composition (Buendi et al., 2014), cancer detection (Teixeira et al., 2018), ischemia monitoring (Guermazi et al., 2014), urea in dialysate measurement (Jensen et al., 2012), tissue analysis (Guermazi et al., 2014), and hemodialysis (Al-Surkhi et al., 2007). Moreover, biological tissues composed of cell clusters and extracellular spaces may be modeled using a simplified but effective distributed parameters lumped circuit approach and fractional calculus models (Freeborn, 2013; Vosika et al., 2013), which result in systems of ODEs (Freeborn, 2013; Ivorra et al., 2004) that may be solved using circuit simulators, for example, SPICE (Nagel and Pederson, 1973; Nagel, 1975). Other permittivity models—single-dispersion RC model, extended single-dispersion and double dispersion Cole models, fractional and multi-scale models—are quoted and referred in the work by Naranjo-Hernández et al. (2019).

The macroscopic, continuous media modelization introduces quantities that average processes and quantities at cellular scale. At the living matter level, tissues are composed of cells with thin membranes of high electric resistivity that behave as capacitors (Grimnes and Martinsen, 2008). For harmonic electric excitation of higher frequency, the electric current flows through tissue and liquids both inside (displacement current) and outside the cells (conduction current). At low frequencies the current circulates only through the liquids outside the cells, a conduction current.

The conduction electric current prevails in the electrolyte solutions of the soft tissues (organs, muscles, etc.) and the body fluids (blood, interstitial fluid, lymph, etc.) (Grimnes and Martinsen, 2014). The osseous and the adipose tissues, the gases in the breathing paths and lungs are permeable to the displacement current, which depends on the frequency of the incident field and the permittivity of the medium (Gabriel et al., 1996)—an electric property that is explained, at the living matter level, by the dipolar polarization of the biologic medium (Chapter 1: Physical, Mathematical, and Numerical Modeling).

The relaxation effects related to the dipoles depend on the frequency of the incident electric field. The higher the frequency is, the larger the time lag of the response to the stimulus is, which leads to an increase in the internal energy. Depending on the frequency of the incident electric field, three groups of processes are observed (Amini et al., 2018; Bhardwaja et al., 2018; Guermazi et al., 2014): (i) in the range 10 Hz to 10 kHz, named "α dispersion region," ionic diffusion through the cell membrane and the counterion processes occur and generate space charge (interfacial) polarization; (ii) in the range 10 kHz to 100 MHz, labeled "β dispersion region," the polarization of cell membranes, proteins, and other organic macromolecules are produced, and (iii) in the range of gigahertz (called "γ dispersion region") the polarization of water molecules happens; both β and γ dispersion regions are characterized by orientational polarization.

To cope with this wide spectrum of frequency-related processes, several macroscopic models of permittivity were proposed. For example, the single-dispersion *Cole–Cole relaxation model* (Cole and Cole, 1941) presents the complex relative permittivity as frequency dependent function for three tissues of the body (skin, muscle, fat) as defined by Gabriel et al. (1996) and Naranjo-Hernández et al. (2019)

$$\varepsilon = \varepsilon_{\infty} + \sum_{n=1}^{3} \frac{\Delta\varepsilon_n}{1 + (j\omega\tau_n)^{\alpha_n}} + \frac{\sigma}{j\omega\varepsilon_0}, \tag{5.2}$$

where $\omega = 2\pi f$ is the angular velocity, f is the frequency, σ is the electrical conductivity, $\Delta\varepsilon_n$ and ε_{∞} are the "static" and "infinite frequency" dielectric constants. $\Delta\varepsilon_n$, τ_n, and $\alpha_n \in (0, 1]$ are the amplitudes, the time constants, and the distribution parameters of the time constants that characterize the three dispersion ranges ($n = 1, 2, 3$ for α, β and γ, respectively).

Based on the variability of dielectric properties, from tissue to tissue, the electrical impedance measurements may provide for global compositional information on the volume of interest. Because dielectric constant and electric conductivity may distinguish between tissues and their physiological states, the impedance spectroscopy is a suitable method for their characterization in what concerns normality and pathogenesis, hydration state, viability, etc. The dependence of these properties with the frequency helps to increase the accuracy of the method.

5.3 The electrical impedance in noninvasive hemodynamic monitoring

Bioimpedance techniques are used to noninvasively monitor the normal or pathologic status of the vascular system, which is of great clinical interest, and its usage dates back to the epoch of the plethysmogram (from Greek *pletusmos*—increase).

The plethysmogram

The fluctuation of the impedance of the body synchronously with the heart is at its origins (Malmivuo and Plonsey, 1995; Segen, 2005). Pathologic changes in the blood volume may be reflected indirectly by small changes in the electrical impedance of the chest or extremities regions: arterial (sclerotic diseases), venous (thrombotic condition), acute deep vein thrombosis (PalmSens, 2019). The impedance plethysmogram (IPG) amplitude, which is about 1/1000 to 1/100 of the average value of the measured impedance, was attributed to the change in the electrical resistivity of the tissues produced by the pulsation of the blood content (Nyboer et al., 1950; Vedru, 1994), and the considerable change in the electrical resistivity of blood that occurs with its velocity (Moskalenko and Naumenko, 1959; Visser, 1989; Visser, 1992). Within the assumptions and limits of the parallel conductor theory (Shimazu et al., 1982) for an equivalent cylindrical region of the body, for instance arm or foot of length L (the distance in between the voltage electrodes), the decrease of the impedance, ΔZ, is related to the blood volume increase, ΔV, through

$$\Delta V(t) = -\frac{\rho L^2}{\overline{Z}^2}\Delta Z(t), \tag{5.3}$$

where ρ is the electrical resistivity of the blood, and $\overline{Z}$ is the per cycle time-averaged value.

The change in $\overline{Z}(t)$ is supposedly produced only by the change in the blood volume within the measurement region, which is equated to an equivalent cylindrical region that is connected, electrically, in parallel with the body segment [forearm in the work by Shimazu et al. (1982)], and modeled as a series of parallel conductors that represent artery, tissue, and vein. The time scale of the current source (typically for IPG 1 mA at 50 kHz) is much smaller than that of the arterial blood flow (60–80 bpm), with which ΔV and ΔZ fluctuate. The initial splitting of the impedance into two parts (Nyboer et al., 1950)

$$Z(t) = \overline{Z}(t) + \Delta Z(t), \tag{5.4}$$

presents its dynamics as the result of a quasi-periodic drift at the heart frequency, $\Delta Z(t)$, about an average value at the breathing frequency, $\overline{Z}(t)$, with different magnitudes and spectra, which bodes the basis to the evaluation of the stroke volume, SV.

The SV usually refers to the left ventricle. It is calculated by subtracting the blood end-systolic volume (ESV) from its end-diastolic volume (EDV) just preceding the beat, in the ventricle. The SVs of the ventricles are usually equal, $\sim$ 70 mL in a healthy 70 kg man. The SV correlates with the cardiac function: the cardiac output (CO) is calculated by multiplying SV and heart rate (HR), and the ejection fraction (EF) results by dividing SV through EDV (Maceira et al., 2006).

Bioimpedance methods and models

On the basis of the plethysmography, Kubicek et al. (1966) proposed a thoracic bioimpedance method, which evolved later to the ICG. ICG uses four special band electrodes around the body and utilizes a signal-processing model to estimate SV (Kubicek et al., 1966, 2006; Vedru, 1994; Woltjer et al., 1997). Later proposed, the integral rheography of the body (IRB) is considered to be a reliable, not expensive, technically simple method for the assessment of overall hemodynamic features of the body. Its efficiency was proved in massive examinations of large-scale cohorts of people. It was intensively used in many fields of medicine, for example, surgery, resuscitation, oncology, and dentistry (Misiura, 2017). For modeling a perfect contact between the band electrodes and the body, the cylindrical conductor model SV reduces to Eq. (5.3), which was boldly applied to regions as complex as the thorax.

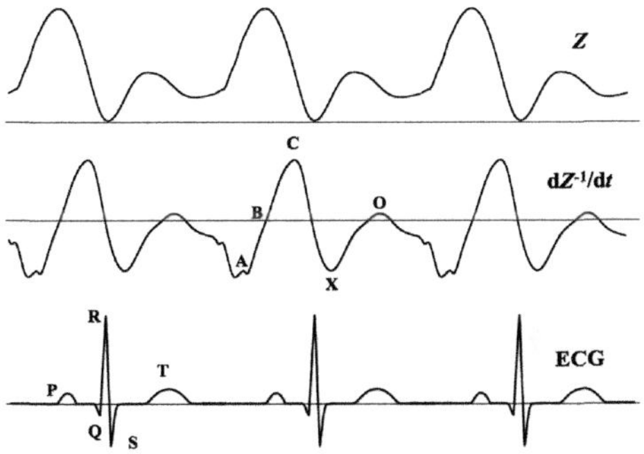

Figure 5.2 Qualitative images for the impedance cardiogram of an ordinary healthy man (top) and its time derivative (middle). The bottom curve is the ECG (Section 5.5).

Fig. 5.2 shows the impedance cardiogram $\Delta Z(t)$ and its time derivative for a normal healthy man, using the traditional inverse rendering (in fact, an admittance).

If the decrease in the impedance corresponds to SV.

$$\Delta Z_{SV} = \min_{t \in \text{Cycle}} \frac{dZ(t)}{dt} \mathrm{T_E}, \tag{5.5}$$

where T_E is the duration of heart ejection (phonocardiography may help to evaluate it), then

$$\Delta \mathrm{V_{SV}} = -\frac{\rho L^2}{\overline{Z}^2} \min_{t \in \text{Cycle}} \frac{dZ(t)}{dt} \mathrm{T_E}. \tag{5.6}$$

The breathing intrinsic component is notably eliminated in the $dZ(t)/dt$ signal because differentiation tends to discard the lower frequencies of spectrum, in contrast to other methods that may use the IPG $Z(t)$.

5.4 Thoracic bioimpedance methods and models

The thoracic electrical bioimpedance

The bioimpedance technology (Sramek, 1986), or thoracic electrical bioimpedance (TEB), replaces the band electrodes with spot, ECG-like ones. These are positioned in eight points, on the band electrodes outlines (Fig. 5.3)—Sramek (1986) cited by Choudari and Panse (2013) and Vedru (1994). This method accounts for the resistivity of blood (or hematocrit) by introducing an individual parameter called the volume of the electrical participating tissue ($\mathrm{V_{EPT}}$), which is defined empirically, based on statistical and experimental data

$$\mathrm{V_{EPT}} = -\frac{(0.17\mathrm{H})^3}{4.2} \frac{\mathrm{P}}{\mathrm{P_{IDEAL}(H)}}, \tag{5.7}$$

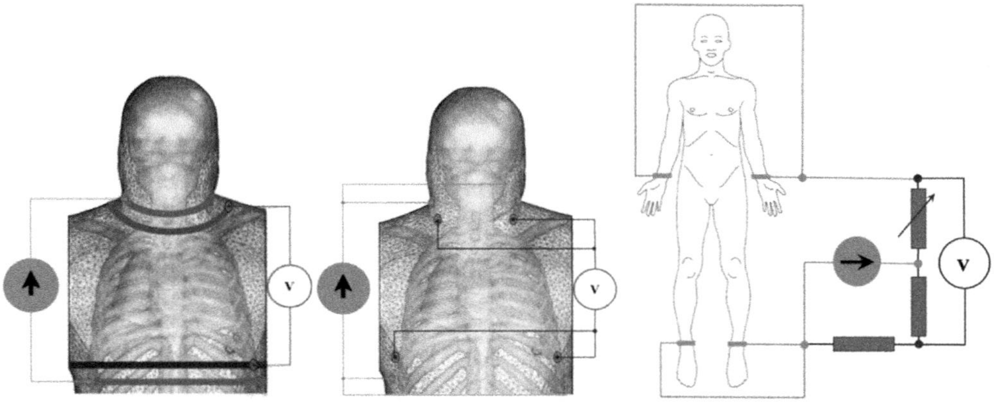

Figure 5.3 Electrodes and arrangement for the measurement of the electrical impedance in SV estimation: (Left) Kubicek bioimpedance (Kubicek et al., 1966), (Middle) Sramek bioimpedance (Sramek, 1986), (Right) IRB (Tishchenko, 1973).

where P is the patient's weight, $P_{IDEAL}(H)$ is the "ideal weight," defined for women and men, which is function of the patient's height, H. Equations (5.6) and (5.7) yield

$$\Delta V_{SV} = -\frac{V_{EPT}}{\overline{Z}} \min_{t \in \text{Cycle}} \frac{dZ(t)}{dt} T_E. \tag{5.8}$$

which is the SV evaluated in the Kubicek bioimpedance technology.

TEB is used for cardiac investigation and diagnosis in a clinical environment (Bernstein, 2010; Sathyaprabha et al., 2008) to continuously monitor the aortic blood flow (Funk et al., 2009; Truijen et al., 2012). TEB dynamics is related to, and might be synchronized and related with the respiratory and ECG vital signs, which offer consistent data for the evaluation of SV, CO, as well as the peak aortic acceleration of blood (PAA), systemic vascular resistance (SVR), velocity of the blood flow (VBF), and flow time (FT) (Osypka, 2009; Stevanović et al., 2008; Ipate et al., 2012).

NASA developed TEB, and introduced the ICG in 1967. In the 1980s, at BioMed Medical Manufacturing Ltd., Sramek developed the apparatus NCCOM3, which brought significant improvements to the clinical accuracy of the method. In 1992, the company was renamed to CDIC and the product was called "BioZ". The bioimpedance measurement of CO (Kubicek et al., 1966; Sramek, 1986) assumes that the blood ejection within the thorax increases the electrical admittance of the thorax as a result of changes in the intrathoracic hemodynamic.

The electrical velocimetry model and the cardiometry method

More sophisticated TEB algorithms have since been proposed. In this group, a new model, the electrical velocimetry™ (EVM) (Bernstein et al., 2015), which introduces a new method named electrical cardiometry™ (ECM) (Bernstein and Osypka, 2003; Norozi et al.,

2008; Osypka, 2009, Grollmuss et al., 2012; Henry et al., 2012) is educed from general cardiac output observances and injury espial of myocardial ischemic (Mellert et al., 2011). EVM model analyses the change in impedance of the aortic blood soon after aortic valve opening, due to the red blood cells (erythrocytes) alignment with the flow.

The observed EVM/ECM bioimpedance signal bears also the concurring effects of respiration and fluctuations in the cardiac cycle: $Z(t) = Z_0 + \Delta Z_R + \Delta Z_C$, where Z_0 is the "quasi-static" part, referred to as the base impedance. ΔZ_R accounts for the effects of respiration and it is suppressed to the estimate SV. ΔZ_C is caused by changes due to the cardiac cycle, produced by the thoracic fluids, the thoracic blood volume included.

In TEB, the stroke volume is calculated using (Osypka, 2009)

$$\mathrm{SV_{TEB}} = \mathrm{C_p} \cdot \overline{\mathrm{v}}_{\mathrm{FT}} \cdot \mathrm{FT}, \tag{5.9}$$

where $\mathrm{C_p}$ [mL] is a patient constant, $\overline{\mathrm{v}}_{\mathrm{FT}}$ [s^{-1}] is the mean velocity index measured during the flow time, FT [s]. EVM introduces the peak aortic acceleration

$$\mathrm{ICON} = \frac{\left| \left(\frac{dZ(t)}{dt} \right)_{\min} \right|}{Z_0} \times 1000, \tag{5.10}$$

and uses it to calculate $\overline{\mathrm{v}}_{\mathrm{FT}}$ yielding

$$\overline{\mathrm{v}}_{\mathrm{FT,EVM}} = \sqrt{\frac{\left| \left(\frac{dZ(t)}{dt} \right)_{\min} \right|}{Z_0}}, \tag{5.11}$$

and "corrects" the flow time rate, which is calculated using the left-ventricular ejection time, LVET and the ECG R-R interval, $\mathrm{T_{RR}}$, yielding

$$\mathrm{FT_C} = \mathrm{LVET}/\sqrt{\mathrm{T_{RR}}}, \tag{5.12}$$

Hence, using the body mass index, $\mathrm{V_{EPT}}$, the stroke volume by ECM yields (Osypka, 2009)

$$\mathrm{SV_{EVM}} = \mathrm{V_{EPT}} \cdot \overline{\mathrm{v}}_{\mathrm{FT,EVM}} \cdot \mathrm{FT_C}, \tag{5.13}$$

and the cardiac output is

$$\mathrm{CO} = \frac{\mathrm{SV_{EVM}}}{1000} \times \mathrm{HR}, \tag{5.14}$$

where HR is the measured heart rate.

To distinguish between ICG and EVM, ICON, Eq. (5.10), in the ICG model (Woltjer et al., 1997) is an index of peak velocity. In contrast, EVM considers it an

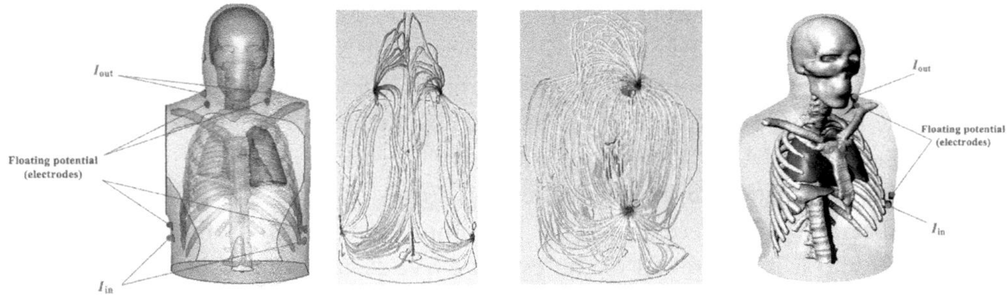

Figure 5.4 Thoracic electrical bioimpedance methods: ICG (left) and ECM (right) (Dobre, 2012).

index of peak acceleration. The EVM calculates the SV using an average velocity index that is obtained by applying a nonlinear transformation to the peak acceleration index.

ECM method uses two pair electrodes—one current electrode and one voltage electrode in each pair. One pair is affixed on the lower left part of the thorax, and the other one to the left side of the neck respectively (Fig. 5.4). The current (outer) electrodes are used to apply a low amplitude (< 0.5 mA) and low frequency (< 1 MHz) electric current $i(t)$. The voltage (inner) electrodes measure a voltage drop. The bioimpedance—the voltage drop, $u(t)$, measured at the voltage electrodes divided through $i(t)$—and its first-order time derivative are measured in real time. The inverse of the bioimpedance, $1/Z(t) = i(t)/u(t)$, so-called derived bioimpedance (in fact, an admittance), is commonly associated with TEB as a measure-of-conductivity. Its derivative, $dZ(t)/dt$, is considered a change-of-conductivity.

The bioimpedance measurement is sensitive to the cyclic variation of the electrical conductivity of the aortic blood, produced by the periodic fluctuations in the stages of the blood flow, which are characterized by adaptations in the orientation of red blood cells (RBC). The synchronized ECG, $Z(t)$, $dZ(t)/dt$, and pulse waveforms in Fig. 5.5 show off the correlations between the cardiovascular indices. The switching in the orientation of the RBCs and their deformation produced by the aorta flow are the main reason for the fluctuation of the blood electrical resistivity. The diastolic period concurs with low electrical conductivity (left, the RBCs are randomly distributed just before aortic valve opens), and the systolic period corresponds to high electrical conductivity. Fig. 5.5 (right) sketches the RBCs streamwise aligned, shortly after the aortic valve opens. Their shape and orientation, the result of the interaction with the viscous hemodynamic flow, also favors the electric current streamwise conduction.

The same principle is applied for the CO evaluation through brachial artery cardiometry measurements (Henry et al., 2012; Dobre et al., 2017; Morega et al., 2018). The interactions between a Hagen–Poiseuille flow for blood and its electrical conductivity have been investigated (see, e.g., Hoetink et al., 2004). These experimental and analytic grounds, which may be implemented on more realistic computational domains, pave the paths for more advanced, realistic numerical models.

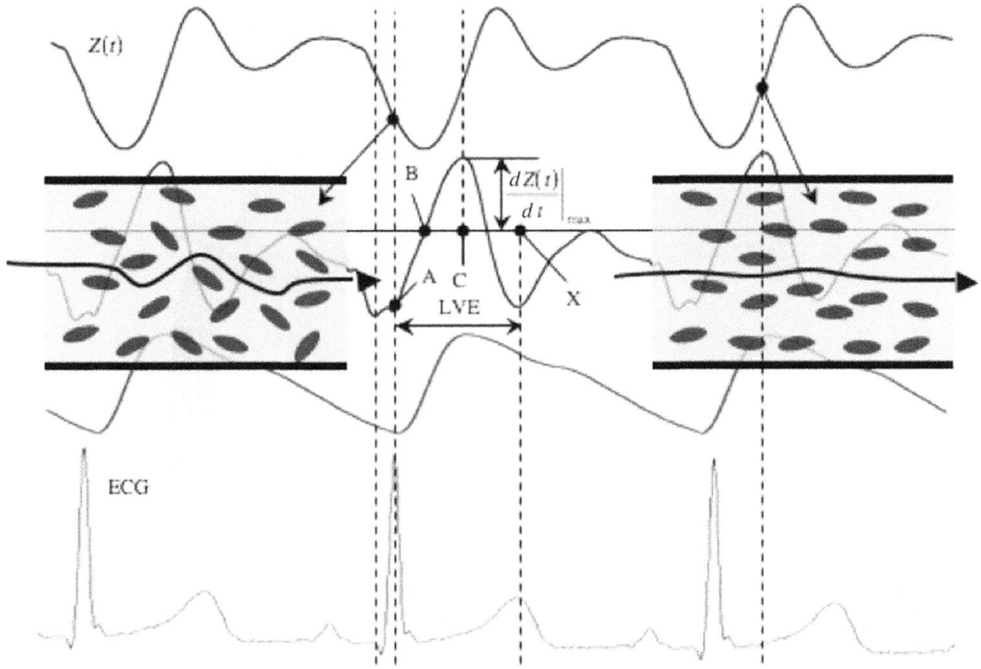

Figure 5.5 The waveforms of ECG, pulse, *Z(t)*, and *dZ(t)/dt*, correlated with the RBCs shape and orientation in the aorta before and shortly after the aortic valve opens (Morega et al., 2016).

5.5 The electrical cardiometry—electrical velocimetry

An ECM numerical model has to take account for the dynamic aortic flow and for the incumbent fluctuations of the blood electrical resistivity: TEB signal is produced by the cyclic change of the blood electrical conductivity that is related to the blood flow dynamic. Moreover, to account for the complex path of the current between the (current) electrodes, through the thorax, numerical simulations have to envisage computational domains that convincingly represent the anatomy of the upper half of a human body, which has an important heterogeneous structure. To this end, realistic computational models are currently used for getting meaningful insights to medical problems (Chapter 3: Computational Domains). Scanners used in computing tomography (CT) and magnetic resonance imaging (MRI) yield patient-personalized image datasets (see, e.g., Cruz-Roa, 2019). The accurate reconstruction is crucial for the relevance of numerical simulations, and it may also provide for and absolute path for a patient-centered therapeutic approach.

Fig. 5.6 shows the computational domain for the upper human body, used to simulate the ECM. A portrayal of the upper body (above the waist) may be obtained using a MRI data set. Based on specialized software (see, e.g., Slicer, 2019; Simpleware, 2010), image processing is applicable to any personalized, quality medical image data set, an important factor

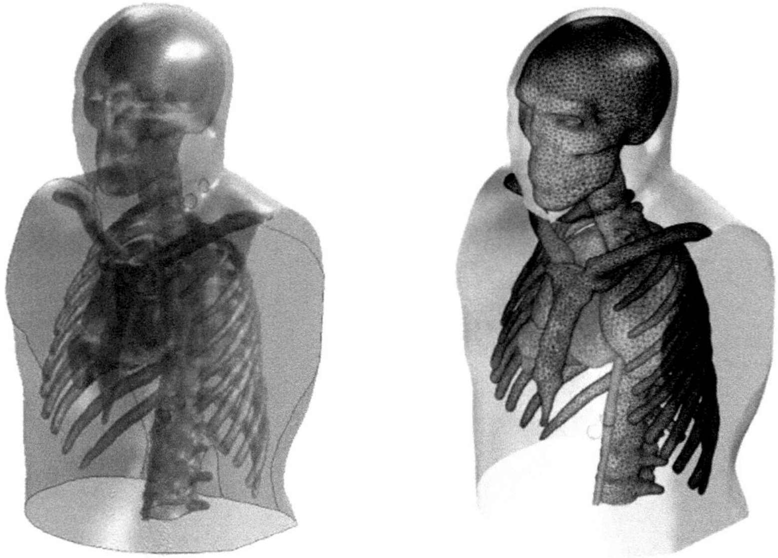

Figure 5.6 The computational domain reconstructed from MRI images (left) and the FEM mesh (right) (Morega et al., 2016). The electrodes are seen on the left side. The mesh is made of ~565,000 tetrahedral, quadratic elements.

for the accuracy of the bioimpedance diagnosis. The 3D model here is based on *ca.* 500 high-resolution MRI images (Digital Imaging and Communications in Medicine) (Visual Human Project, 2015). A FEM suitable mesh was then constructed (Morega et al., 2013).

The electrical conductivity of the blood

The blood in the aorta is regarded as continuous medium, and it is presumed that the RBCs are a dilute suspension of deformable, ellipsoidal particles in plasma (Visser, 1992; Gaw et al., 2008; Hoetink et al., 2004). The classical Maxwell–Fricke model (Visser, 1992; Jaspard et al., 2003) may be used to derive the electrical conductivity of this aggregate structure, which is subjected to a time-harmonic, relatively low frequency electric field.

The pulsatile blood flow (viscous, shear) in the aorta is a complex process (Chandran, 2001; Shahcheraghi et al., 2002; Morris et al., 2005), marked by the synchronism between $Z(t)$ and the pending changes that occur in the orientation of ellipsoids during flow acceleration, and an exponential decay observed during flow deceleration with a relaxation time, depending on the hematocrit, of the order of 0.21–0.29 s. These hypotheses were introduced and tested in the study by Morega et al. (2012), while Hoetink et al. (2004) clearly outlines the blood resistivity changes in transbrachial EVM.

The electrical conductivity of the blood suggested by Gaw et al. (2008) is

$$\sigma_{\mathrm{b}} = \sigma_{\mathrm{pl}} \frac{1 - H}{1 + (C - 1)H}, \tag{5.15}$$

where σ_{pl} and σ_{b} are plasma and blood conductivities, respectively, and H is the hematocrit. Equation (5.15) introduces a nondimensional aspect factor, C. In the round duct model, C is a function of the tube radius, r_0

$$C(r_0) = f(r_0) \cdot C_b + [1 - f(r_0)] \cdot C_{\mathrm{r}}, \tag{5.16}$$

$$C_r = (C_a + 2C_b)/3, \; C_a = 1/M, \; C_b = C(r_0) = 2/(2 - M). \tag{5.17}$$

In the above, $f(r_0)$ is the orientation rate of the RBCs, r_0 is the tube radius, and M is a function the RBCs shape, presumed ellipsoids with a and b ($a < b$) axes

$$M = \cos\varphi \cdot (\varphi - \sin 2\varphi)/\sin^3\varphi, \quad \cos\varphi = a/b. \tag{5.18}$$

Setting the average value $a/b = 0.38$, used for RBCs, yields the simplified expression $M \approx a/b$. An expression that approximates the orientation rate is

$$f(r) = \frac{n}{n_0} \frac{\theta_0^{-1}(r)}{\theta_d^{-1}(r) + \theta_0^{-1}(r)}, \tag{5.19}$$

with r is the radius of the cylindrical duct, n is the volume density of the RBCs fraction that are stably aligned with the flow, n_0 is the volume density of RBCs, θ_0 is the cell orientation from random to lined up with the flow time constant, and θ_d is the cells randomization (or cell disorientation) time constant. In our model $r = r_0$, Morega et al. (2012, 2016), θ_0 scales inverse proportionally with the shear rate, and θ_d scales inverse proportionally with the inverse of the square root of the shear rate.

Hemodynamic of larger vessels

The RBCs deformation is due to shear stress. In round tubes fully developed flows, the average friction factor (Bejan, 1993) is a measure of the shear rate, τ_{w} [N/m^2]

$$4\eta \frac{U}{r_0} = \tau_w = \mu \left(-\frac{du}{dr} \right) \Big|_{r=r_0}, \tag{5.20}$$

where u is the velocity, U the average velocity, and η the dynamic viscosity.

These important results may not be applied directly because, when more realistically rendered through reconstruction, the aorta departs from the tube geometry. To circumvent this difficulty, an aorta-equivalent round tube is introduced. The cylinder radius r_0 is computed out of the average cross-section area of the aorta segment considered in the model.

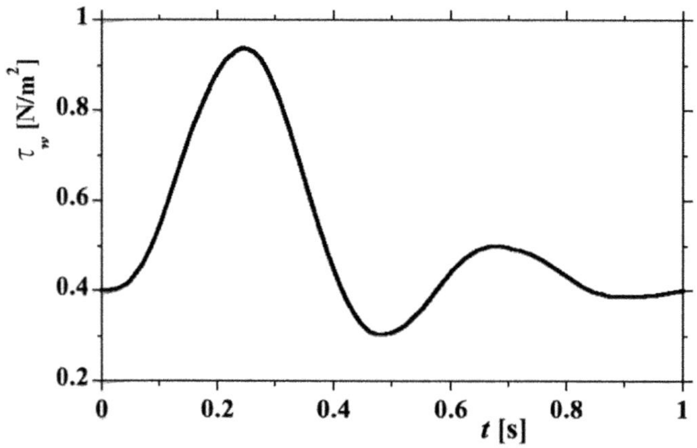

Figure 5.7 The equivalent shear rate for blood flowing through aorta.

In sequel, the tube length can be calculated as the aorta volume divided through its average cross-section area. Mass flow rate can then be used to calculate the average velocity U.

Using this approach, an equivalent shear rate τ_w is computed (Fig. 5.7). The shear rate scale obtained ($\tau_w > 0.1\ \mathrm{N/m^2}$) confirms the mathematical model predicted by Hoetink et al. (2004) for the electrical conductivity of blood, Eq. (5.15).

Essentially, two physical phenomena are coupled: the pulsatile aortic blood flow (a thoracic section consistent with ECM) and the electric current that flows within, between the current electrodes (Fig. 5.4, right).

The aorta is a large artery of "resistive" type (Chapter 1: Physical, Mathematical and Numerical Modeling). The incompressible, pulsatile, laminar flow of the aortic blood—presumed to be a Newtonian fluid—is described by the momentum balance

$$\rho\left[\frac{\partial \mathbf{u}}{\partial t} + (\mathbf{u}\cdot\nabla)\mathbf{u}\right] = \nabla\left[-p\mathbf{I} + \eta\left(\nabla\mathbf{u} + (\nabla\mathbf{u})^T\right)\right], \tag{5.21}$$

and the mass conservation law

$$\nabla\cdot\mathbf{u} = 0. \tag{5.22}$$

In the numerical simulations presented next, $\rho = 1050\ \mathrm{kg/m^3}$, and η is given by (Hoetink et al., 2004) ($\eta_{\mathrm{pl}} = 1.35$ mPa·s is plasma dynamic viscosity, H is hematocrit)

$$\eta = \eta_{\mathrm{pl}}\left(1 + 2.5H + 7.37\times10^{-2}H\right). \tag{5.23}$$

The arterial wall is considered rigid due to its low, negligibly small deformations under the blood flow pressure (Dobre et al., 2010; Morega et al., 2010). The hemodynamic and

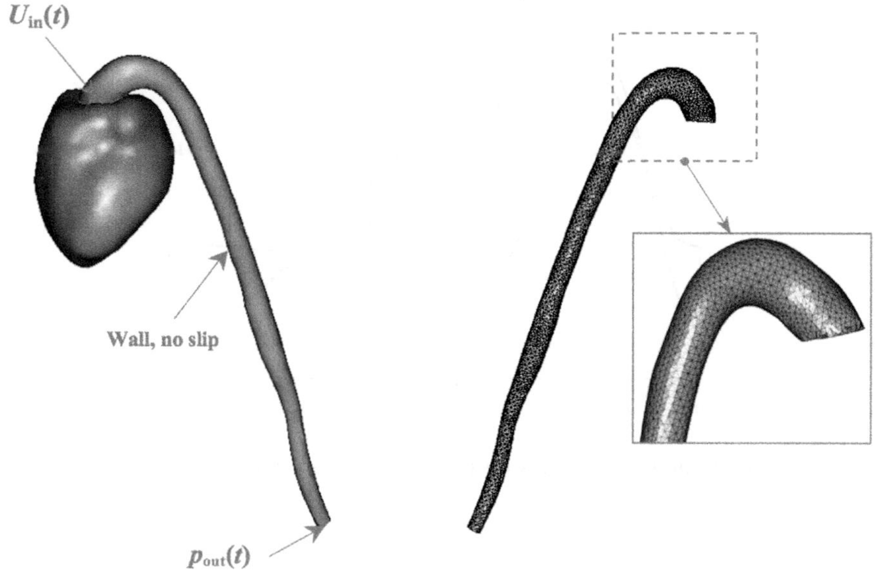

Figure 5.8 The flow boundary conditions for the aorta hemodynamic problem (left) and the FEM mesh used in numerical simulations made of ~29,000 tetrahedral, quadratic elements (right) (Dobre, 2012).

mechanical activity of the heart (its volume is filled with blood) is not included in this study. Fig. 5.8 presents the boundary conditions.

A periodic, uniform profile for the inlet velocity, derived from the Womersley theory (Taylor et al., 1998; Dobre et al., 2011), is set for the aorta inlet, and a periodic, uniform pressure profile is set at the outlet (Fig. 5.9). The time-variation of the inlet velocity reproduces the shear rate profile, as predicted by Hoetink et al. (2004), found through numerical simulation.

The velocity solution is used in sequel to evaluate the dynamic electrical conductivity that is used in the electric field problem.

The electromagnetic field

First, an equivalent electrokinetic (DC) model is used to obtain the data for the $Z(t)$ and $dZ(t)/dt$ waveforms calculations. At frequencies of the order $\sim 10^2$ kHz the tissues electric impedances are dominantly resistive (they are relatively good electric conductors), that is, conduction electric current density prevails the displacement current density. The electrical conductivity of the aorta blood, higher than within the surrounding regions, fluctuates with the blood flow (acceleration, deceleration), which rules the dynamics of $Z(t)$.

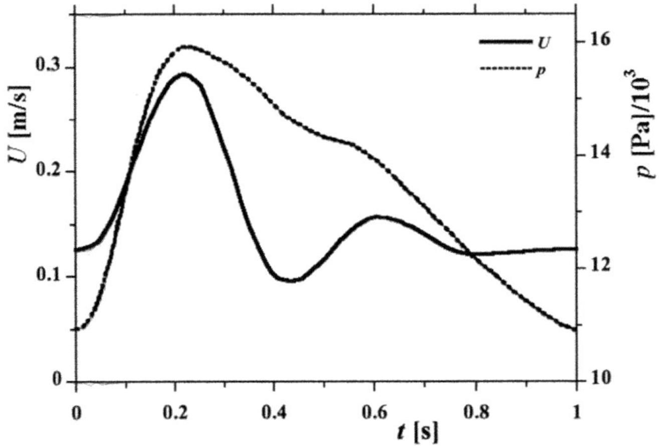

Figure 5.9 Inlet velocity and outlet pressure profiles boundary conditions used for the hemodynamic problem.

The electric field problem is then solved, first, for stationary conditions (electrokinetic regime, DC currents). Next, quasi-stationary (AC) conditions are considered (the actual procedure). The DC model is consistent with the discrepancy between the characteristic time scales for blood flow [slow, $O(s^{-1})$], and electromagnetic field time scale [fast, $O(s^{-5})$]. The DC and AC models are r.m.s. (root mean square) equivalent. Furthermore, the companion DC model is valuable because its specific discrete algebra involves scalar, real quantities, as compared to the vector, complex quantities required by the AC model.

The DC electric field strength is potential hence the mathematical model is described by

$$\Delta V = 0. \tag{5.24}$$

Four electrodes, in ECM arrangement, are electric terminals: floating potential is assumed for the inner electrodes (voltage, measurement), inward current (bottom), and ground (top) for the other pair of electrodes (current, excitation). The surface of the thorax and the cross-sectional cuts that delimit the computational domain are presumed electrically insulated ($\mathbf{n} \cdot \mathbf{J} = 0$).

Next, an AC problem models a 500-kHz ECM setup. The penetration depth for tissues with $\sigma \sim$ 1-100 mS/m, $\delta = 1/\sqrt{\mu\sigma\pi f}$, shows off values larger than the characteristic lengths of the anatomical domains. Table 5.1 presents the electrical properties and the corresponding penetration for the tissues considered in the model.

The electric field is actually irrotational because the penetration depth exceeds local length scales of the anatomic regions, hence $\underline{\mathbf{E}} = -\nabla \underline{V}$ (complex quantities are underlined). Using the electric charge conservation law, $\nabla \cdot \underline{\mathbf{J}} = -j\omega\rho_v$, where

Table 5.1 Electrical properties (compiled from Andreuccetti et al., 1997) and penetration depths for anatomical regions at 500 kHz.

Region	σ [S/m]	ε_r	δ [m] at 500 kHz
Brain (averaged)	0.110	1050	2.147
Thorax (averaged)	0.044	3000	3.395
Liver	0.148	2770	1.851
Lungs	0.123	1025	2.030
Heart	0.281	3265	1.343
Blood	0.748	4189	0.823
Bone (averaged)	0.006	200	9.193

$\mathrm{j} = \sqrt{-1}$, ρ_V is the electric charge density, and the electric flux law, $\nabla \cdot \underline{\mathbf{D}} = \rho_v$, where **D** is the electric flux density and yields the quasi-stationary diffusion model for the EMF

$$-\nabla\left[(\sigma + j\omega\varepsilon_0\varepsilon_r)\nabla\underline{V} - \underline{\mathbf{J}}^e\right] = 0, \tag{5.25}$$

where $\mathbf{J}^e$ is the external electric current density, ε_0 is the permittivity of the free space, and ε_r is the relative permittivity.

The hemodynamic problem Eqs. (5.21)−(5.23) was solved first and then the electric field problem (DC and then AC). To attain the periodic pulsatile flow structure, several cardiac cycles were simulated, and the friction coefficient was used to ascertain flow periodicity. The last hemodynamic cycle was used then to calculate the dynamic conductivity of blood Eqs. (5.15)−(5.19) and the electric field Eqs. (5.24), (5.25). Quadratic, Lagrange, $P1-P2$ elements were used to integrate the flow problem, Lagrange linear elements for the electrokinetic problem, and first-order vector elements for the quasi-steady electric field problem.

Fig. 5.10 presents the simulation results at the peak flow rate (Dobre, 2012). The aorta is the current path with the highest electric current density, which suggests that the flow dynamics should be echoed by any fluctuation in the electric conductivity of the aortic blood. This process is reflected by the bioimpedance, $Z(t)$.

Fig. 5.11 displays experimental results (Woltjer et al., 1997) and the derived bioimpedance and its time derivative obtained by numerical simulation.

The waveforms obtained by numerical simulation have features that fairly resemble with those acquired using experimental setups. The morphology of $Z(t)$ and $dZ(t)/dt$ bear the expected trends, and confirms that the ECM method presents the aorta flow. Several particular cardiovascular moments, outlined and analyzed by Taylor et al. (1998), and evidenced through the numerical simulation here too: X—aortic valve closure; B—start of blood ejection, left ventricle; C—major upward systole deflection; O—diastolic upward deflection, LVET and systolic dZ/dt_{max}.

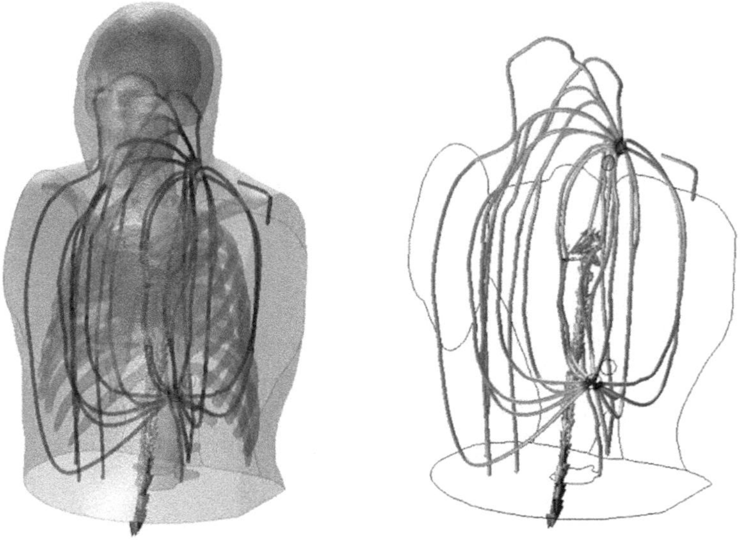

Figure 5.10 Electric potential, through surface gray map, and electric current flow through streamline tubes (left); electric current density, through streamlines, and the blood flow, through arrows (right).

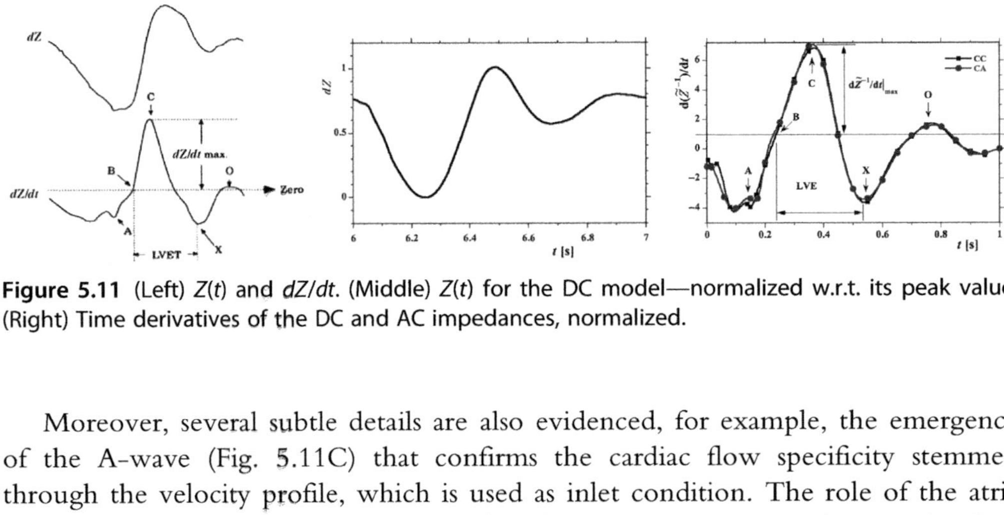

Figure 5.11 (Left) $Z(t)$ and dZ/dt. (Middle) $Z(t)$ for the DC model—normalized w.r.t. its peak value. (Right) Time derivatives of the DC and AC impedances, normalized.

Moreover, several subtle details are also evidenced, for example, the emergence of the A-wave (Fig. 5.11C) that confirms the cardiac flow specificity stemmed through the velocity profile, which is used as inlet condition. The role of the atria dynamics in the A-wave is still to clarify. The A-wave appears to be associated to their contraction, although several studies indicate that the wave of blood reflected from the atria to the central veins might produce it. The ejection in the left atrium could be one major source of the reflected wave, hence the pending ejection portion is cogently related to the A-wave magnitude (Taylor et al., 1998).

The AC and DC models outcome resembles $Z(t)$ and $dZ(t)/dt$ profiles (Fig. 5.11C) and it is expected to produce similar cardiovascular indices. It may be inferred that, without loss of accuracy, the AC model may be replaced with a simpler DC model. This may be explained by the irrotationality of the electric field and the apparent disparity between the operating frequency of the ECM and the hemodynamic of the time scale.

A key problem of the electromechanic coupling is the blood electrical conductivity. This is solved here with an equivalent quantity, calculated out of analytical expressions using averaging methods. However, some electrophysiology effects are not evidenced when embracing the numerical model, for example, the nonlinear change in the blood electrical conductivity that happens for increasing and decreasing the flow rates. Even so, the sensitivity of the solution to blood flow pulsations is evidently outlined.

The ECM impedance and its derivative with respect to time found by numerical modelization fit fairly well with the experimental results. Significant cardio-hemodynamic indices, consistent with experimental findings, are evidenced by the numerical results. Several cardiovascular indices of importance in medical diagnosis become thus mathematically tractable, for example: the start of blood ejection by the left ventricle, the systolic major upward deflection, the aortic valve closure, the diastolic upward deflection, the left-ventricular ejection time, and dZ/dt_{max}.

5.6 The ECM brachial bioimpedance

A localized version of EVM model for the arm, called transbrachial electrical bioimpedance velocimetry (TBEVM), was proposed to compute SV (Henry et al., 2012). Later, the (BCVI) introduced by Dobre et al. (2017) and Morega et al. (2018) was modeled to elicit brachial cardiovascular indices at the arm level, which is the traditional place for blood pressure measurement (Fig. 5.12).

This particular location is favored by the presence and accessibility of the brachial artery, close to the aortic arch, and relatively close to the surface of the arm. It may be inferred that, although may be not all hemodynamic indices are available as compared to the ECM, BCVI is a relevant, useful, and easily accessible cardiovascular monitoring technique, complemented by blood pressure monitoring.

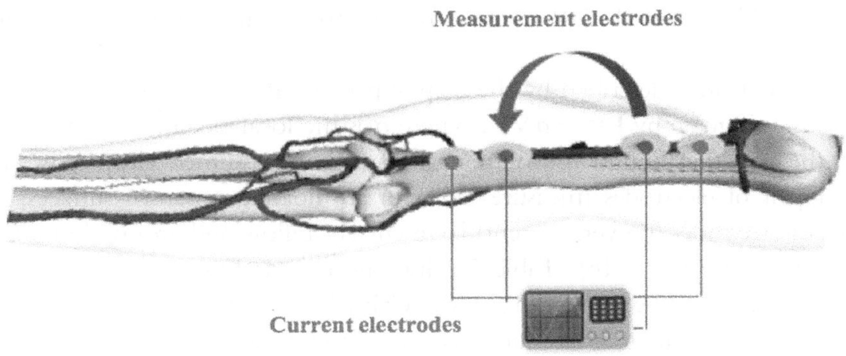

Figure 5.12 The BCVI implementation.

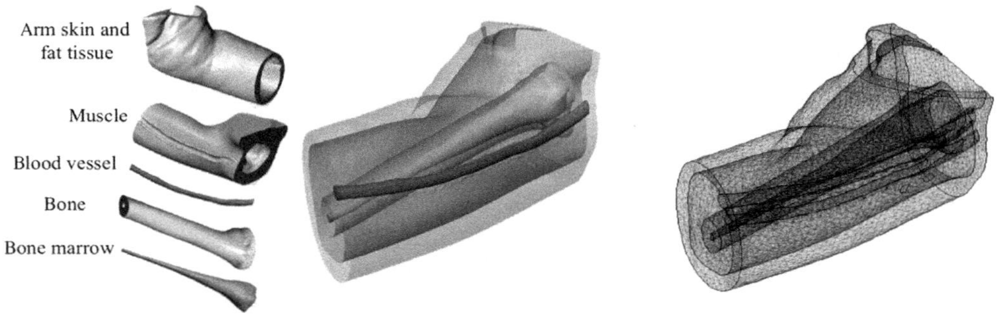

Figure 5.13 The model of the upper arm and the FEM mesh used to numerically simulate the BCVI: (Middle and left) The arm and its inner anatomical regions; (Right) Finite element mesh made of ~130,000 tetrahedral, quadratic elements.

Mathematical modeling and numerical simulations may be utilized to validate this idea. Previously, it has been shown that a DC model can be used to probe the relevance of ECM. A more realistic human anatomical model (upper arm) with a heterogeneous was constructed using high-resolution MRI images of the arm (VHP, 2019; Slicer, 2019; Simpleware, 2010; Morega et al., 2010; Morega et al., 2013; Dobre et al., 2017; Morega et al., 2018) (Fig. 5.13).

As for the thoracic ECM model, two coupled phenomena occur, namely the brachial hemodynamic of the artery and the electric current flow.

The brachial artery wall deformation is negligibly small (Dobre et al., 2010; Morega et al., 2010). Moreover, this artery is a relatively large vessel (Morega et al., 2016), hence the rheological model of the blood is Newtonian, with constant properties. Its flow is pulsatile, laminar, incompressible, described by Eqs. (5.21) and (5.22). The dynamic viscosity η depends on the viscosity of the plasma and on the volume of hematocrit, H, Eq. (5.23). The deep brachial vein is included in the model but the electrical conductivity of the venous blood does not change with the flow, which is stationary.

The upstream flow is modeled by an inlet cyclic, uniform boundary velocity profile, Fig. 5.14 (Morega et al., 2018). The pressure profile, at the outlet, is assumed uniform (Fig. 5.15).

The electric field is described by the elliptic partial differential Eq. (5.1). The electric current density is given by $\mathbf{J} = -\sigma \nabla V$, where σ is the local electrical conductivity. The skin and the cuts that bound the numerical domain are assumed electrically insulated. The inner pair of electrodes (measure, voltage) are floating potential surfaces, and the outer pair of electrodes (power, current) have current inflow and ground boundary conditions, respectively (Fig. 5.16). Table 5.2 lists the electrical conductivity for the main anatomic regions accounted for in this study (Gabriel et al., 1996).

More elaborated theories are devoted to blood flow—conductivity interactions are available (see, e.g., Hoetink et al., 2004), but they are consistent with a

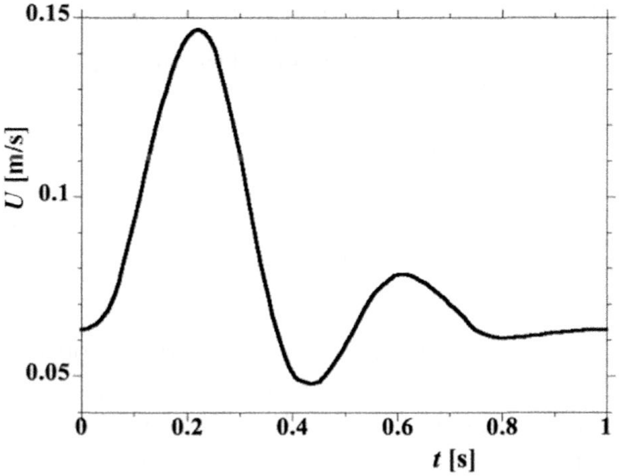

Figure 5.14 The inlet velocity profile.

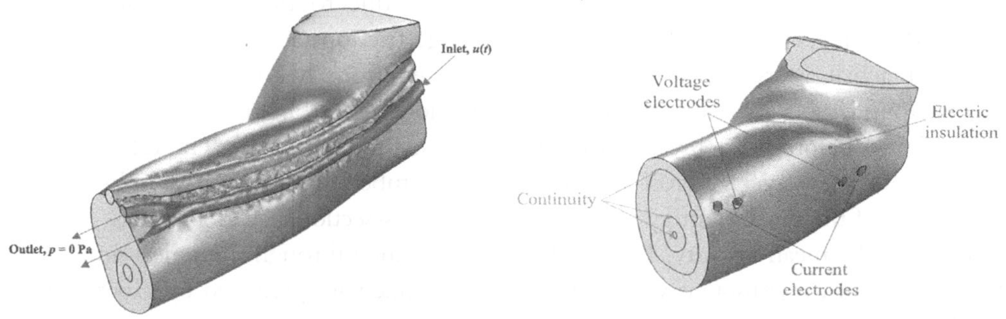

Figure 5.15 The hemodynamic (left) and the electric field boundary conditions (right) for the BCVI problem.

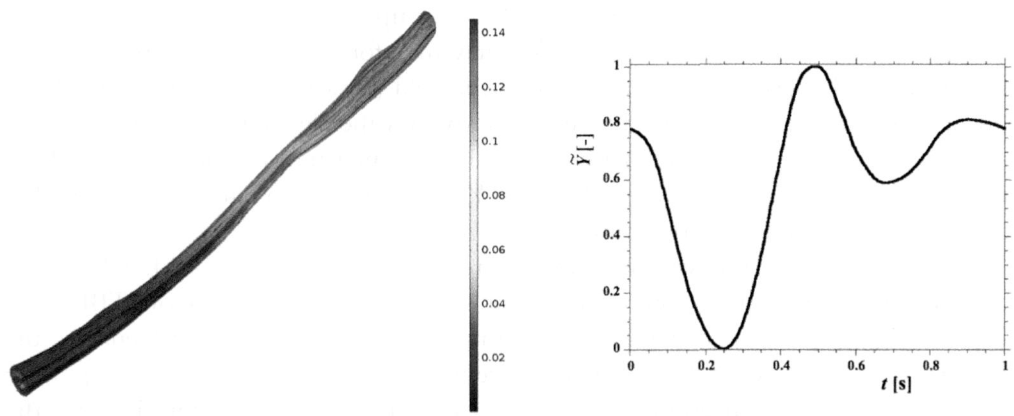

Figure 5.16 (Left) The hemodynamic flow—values are in m/s; (Right) The electrokinetic field—values are in volts. Numerical simulation results at $t = 0.22$ s, during the maximum diastolic flow rate (Fig. 5.14). Dimensions are in meters.

Table 5.2 The electrical conductivity of the different anatomic regions (low frequency).

Tissue	Electrical conductivity [S/m]
Blood	0.66
Bone	0.006
Muscle	0.355
Marrow	0.00247
Arm	0.17

Hagen-Poiseuille flow, hence they may be less adequate for arterial hemodynamic. Here an equivalent electrical conductivity for the arterial blood is used, which assimilates the RBCs within plasma with a dilute suspension of ellipsoidal globules.

To compute the parameters that are used to calculate the electric conductivity of the blood, as for the ECM model, the brachial artery is treated as an equivalent circular, straight, cylindrical tube. The flow dependent electrical conductivity is described by Eqs. (5.15)–(5.19). The deformation of RBCs due to the viscous shear flow is measured in Hagen–Poiseuille flow by the average friction factor.

The brachial artery in our study is reconstructed from MRI slices and it is not a straight, round tube, therefore the mathematical model Eqs. (5.15)–(5.19) is not readily applicable. As before, an equivalent round tube is used for the artery. The tube radius, r_0, is that of the artery average cross-sectional area. The tube length is calculated by dividing the volume of the brachial artery through its mean cross-sectional area. The mass flow rate provides for the mean velocity, U. The shear rate obtained through numerical simulations, $\tau_w > 0.1$ N/m^2, is consistent with the electrical conductivity predicted by Hoetink et al. (2004).

The FEM solution to the BCVI is divided into two steps: the brachial flow is integrated first; then, the DC problem is solved for each time step (saved flow) for one cycle, and the electrical conductivity of the blood is updated (Morega et al., 2018).

Fig. 5.16 shows the brachial flow (surface gray map for pressure, and streamline tubes and arrows for the velocity) and the electrokinetic field (surface gray map for the voltage, constant potential surfaces, and field tubes and arrows for the electrical current density).

Fig. 5.17 graphs the nondimensional derived impedance of the brachial blood where $\tilde{Y} = (Y - Y_{min})/(Y_{max} - Y_{min})$. Here, $Y_{\max} = 8.1481$ mS and $Y_{\min} = 8.1414$ mS, obtained through numerical simulation.

Apparently, BCVI follows the velocity profile depicted in Fig. 5.14, which is a new numerical simulation experiment result since the reported data on TBEV are concerned with experimental works. The time derivative of the BCVI enables the characterization of several hemodynamic events: B—start of left ventricle ejection; O—diastolic upward deflection; C—the major upward systole deflection; LVET—the left-ventricular ejection time; X—aortic valve closure, and $d(dZ)/dt_{\max}$ the maximum change during the systole phase.

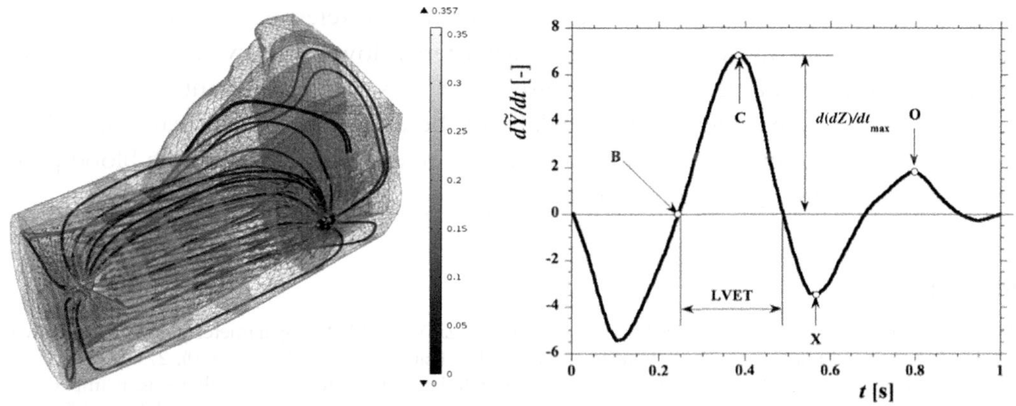

Figure 5.17 The hemodynamic flow and the EMF in the BCVI (left), non-dimensional time derivative of the BCVI (right).

5.7 Some comments on numerical modeling results

The values of the cardiovascular found through numerical modeling may differ from those of their counterparts measured using the ECM because it seems reasonable to expect discrepancies when using impedance measurements on different arterial branches, situated farther to the heart.

Details such as the A-wave, revealed experimentally and numerically by the ECM, Fig. 5.11C, are not identified in the numerical simulation of the BCVI. It remains to investigate experimentally if this is a limitation concerning the BCVI with respect to the ECM.

As expected, the usage of non-deformable computational domains discards the direct influence of "artifacts" in the bioimpedance signal that are produced by the displacements caused by factors such as respiration and heart mechanics. It is not possible to correlate it with the respiration. The "good" part is that $Z(t)$ does not need specific filtering, signal-processing. In what concerns the heart mechanics, the inlet velocity profiles used in the numerical modeling account for their effects upon the blood flow, therefore $Z(t)$ numerical simulation results are close to experimental data.

On closing this discussion, a major problem in passing from idealized laboratory experiments performed to characterize the blood electrical conduction or from analytic solutions to the real conditions of the EBI procedure is their consistency with the computational domains. Anatomically realistic computational domains used to compute arterial flows may raise concerns in using experimental results on the electrical conductivity of the arterial blood. This difficulty may be overcome by using an equivalent electrical conductivity of the arterial blood (function of flow rate), based on known analytic and experimental results. Several nonlinear electrophoresis effects (e.g., the delayed, hysteretic change of the electrical conductivity of blood for decelerating and accelerating flows), not addressed here, are approachable too, but—within the limits of the assumptions that have been presented—the relevance of BCVI to generate a metrics for the dynamics of the brachial arterial flow is clearly sustained. As presented, its time derivative enables the assessment of several important hemodynamic indices, and it may be inferred that BCVI may be a relevant, useful, easily accessible cardiovascular monitoring technique, which may be complemented by the blood pressure monitoring, using a unique apparatus.

References

Ahmad, S., Batkin, I., Kelly, O., Dajani, H.R., Bolic, M., Groza, V., 2013. Multiparameter physiological analysis in obstructive sleep apnea simulated with maneuver. IEEE Trans. Instrum. Meas. 62 (10), 2751–2762.

Alexander, F.A., Price, D.T., Bhansali, S., 2013. From cellular cultures to cellular spheroids: is impedance spectroscopy a viable tool for monitoring multicellular spheroid (MCS) drug models? IEEE Rev. Biomed. Eng. 6, 63–76.

Al-Surkhi O.I., Riu P.J., Vazquez F.F., Ibeas J., 2007. Monitoring Cole–Cole parameters during haemodialysis (HD). In: Proceedings of the 2007 29th Annual International Conference of the IEEE Engineering in Medicine and Biology Society, Lyon, France, pp. 2238–2241, August 2007.

Amini, M., Hisdal, J., Kalvøy, H., 2018. Applications of bioimpedance measurement techniques in tissue engineering. J. Electric Bioimp. 9, 142–158.

Andreuccetti, D., Fossi, R., Petrucci, C., 1997. An Internet Resource for the Calculation of the Dielectric Properties of Body Tissues in the Frequency Range 10 Hz–100 GHz. IFAC-CNR, Florence (Italy), Based on data published by C. Gabriel et al. in 1996. <http://niremf.ifac.cnr.it/tissprop>. (accessed 30 January, 2015).

Bejan, A., 1993. Heat Transfer. Wiley, New York, pp. 295–298.

Bera, T.K., 2014. Bioelectrical impedance methods for noninvasive health monitoring: a review. J. Med. Eng. Article ID 381251, 28 pages.

Bernstein, D.P., 2010. Impedance cardiography: Pulsatile blood flow and the biophysical and electrodynamic basis for the stroke volume equations. J. Electric. Bioimp. 1 (1), 2–17.

Bernstein, D.P., Henry, I.C., Lemmens, H.J., Chaltas, J.L., DeMaria, A.N. Moon, J.B., et al., , 2015. Validation of stroke volume and cardiac output by electrical interrogation of the brachial artery in normals: assessment of strengths, limitations, and sources of error, J Clin Monit Comput 29 (6), 789–800.

Bernstein D.P., Osypka M.J., *Apparatus and method for determining an approximation of stroke volume and cardiac output of the heart*, US Patent No. 6,511,438, Jan. 28, 2003, https://patentimages.storage.googleapis.com/d4/a5/57/351bc704baed48/US6511438.pdf.

Bhardwaja, P., Raia, D.V., Garga, M.L., Mohanty, B.P., 2018. Potential of electrical impedance spectroscopy to differentiate between healthy and osteopenic bone. Clin. Biomech. 57, 81–88.

Buendi, R., Seoane, F., Bosaeus, I., Gil-Pita, R., Johannsson, G., Ellegård, L., et al., 2014. Robustness study of the different immittance spectra and frequency ranges in bioimpedance spectroscopy analysis for assessment of total body composition. Physiol. Meas. 35 (7), 1373–1395.

Chandran, K.B., 2001. *Chapter 5: Flow Dynamics in the Human Aorta: Techniques and Applications*, in *Cardiovascular Techniques*, II. CRC Press LLC.

Choudari P., Panse M.S., 2013. Measurement of cardiac output using bioimpedance method. In: International Conference on Communication Technology 2013, Guilin, China, pp. 28–33. IJCATM: www.ijcaonline.org.

Cichoż-Lach, H., Michalak, A., 2017. A comprehensive review of bioelectrical impedance analysis and other methods in the assessment of nutritional status in patients with liver cirrhosis, Hindawi. Gastroenterol. Res. Pract. 2017, Article ID 6765856, 10.

Cole, K.S., 1940. Permeability and impermeability of cell membranes for ions. Proc. Cold Spring Harb. Symp. Quant. Biol 8, 110–122.

Cole, K.S., Cole, R.H., 1941. Dispersion and absorption in dielectrics. I. Alternating current characteristics. J. Chem. Phys. 9, 341–351.

Cruz-Roa A., 2019. <https://sites.google.com/site/aacruzr/image-datasets> (accessed 2019).

Dai, T., Adler, A., 2009. In vivo blood characterization from bioimpedance spectroscopy of blood pooling. IEEE Trans. Instrum. Meas. 58 (11), 3831–3838.

Di Vincenzo, O., Marra, M., Scalfi, L., 2019. Bioelectrical impedance phase angle in sport: a systematic review. J. Int. Soc. Sports Nutr. 16 (49), 12.

Dobre A.A., 2012. Investigation methods for the analysis of coupled phenomena specific to the medical engineering field (Doctoral Thesis), University Politehnica of Bucharest, Bucharest, Romania.

Dobre A.A., Morega A.M., Morega M., 2010. The investigation of flow—structural interaction in an arterial branching by numerical simulation. In: IFMBE Proceedings of 10th IEEE International Conference on Information Technology and Applications in Biomedicine (ITAB), Corfu, Greece.

Dobre, A.A., Morega, A.M., Morega, M., Ipate, C.M., 2011. Numerical simulation in electrocardiography. Rev. Roum. Sci. Techn. Électrotechn. Énerg. 56 (2), 209–218.

Dobre A.A., Morega M., Morega A.M., Veli Y., Săndoiu A.M., 2017. System for hemodynamic indices monitorization through bioimpedance techniques (HEARTACT), (in Romanian) Research Report UPB-GEX-2017, ET-02-17-03, Univ. Politehnica of Bucharest, Romania.

Einthoven's Triangle & Cardiac Monitoring, Clinical Skills Education LLC, https://ekg.academy/learn-ekg?courseid = 324&seq = 6. (accessed in 2020).

Freeborn, T.J., 2013. A survey of fractional-order circuit models for biology and biomedicine. IEEE J. Emerg. Sel. Top. Circuits Syst. 3 (3), 416–424.

Funk, D.J., Moretti, E.W., Gan, T.J., 2009. Minimally invasive cardiac output monitoring in the perioperative setting. Anesth. Analg. 108 (3), 887–897.

Gabriel, S., Lau, R.W., Gabriel, C., 1996. The dielectric properties of biological tissues. III. Parametric models for the dielectric spectrum of tissues. Phys. Med. Biol. 41 (11), 2271–2293.

Gaw, R.L., Cornish, B.H., Thomas, B.J., 2008. The electrical impedance of pulsatile blood flowing through rigid tubes: a theoretical investigation. IEEE Trans. Biomed. Eng. 55 (2)), 721–727.

Grimnes, S., Martinsen, G.Ø., 2008. Bioimpedance and Bioelectricity Basics, 2nd ed. Academic Press, 3rd ed., 2014, Elsevier.

Grollmuss, O., Demontoux, S., Capderou, A., Serraf, A., Belli, E., 2012. Electrical velocimetry as a tool for measuring cardiac output in small infants after heart surgery. Intensive Care Med. 38, 1032–1039.

Guermazi, M., Kanoun, O., Derbel, N., 2014. Investigation of long time beef and veal meat behavior by bioimpedance spectroscopy for meat monitoring. IEEE Sens. J. 14 (10), 3624–3630.

Henry I.C., Bernstein D.P., Banet M.J., 2012. Stroke volume obtained from the brachial artery using transbrachial electrical bioimpedance velocimetry. In: Proceedings of 34th International Conference on IEEE-EMBS, San Diego, California USA, pp. 142–145.

Hoetink, A.E., Faes, T.J.C., Visser, K.R., Heethaar, R.M., 2004. On the flow dependency of the electrical conductivity of Blood. IEEE Trans. Biomed. Eng. 51 (7), 1251–1261.

Hong, J., Kandasamy, K., Marimuthu, M., Choi, C.S., Kim, S., 2011. Electrical cellsubstrate impedance sensing as a non-invasive tool for cancer cell study. Analyst 136 (2), 237–245.

Huertas, G., Maldonado, A., Yufera, A., Rueda, A., Huertas, J.L., 2015. The bio-oscillator: a circuit for cell-culture assays. IEEE Trans. Circuits Syst. II: Express Briefs 62 (2), 164–168.

Ipate, M.C., Dobre, A.A., Morega, A.M., 2012. The stroke volume and the cardiac output by the impedance cardiography. UPB Sci. Bull. (Ser. C) 74 (3), 219–232.

Ivorra A., Gomez R., Aguilo J., 2004. A SPICE netlist generator to simulate living tissue electrical impedance. In: Proceedings of International Conference on Electrical Bioimpedance, Gdansk, Poland, pp. 317–320.

Jaspard, F., Nadi, M., Rouane, A., 2003. Dielectric properties of blood: an investigation of haematocrit dependence. Physiol. Meas. 24, 137–147.

Jensen, P.D., Meaney, P.M., Epstien, N.R., Paulsen, K.D., 2012. Cole–Cole parameter characterization of urea and potassium for improving dialysis treatment assessment. IEEE Antennas Wireless Propag. Lett. 11, 1598–1601.

Kassanos, P., Constantinou, L., Triantis, I.F., Demosthenous, A., 2014. An integrated analog readout for multi-frequency bioimpedance measurements. IEEE Sens. J. 14 (8), 2792–2800.

Križaj, D., 2018. Basics of numerical simulations of bioimpedance phenomena. Bioimpedance in Biomedical Applications and Research. Springer International Publishing, pp. 101–116.

Kubicek, W.G., Karnegis, J.N., Patterson, R.P., Witsoe, D.A., 1966. Development and evaluation of an impedance cardiac output system. Aerospace Med. 37, 1208–1212.

Kubicek, W.G., Karnegis, J.N., Patterson, R.P., Witsoe, D.A., 2006. Impedance cardiography as a noninvasive method of monitoring cardiac function and other parameters of the cardiovascular system. Ann. N Y Acad. Sci. 170 (2), 724–732.

Kyle, U.G., Bosaeus, I., De Lorenzo, A.D., Deurenberg, P., Elia, M., Gómez, J.M., et al., 2004. Bioelectrical impedance analysis. I. Review of principles and methods. Clin. Nutr. 23 (5), 1226–1243.

Lee, W., Cho, S., 2015. Integrated all electrical pulse wave velocity and respiration sensors using bioimpedance. IEEE J. Solid-State Circuits 50 (3), 776–785.

López-Gómez, J.M., 2011. Evolution and applications of bioimpedance in managing chronic kidney disease. Nefrologia 31 (5), 537–544.

Maceira, A.M., Prasad, S.K., Khan, M., Pennell, D.J., 2006. Reference right ventricular systolic and diastolic function normalized to age, gender and body surface area from steady-state free precession cardiovascular magnetic resonance. Eur. Heart J. 27, 2879–2888.

Majumder, S., Mondal, T., Deen, M.J., 2019. A simple, low-cost and efficient gait analyzer for wearable healthcare applications. IEEE Sens. J. 19 (6), 2320–2329.

Malmivuo, J., Plonsey, R., 1995. Bioelectromagnetism. Principles and Applications of Biomagnetic Fields. Oxford Press.

Mellert, F., Winkler, K., Schneider, C., Dudykevych, T., Welz, A., Osypka, M., et al., 2011. Detection of (reversible) myocardial ischemic injury by means of electrical bioimpedance. IEEE Trans. Biomed. Eng. 58 (6), 1511–1518.

Misiura, K.V., 2017. Specific features of hemodynamics in individuals with different body mass: correlation with composition of the body and insulin resistance. Endokrynologia 22 (4), 315–325.

Morega A.M., Dobre A.A., Morega M., 2010. Numerical simulation of magnetic drug targeting with flow-structural interaction in an arterial branching region of interest. In: Proceedings of Comsol Users Conference, Paris, France, pp. 17–19.

Morega A.M., Dobre A.A., Morega M., 2012. Numerical simulation in electrical cardiometry. In: Proceedings of13th International Conference on Optimization of Electrical and Electronic Equipment (OPTIM), Brasov, Romania, pp. 1407–1412.

Morega, A.M., Dobre, A.A., Morega, M., 2013. Blood flow indices assessment by electrocardiometry. In: Ion, S., Popa, C. (Eds.), Topics of Mathematical Modeling of Life Sciences Problems. MatrixRom, Bucharest, RO, pp. 43–66.

Morega, A.M., Dobre, A.A., Morega, M., 2016. Electrical cardiometry simulation for the assessment of circulatory parameters. Proc. Romanian Acad Ser. A 17 (3), 259–266.

Morega, A.M., Dobre, A.A., Morega, M., 2018. The brachial electrical bioimpedance as a localized cardiovascular investigation technique. Rev. Roumaine Sci. Techn. Electrotech. Energ 63 (2), 162–171.

Morris, L., Delassus, P., Callanan, A., Walsh, M., Wallis, F., Grace, P., et al., 2005. 3-D numerical simulation of blood flow through models of the human aorta. J. Biomech. Eng. 127, 767–775.

Moskalenko, Y.E., Naumenko, A.I., 1959. Movement of the blood and changes in its electrical conductivity. Bull. Exp. Biol. Med. 47 (2), 211–215. Available from: https://link.springer.com/article/10.1007/BF00788156.

Nagel, L.W., 1975. *SPICE2: A Computer Program to Simulate Semiconductor Circuits*, Memorandum No. ERL–M520. University of California, Berkeley.

Nagel, L.W., Pederson, D.O., 1973. *SPICE (Simulation Program with Integrated Circuit Emphasis)*, Memorandum No. ERL–M382. University of California, Berkeley.

Naranjo-Hernández, D., Reina-Tosina, J., Mart Min, M., 2019. Fundamentals, recent advances, and future challenges in bioimpedance devices for healthcare applications. Hindawi J. Sens. 2019, Article ID 9210258, **42**.

Norozi, K., Beck, C., Osthaus, W.A., Wille, I., Wessel, A., Bertram, H., 2008. Electrical velocimetry for measuring cardiac output in children with congenital heart disease. Br. J. Anaesth. 100 (1), 88–94.

Nyboer, J., Kreider, M., Hannapel, L., 1950. Electrical impedance plethysmography. A physical and physiologic approach to peripheral vascular study. Circulation II, 811–821.

Osypka, M., 2009. An Introduction to Electrical Cardiometry. Electr. Cardiometry™, pp. 1–10, <https://osypka-asia.com/pdf/technique.pdf> (accessed Dec. 2019).

PalmSens Compact Electrochemical Interfaces, 2019. <https://www.palmsens.com/> (accessed Dec. 2019).

Piuzzi, E., Pisa, S., Pittella, E., Podesta, L., Sangiovanni, S., 2019. Low-cost and portable impedance plethysmography system for the simultaneous detection of respiratory and heart activities. IEEE Sens. J. 19 (7), 2735–2746.

Punj, R., Kumar, R., 2019. Technological aspects of WBANs for health monitoring: a comprehensive review. Wireless Netw. 25 (3), 1125–1157.

Rapin, M., Braun, F., Adler, A., Wacker, J., Frerichs, I., Vogt, B., et al., 2019. Wearable sensors for frequency-multiplexed EIT and multilead ECG data acquisition. IEEE Trans. Biomed. Eng. 66 (3), 810–820.

Roşu-Hamzescu M., 2019. System for the analysis of the dynamics of biointerfaces, Doctoral Thesis, The University of Bucharest, Bucharest.

Sathyaprabha, T.N., Pradhan, C., Rashmi, G., Thennarasu, K., Raju, T.R., 2008. Noninvasive cardiac output measurement by transthoracic electrical bioimpedence: influence of age and gender. J. Clin. Monit. Comput. 22 (6), 401–408.

Segen, J.C., 2005. Concise Dictionary of Modern Medicine. MacGraw-Hill Medical.

Shahcheraghi, N., Dwyer, H.A., Cheer, A.Y., Barakat, A.I., Rutaganira, T., 2002. Unsteady and three-dimensional simulation of blood flow in the human aortic arch. Trans. ASME 124, 378–387.

Shimazu, H., Yamakoshi, K.I., Togawa, T., Fukuoka, M., Ito, H., 1982. Evaluation of the parallel conductor theory for measuring human limb blood flow by electrical admittance plethysmography. IEEE Trans. Biomed. Eng. BME-29 (1), 1–7.

Simpleware v. 4.2, Simpleware Ltd., UK, 2010.

Slicer v. *4.10*, https://www.slicer.org. (accessed 2019).

Sramek, B.B., 1986. BoMed's Electrical Bioimpedance Technology for Thoracic Applications (NCCOM): Status Report. BoMed Ltd, Irvine, pp. 19–21, May 1986 Update.

Stevanović, P., Šćepanović, R., Radovanović, D., Bajec, D., Perunović, R., Stojanović, D., et al., 2008. Thoracic electrical bioimpedance theory and clinical possibilities in perioperative medicine. Signa Vitae 3 (Suppl 1), 22–27. Available from: www.signavitae.com.

Taylor, C.A., Hughes, T.J.R., Zarins, C.K., 1998. Finite element modeling of three-dimensional pulsatile flow in the abdominal aorta: relevance to atherosclerosis. Ann. Biomed. Eng. 26, 975–987.

Tishchenko, M.I., 1973. Measurement of the stroke blood volume by an integral rheogram of the human body. Fizio.l Zh. SSSR, Im. Sechenova I.M. 59 (8), 1216–1224 (in Russian).

Teixeira, V.S., Krautschneider, W., Montero-Rodriguez, J.J., 2018. Bioimpedance spectroscopy for characterization of healthy and cancerous tissues. In: Proceedings of the 2018 IEEE International Conference on Electrical Engineering and Photonics (EExPolytech), Saint Petersburg, Russia, pp. 147–151, October 2018.

Truijen, J., van Lieshout, J.J., Wesselink, W.A., Westerhof, B.E., 2012. Noninvasive continuous hemodynamic monitoring. J. Clin. Monit. Comput. 26 (4), 267–278.

Vedru, J., 1994. Electrical Impedance Methods for the Measurement of Stroke Volume in Man: State of Art. Institute of General and Molecular Pathology Group of Biomedical Engineering, Veski 34, EE2400 Tartu, Estonia.

VHP – Visible Human Project, U.S. National Library of Medicine, National Institutes of Health. <http://www.nlm.nih.gov/research/visible/visible_human.html> (accessed 2019).

Visual Human Project, U.S. National Library of Medicine, National Institutes of Health http://www.nlm.nih.gov/research/visible/visible_human.html. Accessed 30 January 2015.

Visser, K.R., 1989. Electric properties of flowing blood and impedance cardiography. Ann. Biomed. Eng. 17, 463–473.
Visser, K.R., 1992. Electric conductivity of stationary and flowing human blood at low frequencies. Med. Biol. Eng. Comput. 30, 636–640.
Vosika, Z.B., Lazovic, G.M., Misevic, G.N., Simic-Krstic, J.B., April 2013. Fractional calculus model of electrical impedance applied to human skin. PLoS One 8 (4), e59483. Available from: www.plosone.org.
Ward, L.C., 2019. Bioelectrical impedance analysis for body composition assessment: reflections on accuracy, clinical utility, and standardisation. Eur. J. Clin. Nutr. 73, 194–199.
Woltjer, H.H., Bogaard, H.J., de Vries, P.M.J.M., 1997. The technique of impedance cardiography. Eur. Heart J. 18, 1396–1403.
Yu, K., Shao, Q., Ashkenazi, S., Bischof, J.C., He, B., 2016. In vivo electrical conductivity contrast imaging in a mouse model of cancer using high-frequency magnetoacoustic tomography with magnetic induction (hfMAT-MI). IEEE Trans. Med. Imaging 35 (10), 2301–2311.

CHAPTER 6

Magnetic drug targeting

6.1 Introduction

Magnetic drug targeting (MDT) is one of the promising methods aimed to target the delivery of the medication needed to more efficiently treat locoregional malignancies and to enhance the healing process for solid tumors, infections, blood coagula, and similar diseases while minimizing the negative side effects (Plavins and Lauva, 1993; Saiyed et al., 2003; Alexiou et al., 2006a; Shapiro et al., 2014). In MDT cancer therapy, magnetic (superparamagnetic) nanoparticles (MNPs) carrying anticancer agents, for example, chemotherapeutic, radionuclide, cancer, or gene-specific antibodies (Dobre, 2012) are administered into the bloodstream to be delivered to the region of interest (ROI), where they are noninvasively guided, squarely aggregated, and retained using an external magnetic field (Gupta and Hung, 1989; Kaminski et al., 2003; Shamsia et al., 2018).

MDT acts also as a complementary intervention for standard chemotherapy as it has the advantage to reduce the aggressive side effects (Thomas et al., 2013). In another circumstance, MD may be injected into articular cavities, where an external magnetic field helps increasing its retention time at that joint level (Manea et al., 2014; Stanciu, 2016).

Researchers have long focused on a method of concentrating significant amounts of toxic drugs within the area of the diseased tissue with a very small amount being absorbed into the surrounding healthy tissue (Gupta and Hung, 1989; Orekhova et al., 1990; Papisov et al., 1987; Kaminski et al., 2003). Thus the first MDT attempts on patients used a ferrofluid with particles (100 nm in size) constructed to chemically bind a chemotherapeutic drug, for example, the *epidoxorubicin* (Lübbe et al., 1996). Later on, different types of ferrofluids available on the market were tested and the iron oxide core was coated with a polymer layer of starch, to provide for biocompatibility (Xu et al., 2005; Kheirkhah et al., 2008; Price et al., 2017; Lima et al., 2019). For instance, *mitoxantrone* was successfully linked to phosphate groups of starch derivatives, with which female specimens of New Zealand rabbits were inoculated at the medial level of the left arm with squamous cell carcinoma. The tumor completely disappeared after 35 days of treatment, and no metastasis or side effects were observed (Alexiou et al., 2006b).

MDT therapy begins by binding the drug to ferrofluid nanoparticles by chemical methods or by incorporating them into a carrier drug particle (Gupta and Hung, 1989; Kaminski et al., 2003; McBride et al., 2013; Abd Elrahman and Mansour,

Computational Modeling in Biomedical Engineering and Medical Physics
DOI: https://doi.org/10.1016/B978-0-12-817897-3.00006-3

2019). The MNPs are then injected into the bloodstream at a suitable location, directed to the tumor site using an external magnetic field produced by a permanent magnet (PM), and directed to the capillary bed of the diseased tissue (Papisov et al., 1987; Papisov and Torchilin, 1987; Alexiou et al., 2000, 2006a,b). Here they could in turn be activated by an enzyme, pH, or trough temperature (Papisov et al., 1987). However this procedure also has its challenges, largely due to the dynamics of the MNPs within the blood flow, which is not well understood hence mastered at the moment (Kaminski et al., 2003; Babincova and Babinec, 2009; Price et al., 2017). In the attempt to alleviate this difficulty, monitoring and control techniques have been suggested to ensure the MD proper localization (Bai et al., 2016; Chuzawa et al., 2013; Shapiro, 2009).

Despites the efforts the magnetic field "design" (spectrum and intensity) is still a challenge (Gleich et al., 2007; Morega et al., 2013a,b,c; Manea et al., 2014; Shapiro et al., 2014; Sharma et al., 2015; Săndoiu, 2019). To produce the field that ought to guide and immobilize particles, MDT utilizes, usually, widely available, inexpensive and powerful PMs. In vivo and in vitro studies have demonstrated the potency of PMs to immobilize MNPs provided they are placed close enough to the target. Several studies were conducted with the aim of optimizing their size, shape, and orientation for MNPs control (Dobre and Morega, 2010; Săndoiu, 2019; Liu et al., 2019; Liu, 2019). Moreover, Halbach arrays (Riegler et al., 2011; Barnsley et al., 2016), magnetic field concentrator ferromagnetic wires (Iacob and Chiriac, 2004), or transdermal ferromagnetic implants (Avilés et al., 2005; Cregg et al., 2010; Cregg et al., 2012) were alternatively proposed.

Although significant progress was made using PMs to control MNPs, they have several disadvantages. The strength of the magnetic field that they produce decreases fast (power law) with the distance to the source. Consequently it has a rather narrow region of influence on the circulatory system, and it cannot act efficiently upon the MD that is injected deeper into the body. Moreover, the magnetic field created by a PM is fixed and static and can only be changed by the physical displacement of the magnet.

To overcome these limitations, electromagnets have been proposed and experimental, and numerical studies have shown that electromagnetic MDT is feasible (Xu et al., 2005; Morega et al., 2013a,b,c). Several works were published on electromagnet systems specially built for MDT that can produce extremely intense, high gradient magnetic fields, confined in rather narrow regions (Hoskins et al. 1990; Gleich et al., 2007; Li et al., 2018). Their magnetic field may be controlled to guide, to some extent, the MNPs by conveniently adjusting the current. Nevertheless, controlling MNPs flow in micrometric capillaries is unrealistic by cause of the rather high rate of the blood flow (Furlani and Furlani, 2007; Shaw, 2020).

The implant-assisted MDT combines the advantages of PMs with the guiding property of electromagnets, and it involves a stent, wire or cylinder that are invasively inserted near the diseased tissue. At each release of MD, the ferromagnetic implant is magnetized by an external magnetic field, leading to a much more effective trap of MNPs (Iacob and Chiriac, 2004; Avilés et al., 2005; Avilés et al., 2008). Moreover, the PM itself may be implanted right in the patient's tumor region and not at the surface of the body. The invasiveness of this technique is confined to a single surgical intervention to implant either the ferromagnetic insert and, or the magnet, which is an improvement over an ordinary method, where the drug MNPs are injected deep into the tissue. A disadvantage would be the difficulty for the drug to cross the wall of the blood vessel to reach the desired location, but this is a common concern with MDT.

Another, quite new, slightly modified *extra corpus* application of this MDT principle, is the purification of bone marrow cells contaminated with tumor cells (Roots-Weiß et al., 1997), using immunomagnetic particles (Saiyed et al., 2003).

6.2 Magnetic nanoparticles for magnetic drug targeting

MNPs are promised to be used in various biological or medical applications, for early diagnosis of maladies, noninvasive imaging and for drug development, but also for controlled drug delivery systems that have the ability to minimize negative systemic side effects. They comprise polymeric micelles, dendrimers, liposomes, inorganic and polymeric nanoparticles, quantum dots, etc. (Hawk's, 2020). All these were tested preclinically and clinically for targeted medication, for gene delivery or for improving diagnostic imaging. Properties existing only at the nanoscale, such as the enhanced fluorescence emission of semiconductor crystals (as in the case of quantum dots) or the magnetic properties of the MNPs, recommend these materials for medical imaging or targeted medication applications (Neuberger et al., 2005).

Magnetic properties of materials used in designing the magnetic drug targeting medication

Iron oxide MNPs with various coatings are more and more used in in vivo applications for reason that they are small enough to be carried through the circulatory or lymphatic system, and at the same time they can be attached to cells, or even they can be inserted into cells, and then transported altogether. When combined with drugs or genes, these particles can transform the viability of the cell or alter the transcription process (Brown, 2020).

According to their magnetic susceptibility, these materials are classified in *diamagnetic,* with negative susceptibility ($\chi \approx -10^{-5}$, i.e., they slightly repulse an external magnetic field), *paramagnetic* ($\chi \approx 10^{-3}...10^{-4}$), and *ferromagnetic* ($\chi \approx 50...10^{4}$) (Fig. 6.1).

Diamagnetic and paramagnetic properties vanish when the external magnetic field is removed, whereas the ferromagnetic materials properties persist—they make, in fact, the PMs. The maximum value of the magnetization is termed *saturation magnetization*

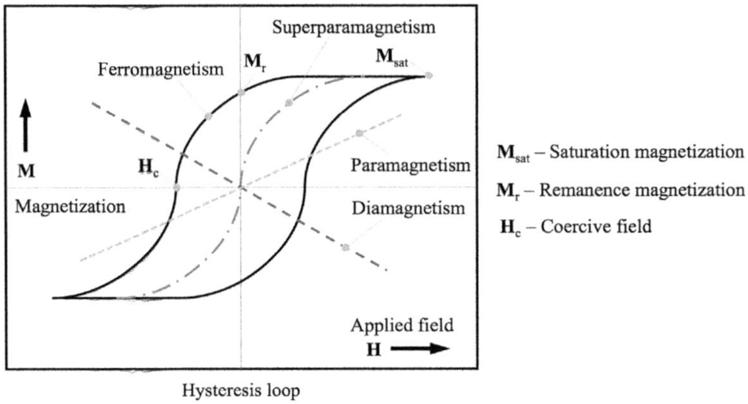

Figure 6.1 Magnetic materials and properties.

(Fig. 6.1). In ferromagnets, a residual magnetization state remains even if an external magnetic field does not exist. That value is called the *remanent magnetization*, M_r. The magnetic field strength that cancels M_r is called *coercive field*, H_c (Mocanu, 1981).

The MNPs used for medical purposes must be biocompatible, nontoxic to cells, and biodegradable. They need to preserve their physical properties after surface treatments, may not affect the characteristics of normal cells, have to be effective for therapeutic doses, and must pose no menace to neighboring tissues (Markides et al., 2012). The modular approach in adding components to MNPs may facilitate the incorporation of particular features and the interchange or combination of functional groups of molecules. Along this line, *ligands* (therapeutic agents and targeting agents, optical dyes, and permeation enhancers) can be conjoined on the surface or inside these nanostructures (Fig. 6.2; Sun et al., 2008).

Superparamagnetic iron oxide nanoparticles

MNPs for biomedical applications have superparamagnetic properties. In this category, the *superpparamagnetic iron oxide nanoparticles* (SPIONs) play a key role. SPIONs have *magnetite* (Fe3O4) or *maghemite* (γ-Fe_2O_3) ferromagnetic kernel, and they are coated with an either organic or inorganic biocompatible polymer, or precipitated into a porous biocompatible polymer. When the radius of their ferromagnetic cores fall below 30 nm the MNPs lose their permanent magnetism and become superparamagnetic (Pankhurst et al., 2003; Neuberger et al., 2005). SPIONs are, so far, the only clinically approved metal oxide MNPs. They may be driven to a certain tissue or organ by an external magnetic field, or they may be used as contrast agents for MRI, for targeted drug delivery or in magnetic hyperthermia (Chapter 8: Hyperthermia and Ablation (Thermotherapy methods)).

The hydrodynamic radii of SPIONs, assumed as elastic spheres of fixed size, are in the range 1–5000 nm (Nacev et al., 2011), their density is 4800–5100 kg/m^3

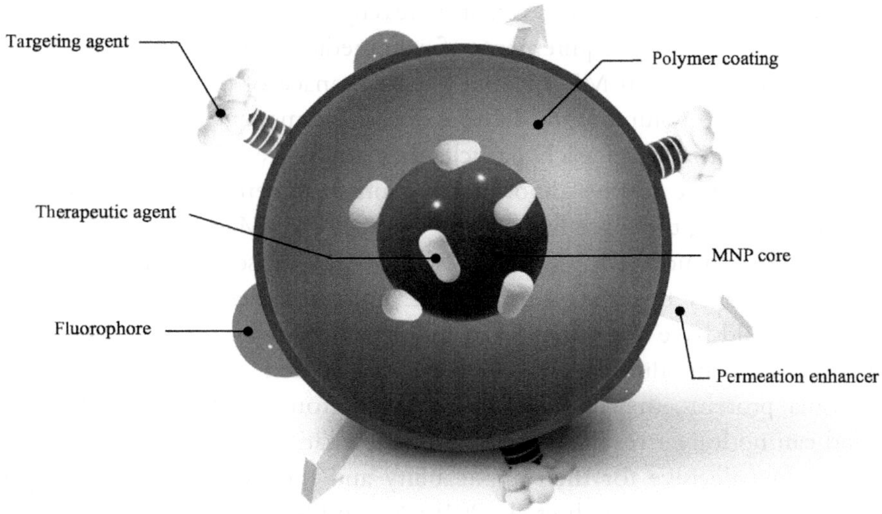

Figure 6.2 MNP (magnetic nanoparticle) possessing various ligands to enable multifunctionality from a single nanoparticle platform.

(Sharma et al., 2015), and their magnetic susceptibility is 20 (Aslibeiki et al., 2012). Considering their hydrodynamic diameter, SPIONs are classified into three categories: *oral* (300 nm to 3.5 μm), *standard* (SSPIO, 50 mm to 150 nm), and *ultrasmall* (USPIO, less than 50 nm), which place them in a range of sizes comparable with those of cells in the human body (10–100 μm), viruses (20–450 nm), proteins (5–50 nm), or genes (2 nm wide and 10–100 nm long). This shows that they are able to come into direct contact with a targeted biological entity. SPIONs with hydrodynamic diameters between 10 and 100 nm in size are considered optimal for intravenous administration, while particles between 200 and 10 nm are either cloistered in the spleen or removed by the kidneys.

SPIONs are suitable for biological applications because they do not attract with each other, so that the peril of clustering in a medical application is minimized. Iron-cored MNPs are biocompatible since the organism is adapted to metabolize *ferritin*, and to use it in subsequent metabolic processes. SPIONs have so far been used as contrast agents in MRI, with products on the market approved by the Food and Drug Agency (FDA), so they could be accepted for other types of applications too (Schütt et al., 1997; Markides et al., 2012).

However, the route of administering and the surface properties of SPIONs are what determine the efficiency of cell absorption, the biodistribution they have, their metabolization and their toxic potential. Magnetite and maghemite are both iron oxides thus they appear naturally as nanometric crystals in the crust of the earth through volcanoes or fires. Therefore it would seem that there are no intrinsic risks associated with them.

The concern that, however, mounts is that increasingly more nanoparticles are produced to meet the rapidly growing requirements of nanomedicine (Singh et al., 2010).

The usage of SPIONs in MDT could be the menace of their aggregation for reason of their surface to volume ratio. In an external, permanent, intense magnetic field, SPIONs get magnetized and each particle may influence (alter) the local magnetic field. This may produce dipole–dipole interactions between particles, so that the particles get aligned along the force lines of the external PM (Markides et al., 2012).

Moreover their nanometric size (large surface area causes increased reactivity and enhances the tendency to diffuse through the biological membranes which can cause cellular stress) could have the potential to induce cytotoxicity, manifested by impaired cell function: mitochondria, nucleus, DNA. In addition, if the particles clump or absorb plasma proteins, they are rapidly removed from the bloodstream by macrophages and can no longer reach the cells aim. The coating and functionalization of the particles are then decisive for their applicability and their degree of biocompatibility with the human organism (Singh et al., 2010; Xu and Sun, 2013).

Superparamagnetic iron oxide nanoparticles synthesis, coating, and functionalization

The iron oxide core of the SPION can be chemically synthesized through various methods: standard synthesis by coprecipitation (Marinin, 2012), reactions in constrained environments (Singh and Lillard, 2009), polyol method (Ali et al., 2016), flow injection synthesis (Salazar-Alvarez et al., 2006), sonolysis (for MNPs used simultaneously to diagnose and treat diseases) (Kudr et al., 2017) or thermolysis (used for formation of iron oxide MNPs from their organometallic precursors) (Lin et al., 2012). Uncoated MNPs poses low solubility and may precipitate (if not small enough) and also have high rates of cloistering in physiological conditions, which may lead to blood vessels clogging in clinical applications. Therefore they are clad with a superficial overlay aimed to ensure efficiency in clinical applications and to improve their biocompatibility, that is, the ability to achieve an appropriate host reaction in a particular situation, and biodistribution, that is, to track the travel of the compounds of interest in an experimental animal or human subject (LexInnova, 2020).

SPIONs coating with polymers for drug delivery preserves their magnetic properties (Mahmoudi et al., 2011) and provides for the most common protection against their oxidation (Faraji and Wipf, 2009; Singh and Lillard, 2009). A major challenge, at this stage, is the uniform size distribution of the coating, a high level of monodispersion, that is, uniformity, and composition homogeneity.

The functionalization of the SPIONs surface using polymer coating with biocompatible molecules, such as dextran, dendrimers, polyethylene glycol, or albumin, aims at binding complex biological molecules, such as drugs, antibodies, hormones, or peptides, with materials such as *silicone*, *dextran*, and *pegylated citrate* (Musielaka et al., 2009), which are used mainly as contrast agents in target organs such as liver, spleen, nodules nodes, and gastrointestinal tract. Commercially available contrast agents are *Feridex*, *Rasovist*, *Survivor*, and *Sinere* (Wang, 2015).

Several SPION systems—starch-coated and methotrexate functionalized; coated in polymeric anhydroglucose and functionalized with epirubicin; starch-coated and functionalized with *epirubicin*; and coated with dextran and functionalized with steptokinase—were successfully used in experiments and in clinical trials (Neuberger et al., 2005; Chomoucka et al., 2010). For instance, the chemotherapeutic drug *gemcitabine* is absorbed on the magnetite MNPs coated with *chitosan* and may release the drug into the cancer cells of the liver, colon and breast (Viota et al., 2013); *dexamethasone* 21-acetate, which is part of the corticosteroid class, may be used with SPION and injected into the joints, where it may be retained with an external magnet for treatment rheumatoid arthritis (Butoescu et al., 2009).

Whatever their specific medical or biological role, or fabrication, in the following the MDT is using MNPs (SIONs) as representative entities, and their spatial distribution is consistent with the continuous media assumptions. As magnetic materials they are superparamagnetic, linear media. The following sections present concerns, stages and results of mathematical and numerical modeling for several MDT application stages—from physical systems and models, to numerical results. The magnetic field sources are either permanent or coils. Certain aspects regarding time and space scales, couplings, which are typical to multiphysics problems, are analyzed, with the aim to produce solvable numerical models consistent with the physics that they represent.

6.3 Several modeling concerns in magnetic drug targeting

MDT therapy utilizes magnetizable particles (SPIONs) as MNPs carriers for the drug. Injected into the vasculature, the MD travels with the blood flow to the targeted location, where an externally produced magnetic field is used to retain and confine it. Magnetization body forces and the gradient driven transfer process (convection through the local vasculature and diffusion) act into hauling it towards the magnetic field source. The MD fabrication and properties (core and coating) (Pankhurst et al., 2003; Berry, 2009; Sun et al., 2008; An et al., 2015; Heidarshenas et al., 2019), the knowledge of its transfer inside the human body (Hobbs et al., 1998; Grief and Richardson, 2005; David et al., 2011), the magnetic field spectrum and the magnets design (Jones, 1995; Schenck, 2000; Preis et al., 1991; Nedelcu, 2016), the deep-body real-time MNPs sensing means and control (Martel et al., 2007; Koch and Josephson, 2009; Martel et al,. 2009; Eckert et al., 2013) are some of the concerns that need to be comprehended and mastered through physical and numerical experiments to the success of MDT. Along this line, MDT traverses several stages: (1) injection and the mixing (with the blood), (2) hemodynamic advection through larger vessels to the

ROI, and (3) transfer through the "leaky" vasculature and the surrounding perfused tissue to the targeted region, Fig. 6.3.

The magnetic field plays a crucial role at stages two and three, and its design is key to the success of the procedure. In general, the magnetic properties of the MNPs core and the incident magnetic field yield the strength and spectrum of the magnetic body forces, whereas their size, shape, and coating properties determine their biocompatibility and spread time (Sun et al., 2008; Veiseh et al., 2010; Nacev, 2013). Despite all the progress in understanding the MD fabrication and transport, the in vivo visualization techniques and the precise measurement of MNPs distribution are yet challenging objectives (Sun et al., 2008; Liu et al., 2019; Liu, 2019; Abd Elrahman and Mansour, 2019).

Numerical modeling may then help and provide for valuable guidelines and insights, unavailable yet through physical experiments. To this aim, consistent mathematical models that rely on available experimental data need to be formulated first for: the fluidic, magnetic and inertia body forces, particle—particle interactions, Saffman lift force in the shear field (Zheng and Silber-Li, 2009; Drochon et al., 2016; Shaw et al., 2017), the drug advection through microvessel by considering the permeability of the microvessel and carrier particle (Furlani and Furlani, 2007; Zheng and Silber-Li, 2009), the rheology of the blood (Newtonian and non-Newtonian) (Misra et al., 1992; Morega et al., 2013a,b,c; Săvăstru, 2016; Shaw et al., 2017), and the aggregate blood MDT flow (single- or two-phase) in larger and microvessels.

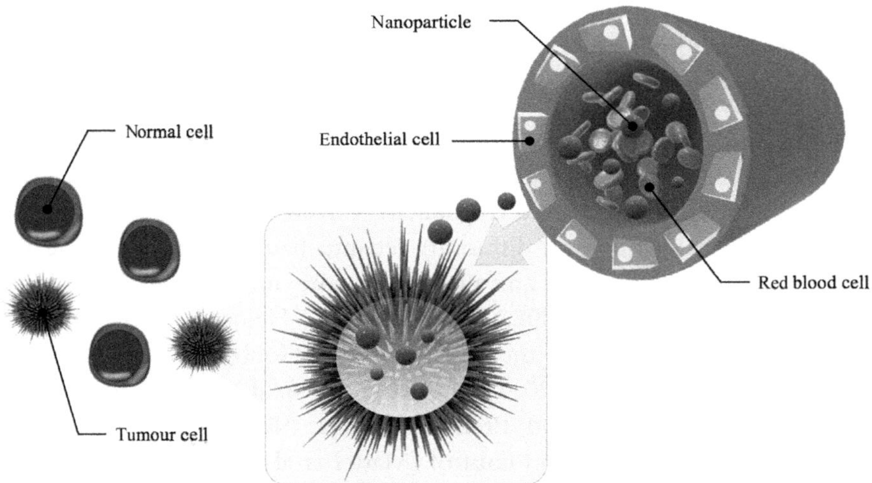

Figure 6.3 The magnetic drug targeted delivery facilitated by "leaky" vasculature. (Left) MNPs receptor-mediated endocytosis internalization and endosome forming. (Middle) Proton pump endosomal acidification increases the osmotic pressure; (right) swelling, and eventual rupture of the endosome releases the MNP and affixed therapeutic agents. *MNP*, Magnetic nanoparticles.

6.4 Magnetic drug mixing

The magnetic drug is injected into the blood stream where it is subjected to a mixing process with the blood and conveyed to the ROI. The degree of mixing depends on many factors (MD dilution, vessel caliber, blood flow rate, MD injection flow rate, etc.) but it is important to find the distance from the injection point where this initially two-phase fluid, blood-MD, turns into a homogeneous, single-phase fluid flow. From that point on, the blood and the MD are a homogeneous magnetizable aggregate fluid (MAF) whose magnetic permeability accounts for the dilution of the MD mixed with the blood.

A simpler, two-dimensional model may help to evidence the underlying features of the two-phase flow produced by injecting the MD into a blood vessel, through a catheter, in the absence of any external magnetic field. The *level set method* (Sussman et al., 1998; Sethian, 1996; Sethian and Wiegmann, 2000) is used to solve this two-phase flow problem that is expressed by the momentum balance and the continuity laws (Chapter 1: Physical, Mathematical, and Numerical Modeling) (Dobre, 2012), which for Newtonian fluids yield

$$\rho\left[\frac{\partial \mathbf{u}}{\partial t} + (\mathbf{u}\cdot\nabla)\mathbf{u}\right] = \nabla\left[-p\mathbf{I} + \eta\left(\nabla\cdot\mathbf{u} + (\nabla\cdot\mathbf{u})^T\right)\right] + \mathbf{f}, \tag{6.1}$$

$$\nabla\cdot\mathbf{u} = 0, \tag{6.2}$$

and the *level set equation*

$$\frac{\partial \Phi}{\partial t} + (\mathbf{u}\cdot\nabla)\Phi = \gamma\nabla\cdot\left[\varepsilon\nabla\Phi - \Phi(1-\Phi)\frac{\nabla\Phi}{|\nabla\Phi|}\right], \tag{6.3}$$

where $\mathbf{f} = \mathbf{f}_{ts}$ stands for the surface tension forces, Φ is the *level set* function (nondimensional, with values in the range 0–1), and γ and ε are stabilization parameters that ensure the existence and uniqueness of the solution. The mass density, ρ, and the kinematic viscosity, η, are provided by

$$\rho = \rho_b + \left(\rho_{MD} - \rho_b\right)\Phi, \quad \eta = \eta_b + \left(\eta_{MD} - \eta_b\right)\Phi, \tag{6.4}$$

where $(\cdot)_b$ and $(\cdot)_{MD}$ denote the blood and the MD, respectively.

This method is useful in studying the dynamics of interfaces and geometric shapes, making it possible to calculate the curvilinear surfaces in the Cartesian coordinate system and does not require the parameterization of the computational domains. Also it facilitates the analysis of the dynamics of the computational domain morphology, when and where its deformation and, or its division into several subdomains occur. On the other hand, there are applications of this method also in the field of image

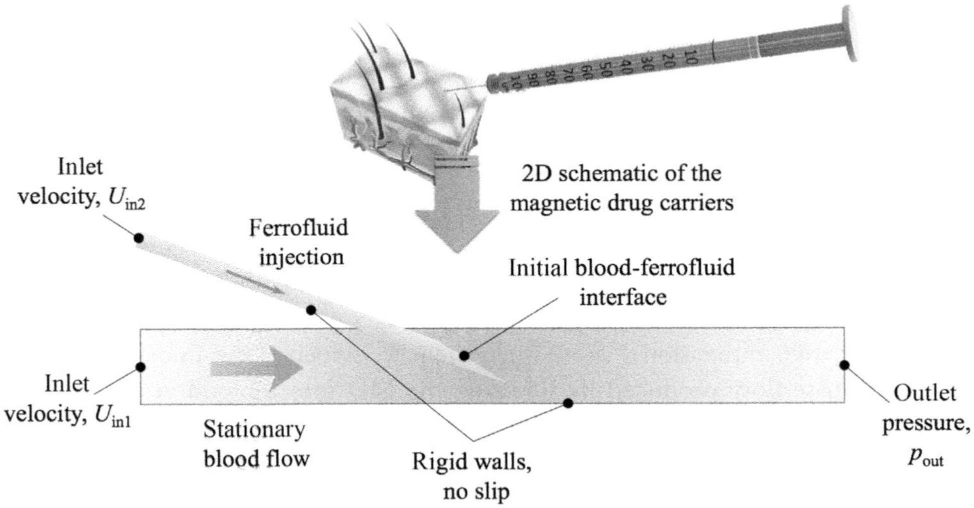

Figure 6.4 The magnetic drug is injected and eventually mixes with the blood. Uniform velocity profiles at inlets, pressure uniform profile at the outlet and no-slip boundary conditions are set.

reconstruction, in the form of segmentation algorithms of ROIs (Osher and Sethian, 1988).

Stationary flow is relevant to the study of this mixing process as the MD injection (at 1 m/s) concerns a vein, Fig. 6.4. Besides, the mixing length is apparently short enough such that in ~0.5 s (half of the cycle for 60 bpm flow dynamics) the MAF is observed to materialize, Fig. 6.5 (Dobre, 2012).

The time scale of this mixing process is related to the size (caliber) of the vessel, the injection flow rate and the blood flow rate. Here, the distance from the injection location to the point where the MAF sets in is 5–6 times the hydraulic diameter of the vessel, and the time scale is less than the cardiovascular period, that is, 1 s for 60 bpm.

More complex models, constructed, for instance, using fused computational domains (Chapter 3: Computational Domains), may be of interest for more realistic, perhaps patient-related evaluation. However, the outlining features of this mixing process are fairly well presented by this 2D analysis, based on which it may be conjectured that the MD (carried on by MNPs) and the blood may be modeled as a MAF in the circulatory system of the ROI.

6.5 Magnetic drug targeting, from the blood vessel to the targeted region

MNPs with usual dimensions less than 400–600 nm can protrude the walls of the blood vessels walls into the adjacent tissue that contains the targeted ROI. MNPs with characteristic size below 25 nm can also be used, but the smaller they are, the harder

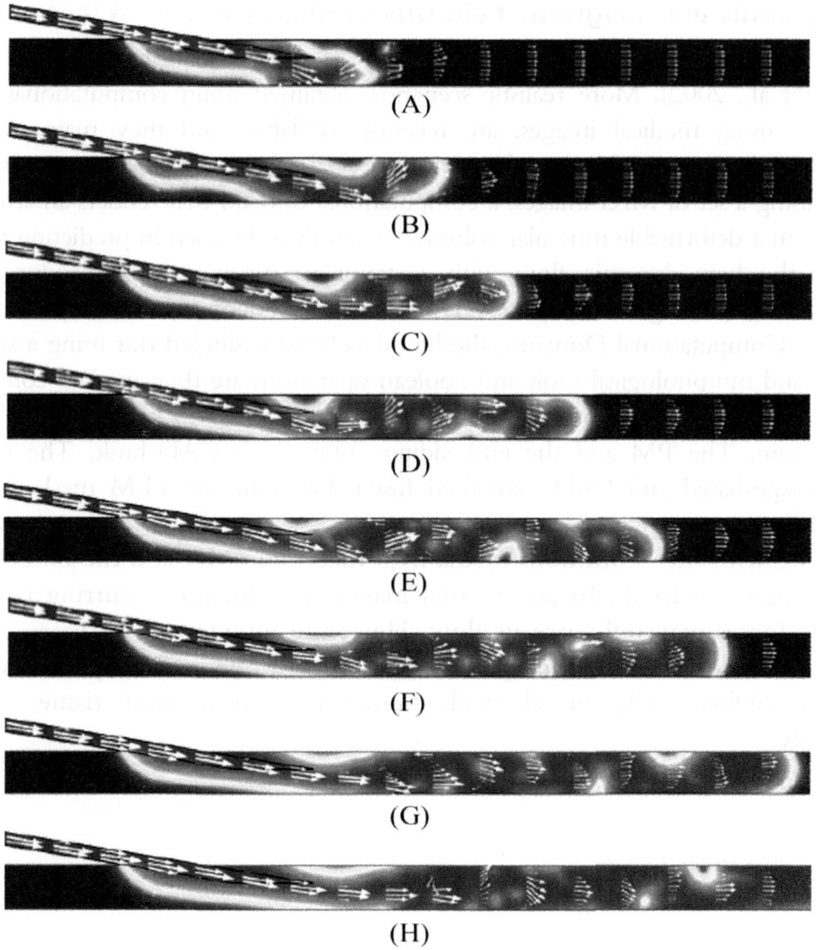

Figure 6.5 The level set function, Φ, shows the two-phase flow mixing progress.

the particles are removed from the circulatory flow and the magnetic forces that may drive them are lower. For instance, forces of the order *O*(pN) are obtained in a magnetic field of about 1 T and with a gradient of about 0.5 T/cm. Numerical simulation may help asserting the dynamics of the MNPs in the magnetic field, otherwise difficult (if possible) to figure out, with the aim to exemplify the MDT optimization, training, planning and intervention.

To substantiate this approach and its merit, we consider an idealized blood vessel through which blood flows with MNPs, and which is exposed to a static magnetic field. Several MNPs mass transfer processes occur in the hemodynamic flow and the external magnetic field: the MNPs travel with the blood flow; they diffuse through the vessel walls, and then travel and diffuse through the adjacent tissue to the contained ROI.

Hemodynamic and magnetic field driven mass transfer in larger vessels

Simplified computational domains are valuable as a first step in the MDT analysis (Voltairas et al., 2002). More realistic scenarios, obtained using computational domains constructed using medical images, are recently available, and they may provide for more thorough, patient-related studies (Morega et al., 2009; Shamloo et al., 2019). For instance, using a set of MRI images, a computational domain that renders an arterial tree, embedded in a deformable muscular volume, was built to be used in predicting the interaction of the hemodynamic flow with a stationary magnetic field produced by an implanted PM (Morega et al., 2010; Dobre, 2012). Following the steps outlined in Chapter 3, Computational Domains, the blood volume is singled out using a segmentation tool, and morphological tools and boolean operations are then used to construct the vascular (arterial) tree. The final 3D solid model is processed to produce the computational domain. The PM and the embedding volume are CAD-built. The two constructs—image-based and CAD—are then fused, Fig. 6.6, and FEM meshed (Morega et al., 2010).

Keeping with the continuous media hypothesis, the MD and the blood, behave as a homogeneous MAF. In an external magnetic field, the occurring magnetization body forces may influence its flow. Here, the interactions between the flow, the vessel wall and the embedding tissue are coupled one-way only: the flow pressure field deforms the vessel wall, hence the embedding tissue, and not reciprocally.

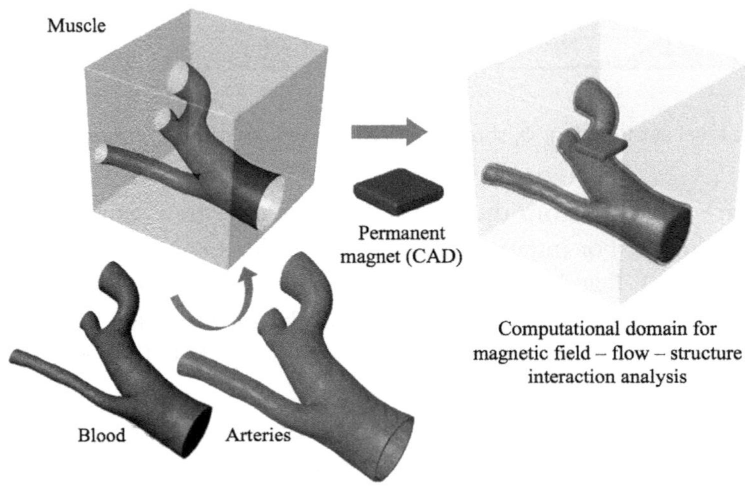

Figure 6.6 The main stages to obtaining a more realistic computational domain constructed using a DICOM dataset and a CAD part.

The magnetic field produced by the PM is described through Ampere's law, the magnetic flux law (Chapter 1: Physical, Mathematical, and Numerical Modeling)

$$\nabla \times \mathbf{H} = 0, \quad \nabla \cdot \mathbf{B} = 0 \tag{6.5}$$

and the material law $\mathbf{B} = \mathbf{B}(\mathbf{H})$, which for linear magnetic media yields $\mu = \mu_0(1 + \chi) = \mu_0\mu_r$

$$\mathbf{B} = \mu_0\mu_{r,mag}\mathbf{H} + \mathbf{B}_{rem}, \quad \text{for the permanent magnet,} \tag{6.1}$$

$$\mathbf{B} = \mu_0\left[\mathbf{H} + \mathbf{M}_{ff}(\mathbf{H})\right], \quad \text{for the aggregate magnetic fluid,} \tag{6.2}$$

$$\mathbf{B} = \mu_0\mathbf{H}, \quad \text{for arterial walls and embedding tissue.} \tag{6.3}$$

In the above $\mathbf{B}$ is the magnetic flux density, $\mathbf{H}$ the magnetic field strength, and μ_0 the magnetic permeability of free space. For the PM, $\mu_r = \mu_{r,mag}$ and $\mathbf{B}_{rem}$ is the remanent magnetic flux density. For the MAF (superparamagnetic medium), $\mathbf{M}_{ff}(\mathbf{H}) = \chi\mathbf{H}$, with χ magnetic susceptibility.

The magnetic vector potential, $\mathbf{A}$ (and the divergence free gauge condition)

$$\mathbf{B} = \nabla \times \mathbf{A}, \quad \nabla \cdot \mathbf{A} = 0, \tag{6.7}$$

may be used to present the mathematical model

$$\nabla \times \left(\mu_{\mathbf{0}}^{-1}\mu_{\mathbf{r}}^{-1}\nabla \times \mathbf{A} - \mathbf{B}_{rem}\right) = 0. \tag{6.8}$$

"Infinite elements" are bordering the computational domain to provide for a boundary that contains the magnetic field within a shorter distance from the magnet, but conveniently sized for the hemodynamic flow and the structural interactions, and where the magnetic field may be verified to be vanishingly small (magnetic insulation, $\mathbf{n} \times \mathbf{A} = 0$). This approach has the advantage of a single and smaller computational domain.

The MAF is assumed Newtonian. Its flow is pulsatile (arterial), incompressible and laminar, described by Eqs. (6.1) and (6.2), with $\mathbf{f} = \mathbf{f}_{mg}$ magnetic body forces (Chapter 1: Physical, Mathematical, and Numerical Modeling) (Rosensweig, 1997; Morega et al., 2010)

$$\mathbf{f}_{mg} = \mu_0(\mathbf{M} \cdot \nabla)\mathbf{H}. \tag{6.9}$$

No-slip conditions are set for the walls and pressure conditions for the inlet and outlet: $p_1 = 13.300$ N/m^2; $p_2 = 13.290$ N/m^2; $p_3 = 13.040$ N/m^2; $p_4 = 13.040$ N/m^2, $p_i = 1 + \mathrm{K} \cdot \sin(t + 3/2)$, with K a factor of the order $O(10^{-1})$ (Fig. 6.7) (Morega et al., 2011).

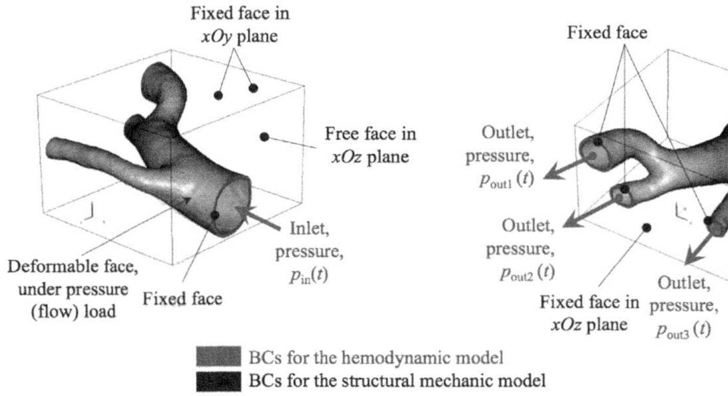

Figure 6.7 Computational domain and boundary conditions for the hemodynamic and structural problems.

The arterial segment is of *resistance type* (Feijóo, 2000) (Chapter 1: Physical, Mathematical, and Numerical Modeling), with relatively large cross section.

The first stage in MDT delivery is the advection of the MD through larger vessels. In view of the one-way coupling magnetic field—flow—structure the magnetic field is solved first. In this "early" stage of MDT, only the vessel volume (MAF) has magnetic properties. The magnetic field produced by a PM ($\mathbf{B}_{rem} = 1.3$ T, e.g., N50NdFeB N42, N45, N48, Ningbo, 2019, or SmCo magnets, Pyrhönen et al., 2009) yields magnetization body forces, Eq. (6.9), which exhibit the expected targeting effect, Fig. 6.8.

Next, the hemodynamic flow and pending the deformation of the embedding tissue and vessels walls, with and without magnetic field, are analyzed. Fig. 6.9 (Morega

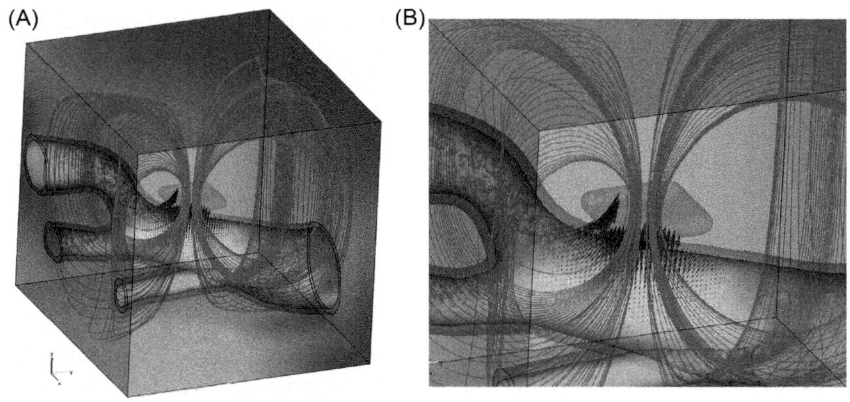

Figure 6.8 Magnetic field and magnetization forces—independent of the flow and structural interactions: (A) numerical simulation results, and (B) enlarged view (Morega et al., 2011; Dobre, 2012).

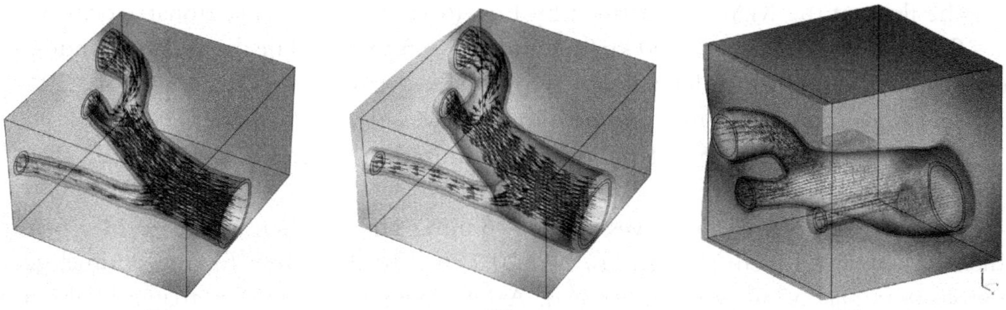
(A) (B) (C)

Figure 6.9 Flow–structure interaction in the absence of a magnetic field. Deformations are amplified by a factor of 122. (A) No flow, no deformation. (B) No magnetic field, maximum flow rate, maximum deformation, 0.5 μm. (C) Magnetic field, maximum flow rate, maximum deformation, 0.52 μm.

et al., 2011; Dobre, 2012) outlines the displacements of the biological tissues, vessel wall and embedding tissue, assuming their structural hyperelastic behavior.

Apparently the deformation is related to the blood flow, therefore in what follows it will be neglected and the vessels will be treated as rigid. The optimization of the magnet—its position, size, shape, magnetic properties, and its field spectrum—is proposed next.

The constructal optimization of the magnetic field source

The magnetic field sources in MDT are PMs and electromagnets, and particular attention is allotted to their design subjected to several optimization criteria (Preis et al., 1991; Hoke et al. (2008); Schenck, 2000). Recently a *constructal* optimization (Chapter 2: Constructal Law Criteria in Morphing Shape and Structure of Systems With Internal Flows) of a PM was suggested (Morega et al., 2018; Săndoiu, 2019), with the aim to enhance the targeting effect of the magnetic field—"shape with a purpose." Starting from the standard, uniformly magnetized parallelepiped bar, different PM configurations may be envisaged. Of these we single out the optimization path where the PM bar is split into several smaller, identical, parallelepiped blocks, with the final aim to optimally cover a specific ROI. The *block width*, SW (in split direction, aligned with the hemodynamic flow) and the spacing between the blocks (the *gap size*, GS) are the design optimization parameters. The total volume of the magnetic material is invariant. Moreover, only the inside vessel volume—a MAF—has magnetic properties.

Instead of Eq. (6.8) we use the magnetic scalar potential V_m, $\mathbf{H} = -\nabla V_m$, which yields

$$\nabla \cdot \left(\mu_0 \mu_r \nabla V_m + \mathbf{B}_{rem}\right) = 0, \tag{6.10}$$

where $\mu_r = \{\mu_{r,mag}, \mu_{r,ff}, 1\}$ for the PM, MAF, and elsewhere, respectively. The boundary condition, magnetic insulation, becomes $\partial V_m / \partial n = 0$ where $\mathbf{n}$ is the outward normal to the boundary.

The flow of the MAF is described by Eqs. (6.1) and (6.2), in stationary form, with $\mathbf{f} = \mathbf{f}_{mg} = \nabla(\mathbf{B}\cdot\mathbf{H})/2$, $\rho = 1000$ kg/m^3 and $\eta = 3.5$ mPa s. The boundary conditions are as follows: uniform inlet velocity profile ($U_{in} = 0.17$ m/s, Quanyu et al., 2017), uniform (zero) outlet pressure, and no-slip at the vessel walls.

A 2D model is used to exemplify the optimization process (Morega et al., 2018), Fig. 6.10.

The diameter of the blood vessel is $d = 6$ mm. The bidimensional size of a 6-mm thick PM of $V = 1.2$ cm^3 volume is, here, 0.2 cm^2. In the search for the optimal geometric *aspect ratio of the permanent magnet*, A (A = height over width) that may provide for best MDT effect, the volume of the PM, *Vol*, is constraint, which further sets the total free PM energy, $W_m = \int_{Vol}(\mathbf{B}\cdot\mathbf{H})/2dv \cong Vol\cdot(B_{rem}\cdot H_c)/2$ (Fig. 6.10). It should be mentioned that the resulting "optimum" elemental cell is rather a compromise between the largest force, its orientation, and its area of action, rather than an *optimum optimorum*.

Fig. 6.11 (top) shows the optimization for the *elemental cell* (a single PM block) as bell-shaped curves for the magnetic forces acting upon the MAF, $F^*_{mg,x}$ (streamwise) and $F^*_{mg,y}$ (orthogonal), where $F^*_{mg(x|y)} = \int_S f_{mg(x|y)}dxdy$, S is the area of the MAF channel. To compare these results with those in the following stages of optimization, SW (the PM block width) is used in the abscissa instead of A. Both curves show off maxima, with different values and locations though. The vertical force acts into attracting the MD towards the PM, whereas the horizontal component may influence (enhance or oppose) the flow. It can be inferred that the PM has to be placed so that the tumor is located between the magnet and the blood vessel carrying the medication. Furthermore, for pulsating flow conditions the action of $F^*_{mg,x}$ is more significant during the minima of the flow rate (Morega et al., 2018; Săndoiu, 2019; Săndoiu et al., 2019).

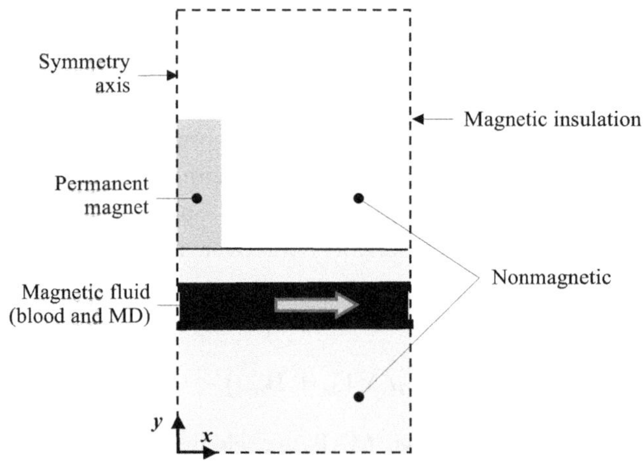

Figure 6.10 The 2D model and the magnetic field boundary conditions. Symmetry is used to reduce the computational domain.

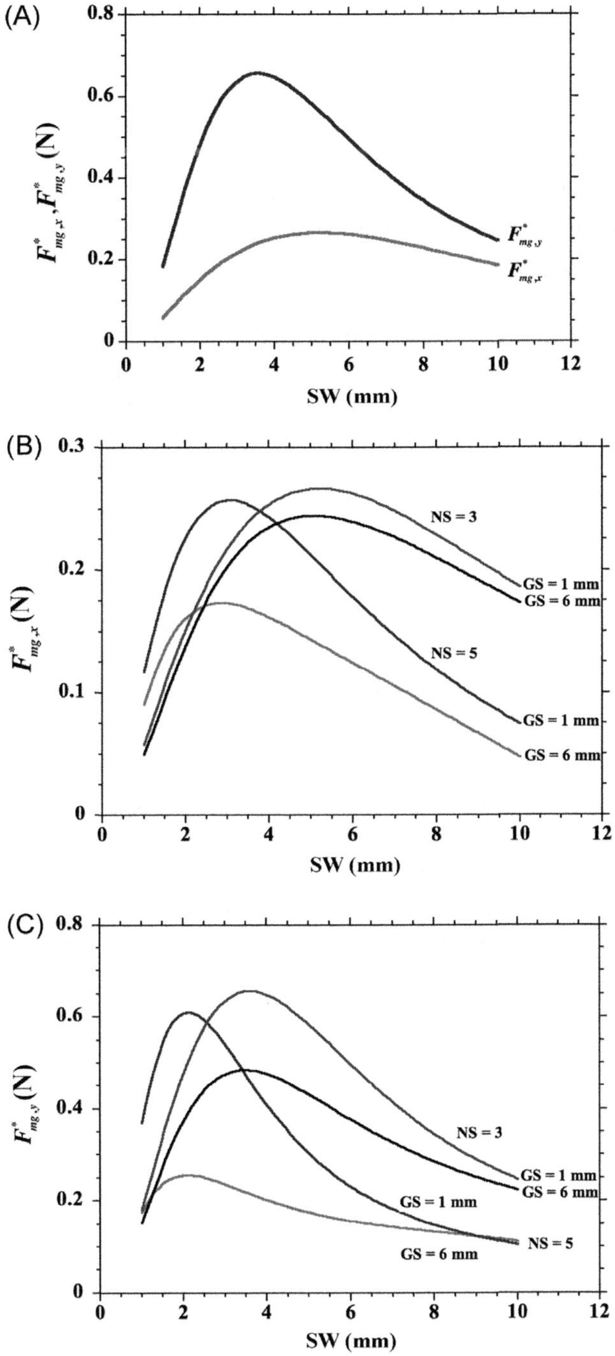

Figure 6.11 The magnetic forces for NS = 1 (A) and NS = 3, 5 (B and C)—2D analysis. (A) Magnetic forces for NS = 1. (B) Horizontal magnetic force. (C) Vertical magnetic force.

The PM may be divided into NS = 2, 3, etc. identical, equally spaced blocks, a sequence which points out that for increasing NS and spacing, GS, in between them, the magnetic blocks act more and more as independent PMs.

The second stage of optimization aims the *first order constructal ensemble*, NS = 2. The block width, SW (7–10 mm) and the spacing between them, GS (1–6 mm) are the optimization parameters. The magnetic volume of each block is *Vol*. Fig. 6.11 (middle and bottom) shows the results in the third stage of optimization when NS = 3, the PMs of volume *Vol*, for two limiting cases, GS = 1 mm and GS = 6 mm. Quasi-independent, noninteracting PMs may result for GS > 6 mm. The maxima of $F^*_{mg,x}$ and $F^*_{mg,y}$ decrease with GS by half (from 6 to 1 mm), for almost the same values of GS. It may be concluded that GS = 1 mm, the smallest GS, provides for an optimum. The optimal configuration was found in the interval SW = 3–4 mm. Depending on the morphology and the location of the targeted tumor volume, the therapist may decide the optimal configuration for the PM array.

The same analysis (GS and SW optimization) was performed for NS = 4, 5, and 6 blocks. Fig. 6.11 depicts $F^*_{mg,x}$ (middle) and $F^*_{mg,y}$ (bottom) for GS = 1 mm and GS = 6 mm, for NS = 3, 5. When comparing with NS = 3, the maximum $F^*_{mg,y}$ for NS = 5 occurs for a thinner (slender) block (SW = 0.002 mm).

Although insightful, the bidimensional analysis may not account for the "duct" type of flow of the MAF and for the third component (perpendicular to the plane) of the magnetic forces. A simple, three-dimensional analysis is shown here, using the computational domain sketched in Fig. 6.12. The *per meter* (in the third dimension) forces may be used, in an order of magnitude sense, to compare with their three-dimensional counterparts by multiplying them with an equivalent out-of-the-plane size of the bidimensional model.

"Infinite" elements are used to border the computational domain and close the magnetic field at a finite distance, and symmetry is applied, leading to a substantial reduction in the mesh size. An unstructured mesh with ~540,000 tetrahedral elements provides for grid independent numerical solutions. The magnetic scalar potential formulation, and quintic Lagrange elements are used. Fig. 6.13 (top)

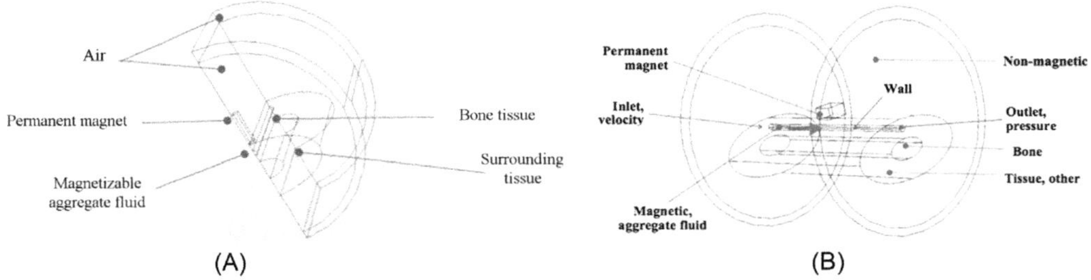

Figure 6.12 The three-dimensional computational domains in the MD problem (Săndoiu, 2019). (A) The magnetic field. (B) The flow.

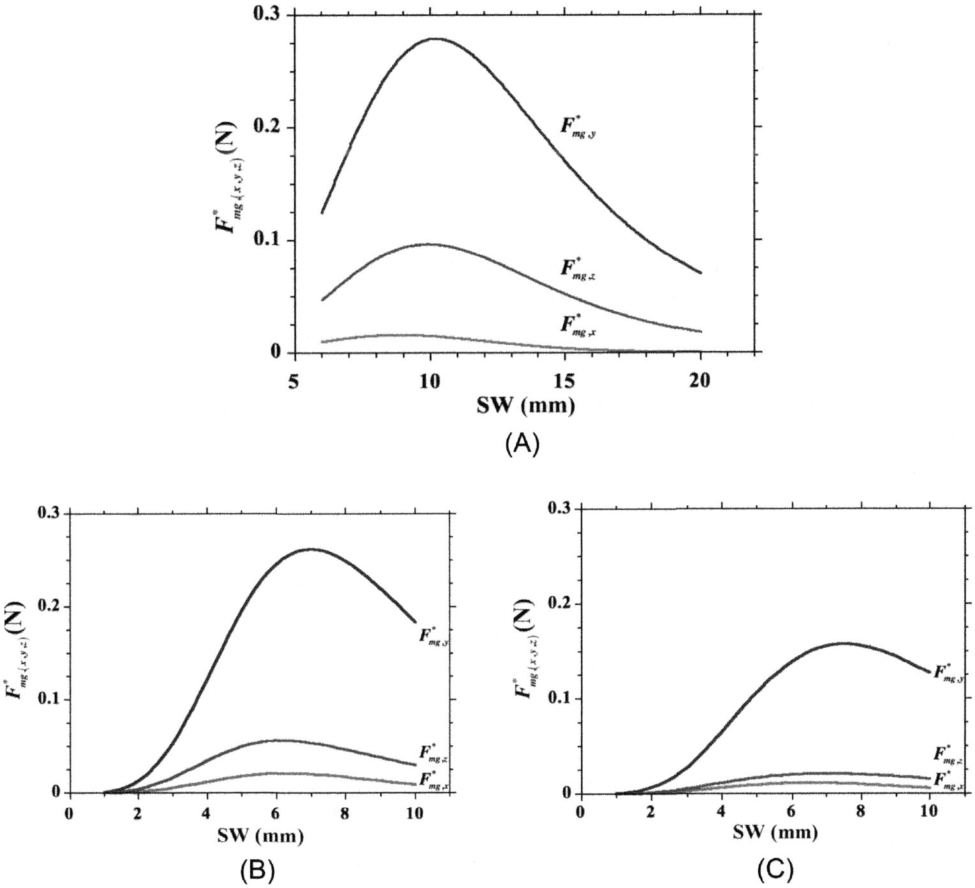

Figure 6.13 The magnetic forces in the 3D analysis (Săndoiu, 2019). (A) NS = 1 (single block permanent magnet). (B) NS = 5, GS = 1 mm. (C) NS = 5, GS = 6 mm.

(Săndoiu, 2019) presents the total magnetic force components for the single block PM. The maxima of $F_{mg,z}$ (streamwise) and $F_{mg,y}$ (collinear with the PM magnetization axis) occur for (almost) same GS, whereas the maximum for $F_{mg,x}$ (lateral, transversal to the PM magnetization axis) is recorded for (slightly) smaller GS.

The optimization performed for NS = 3, 4, and 5 shows that the streamwise magnetic force may play a role if the PM array is to cover a larger surface of the vessel and that $F_{mg,x}$ component acts into transferring the MD from the vessel to the tissue. In general, similar bell-shaped graphics are obtained, and they show that increasing NS (PM splitting) and GS (distancing the PM blocks) leads to maxima decrease, but a larger surface of the blood vessel is covered. This aspect may suggest a tradeoff optimal design solution—a large enough force over a large enough area. Two, twice optimized (w.r.t. GS and SW) MP arrays are shown here, NS = 1 and 5, Fig. 6.13 (bottom).

Table 6.1 (Săndoiu, 2019) lists the optimization results NS = 3, 4, and 5 identical blocks. The GS limiting cases only are given here. SW, related to GS, is the running the optimization parameter.

The three-dimensional results (see the magnetic field spectra in Fig. 6.14) are consistent with the bidimensional ones, smaller but of the same order of magnitude, and the maxima of the magnetic forces decrease when GS increases. The reason is that the extrapolated 2D MAF volume is almost half the three-dimensional one, while the average magnetic flux field tube is shorter. Hence, the bidimensional model is prone to produce a more intense field. Moreover, is a $F_{mg,x}$ force component, orthogonal to the stream and to the PM axis of magnetization. The three-dimensional maxima SW-shift is less apparent but, interestingly, when compared with their bidimensional counterparts, the maxima of $F_{mg,y}$ (attraction to the PM) occur later downstream than those of $F_{mg,z}$ (stream-wise). The design with the highest magnetic forces corresponds to GS = 1 mm (Fig. 6.15).

The MAF stationary flow under the influence of the magnetic field is analyzed using the computational domain presented in Fig. 6.12B, and Fig. 6.14 shows the magnetic field and the flow field for NS = 1, 5. The wider spatial coverage of the magnetic field

Table 6.1 Magnetization forces maxima recorded along the top edge of the blood vessel and the design parameters for different PM arrays.

NS (−)	GS (mm)	SW_{opt} (mm)	$F_{mg,x}$ (N)	SW_{opt} (mm)	$F_{mg,y}$ (N)	SW_{opt} (mm)	$F_{mg,z}$ (N)
3	1	7.4	0.019051	8.6	0.272633	8	0.07751
	6	7.6	0.012235	8.8	0.181689	8.4	0.038709
4	1	8.8	0.015457	10	0.262231	9.8	0.087083
	6	8.8	0.011722	10	0.198195	9.2	0.04515
5	1	6.2	0.02113	7	0.2621	6.2	0.056329
	6	6.4	0.011493	7.4	0.157431	6.8	0.020921

GS, Gap size; *NS*, number of slope; *PM*, permanent magnet.

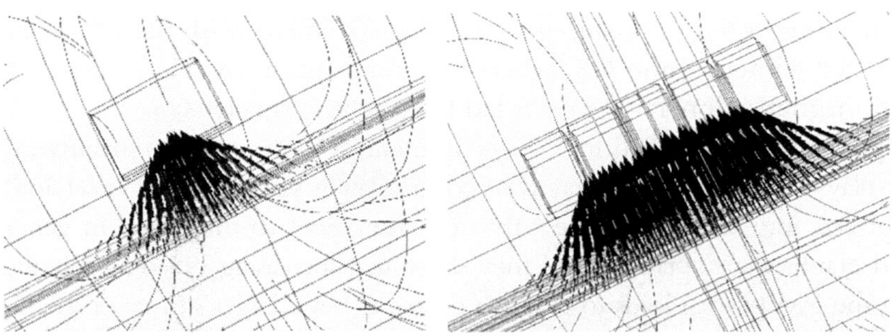

(A) (B)

Figure 6.14 Magnetic flux density (blue lines), velocity (red lines), and magnetic forces (arrows). (A) NS = 1 block. (B) NS = 5 blocks.

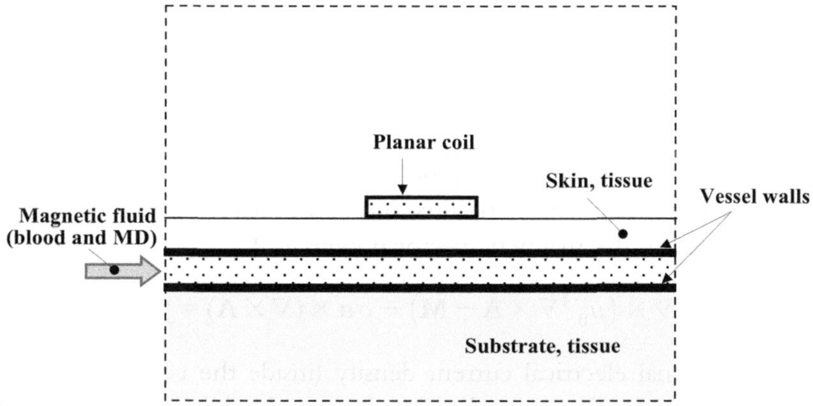

Figure 6.15 The bidimensional computational domain in the MDT controlled using a planar coil.

Table 6.2 The total force that different configurations of the magnetic field source.

Number of slots, NS	1	2	3	5
$F_{mg,y}$ [N]	1.0763	1.0632	1.0512	1.0093
$F_{mg,y}/(\text{AL} \times \text{SW})$ [daN/m^2]	538.15	548.323	468.449	369.707

control (of higher gradient due to its spatial spectrum rather than intensity) for NS $=5$ may supposedly help retaining and "collecting" more efficiently the MNPs.

Table 6.2 (Săndoiu, 2019) lists the maxima of the attraction force. Here AL is the linear size of magnetic array in streamwise direction, computed as $\text{AL} = \text{NS} \times \text{SW} + (\text{NS} - 1) \times \text{GS}$. Apparently the array of NS $=2$ blocks provides for the best magnetic extraction force.

The total stream-wise force is defined as $F_{total} = \int_S pdS + \int_V F_{mg,z} dV$, where S is area of the wetted surface of the vessel, and V is the volume of the MAF. The magnetic term contributes to the mixing of the MAF.

Using electromagnets for magnetic drug targeting

For the magnetic fields to be used in MDT to direct inner body actions, they have to be intense and to exhibit high gradients able to produce forces required to precisely lead the MD to the ROI, thus limiting the toxic drug spreading to unharmed cells (Alexiou et al., 2003, 2006a,b; Li et al., 2018). To this aim current coils may be used too. Their design and control can be tailored to meet specific needs and a companion bidimensional model is useful to prove this concept (Morega et al., 2015). First an analysis derived from the previous, PM study, is conducted to validate this concept, Fig. 6.15.

The vessel volume is the only magnetizable medium. In the linear limit, its magnetization is

$$M_{ff} = \gamma \arctan(\delta H) \approx \gamma \delta H \tag{6.11}$$

where γ [A/m] and δ [m/A] are empiric constants (Morega et al., 2015; Săvăstru, 2016). The magnetic circuit, magnetic flux and constitutive laws yield the mathematical model formulated in the magnetic vector potential, $\mathbf{A}$,

$$\nabla \times \left(\mu_0^{-1} \nabla \times \mathbf{A} - \mathbf{M}\right) - \sigma \mathbf{u} \times (\nabla \times \mathbf{A}) = \mathbf{J}^e, \tag{6.12}$$

where $\mathbf{J}^e$ is the external electrical current density (inside the coil only), the magnetic field source, $\mathbf{u}$ the MAF velocity, σ the electrical conductivity of the MAF. Magnetic insulation boundary condition closes the problem.

An order of magnitude analysis of Eq. (6.12) suggests the balance

$$\frac{A_0}{\mu_0 L^2}, \quad \frac{M_0}{L}, \quad \sigma \mu_0 U_0 L J_0 \sim J_0. \tag{6.13}$$

Using as reference quantities $J_0 \sim 10^6$ A/m^2 (electrical current density scale), $L \sim 0.1$ m (length scale), $H_0 \sim J_0 L$, $A_0 \sim \mu_0 L H_0$, $M_0 \sim \gamma \delta H_0$, $U_0 \sim 0.1$ m/s, rel. Eq. (6.13) yields

$$O(1), \quad O(1), \quad O\left(10^{-7}\right) \sim O(1), \tag{6.14}$$

where $O(\cdot)$ means "order of magnitude." The transport term is much smaller than the others, thus it can be discarded, leading to an important reduction in complexity for reason that the magnetic field problem may be solved independently of the flow, only once, in the beginning.

The arterial section of interest here is of "resistance" type (Feijóo, 2000) and MAF is a Newtonian fluid. The quasisteady, incompressible, laminar, flow is modeled by Eqs. (6.1) and (6.2), where $\mathbf{f} = \mathbf{f}_{mg} = \mu_0 (\mathbf{M} \cdot \nabla) \mathbf{H}$ is the magnetic body force. The boundary conditions are zero at the walls, uniform periodic velocity profile at the entrance (Fig. 6.16) and uniform periodic pressure for the exit (Fig. 6.17). The morphologies and the significance of these profiles are described in Morega et al. (2012). For convenience, the leading pulse is set to 60 bpm.

The electromagnet is coreless, essentially a planar coil made of concentric, with square cross-section turns, which can be powered independently. Fig. 6.18 (Morega et al., 2015) shows the horizontal (left) and vertical (right) magnetic forces at the upper side of the vessel wall that is closer to coil, for several powering schemes.

The grey blocks designate the currents *entering* the plane and the black ones show the currents *exiting* the plane. In general, the *extraction* component exhibits maximum values about the coil axis, whereas the streamwise component has maxima by the coil

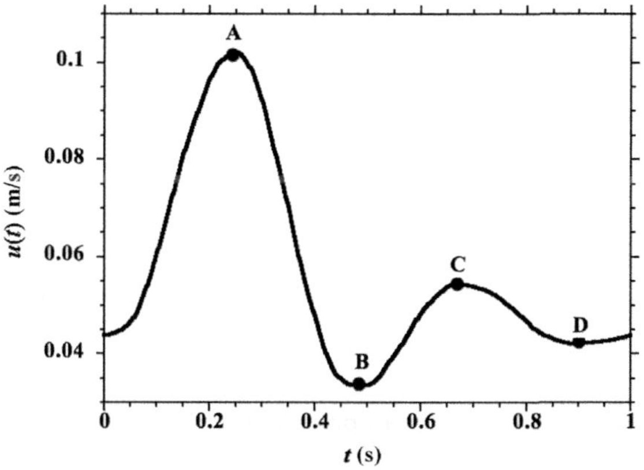

Figure 6.16 Inlet velocity profile, one period at 60 bpm.

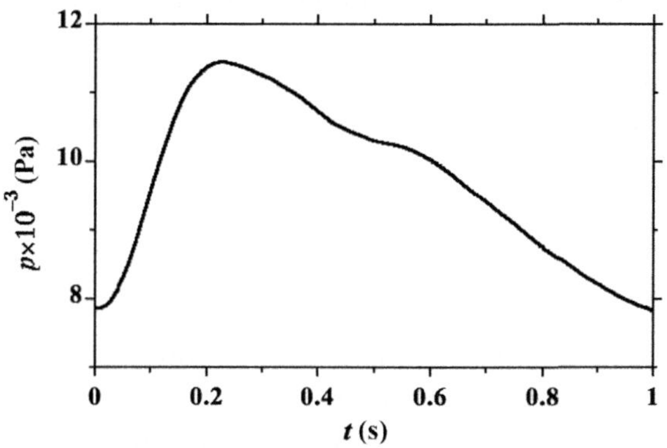

Figure 6.17 Outlet pressure profile, one period at 60 bpm.

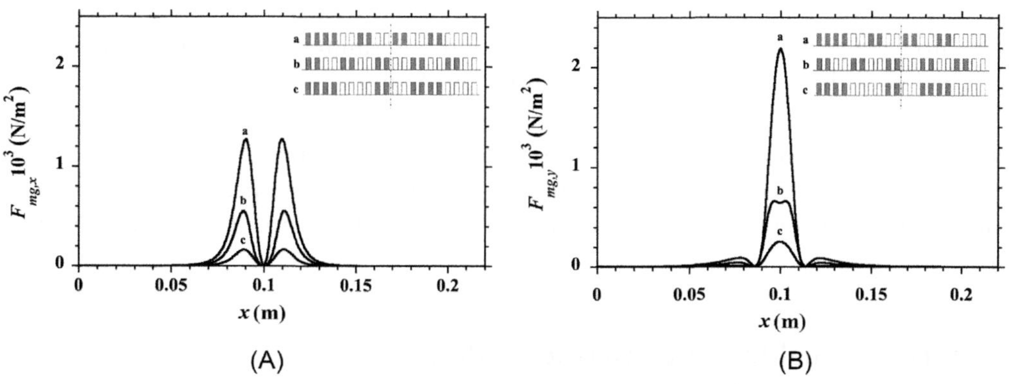

Figure 6.18 The magnetic forces at the upper part of the wall. Other powering schemes are possible. (A) Streamwise magnetic force. (B) Vertical (extraction) magnetic force.

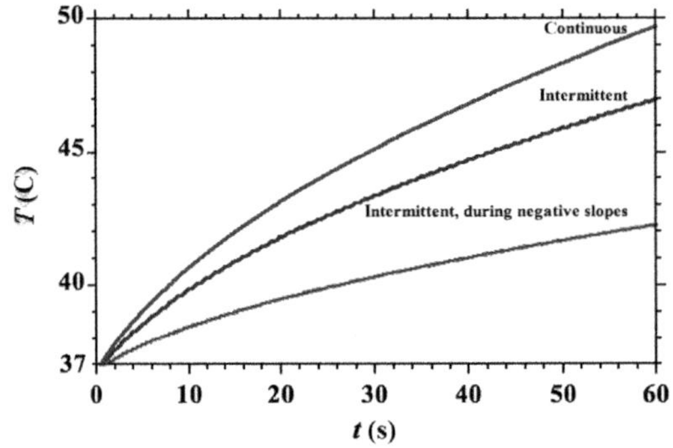

Figure 6.19 Temperature profiles for different powering schemes.

lateral margins. The planar coil splitting (A) provides for larger forces. Different powering schemes are though possible, for instance, keeping the same total number of ampere-turns, with continuous or intermittent powering, with different aspect ratios of the rectangular conductor or the spacing between turns, etc.

Conduction electrical currents are heat sources (by the Joule effect), and they can menace the adjacent tissue, therefore a heat transfer analysis is needed. The mathematical model used to this aim is

$$\rho c_P \left[\frac{\partial T}{\partial t} + (\mathbf{u} \cdot \nabla) T \right] = \nabla (k \nabla T) + Q, \tag{6.15}$$

where c_P is specific heat, T temperature, k heat conductivity, and Q the heat source (produced by Joule effect, in the conductors).

Intermittent and mass flow related powering could reduce heating by suppressing the heat (and the magnetic field) source when the magnetic forces are less effective in the MD control. The time intervals with negative flow slope. Fig. 6.19 (Morega et al., 2015) shows the temperature profiles for continuous powering, intermittent powering (synchronized with the pulse, Fig. 6.16), and intermittent powering respectively, matched with the negative slopes of the flow (A—B and C—D, Fig. 6.16).

The third strategy seems more difficult to implement but is best since it extends the duration of the therapy, within acceptable temperature levels. The same moments for the on/off powering could be set guided by the pressure profile, more convenient from technical point of view. Fig. 6.20 shows the heat that goes with the blood stream.

From conceptual to more realistic models

Inflammatory maladies of the joints are frequent conditions, and for many of them remedies do not exist at present. Intraarticular medication can be a better option for

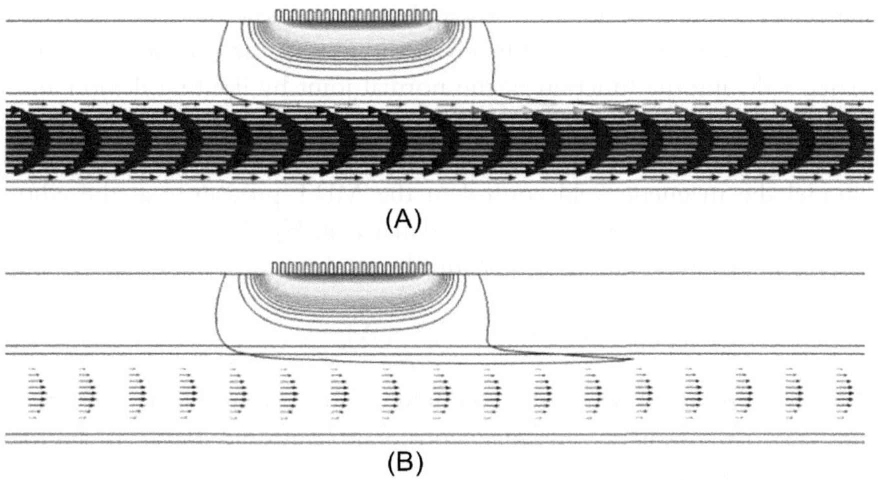

Figure 6.20 Heat entrainment by the MAF flow shown through isotherm profiles—see the moments A, B in Fig. 6.16. *MAF*, Magnetizable aggregate fluid.

the reason that an increased drug quantity is administrated at the major place of swelling while reducing the undesired toxicity. This posology has though the foremost drawback that the drug fast vanishes from the joint nook, requiring recurring injections, which may produce affliction and dysfunctions of the joints. Retard medication is then often recommended but it is constantly linked with negative collateral effects, mainly when the medication is provided orally or systemically.

To alleviate these difficulties while maintaining the requested amount longer could be to instill the medication attached to MNPs and use magnetic fields to increase the retention of the active admixture in the joint nook and to enhance their in situ effect (Schulze et al., 2005; Butoescu et al., 2009). Some difficulties of this MDT protocol, for example, limited penetration depth and distribution of the released drug to the disease site, were evidenced through preclinical studies though, but these can be surpassed by active directing through the attachment of ligands of high affinity onto the MNP surface (Chegini et al., 2018). There are very few studies that address this method though, and significant unmet needs in the joint treatment persist.

The tissues concerned in the joint maladies mainly consist of the synovial membrane, the capsular tissue, the hyaline cartilage, and the subchondral bone, where *synoviocytes*, *chondrocytes*, and *osteoblasts* play a major role. These cells come from the mesoderm level in embryogenesis, and, in compliance to the environment, they transform in their final form. Chronic swelling can lead to dedifferentiation (e.g., fibrous tissue), which is normally associated with a loss of function that is damaging to the overall functionality of the joint. The hyaline cartilage degenerates into fibrocartilage and bone resorption and in addition, osteophyte formation alters the bone structures of the joints. In addition, excessive fibrosis of the fibrous capsule and the inside of the joint occurs, reducing the function of the normal joint by limiting flexion and extension. Therefore the biocompatibility of SPION MNPs, their eventual distribution and elimination are important issue (Shi and Gu, 2008).

To model the magnetic field control in the MDT procedure at the elbow joint level we consider, as magnetic field source planar coil, Fig. 6.21.

Assuming an electrokinetic (DC) system, the mathematical model is

$$\nabla \times \left(\mu_0^{-1}\mu_r^{-1}\nabla \times \mathbf{A}\right) = \mathbf{J}^e, \tag{6.16}$$

where $\mathbf{J}^e$ is the current density in the coil. The magnetization of the SPION-MD (an aggregate, superparamagnetic, homogeneous, and isotropic fluid) is approximated through the linear form

$$M = \alpha\,\mathrm{atan}(\beta H) \approx \alpha\beta H = \chi H, \tag{6.17}$$

where $\alpha = 10^{-4}$ A/m and $\beta = 3 \times 10^{-5}$ m/A are empiric constants (Dobre, 2012). The MNPs mass fraction in the synovial fluid is set to 0.1. The synovial liquid and the MNPs make a MAF.

Fig. 6.22 presents the magnetic flux density spectrum (max. 0.04 T for $J^e = 6$ A/cm^2) and the associated magnetic forces ($\sim$004 N/m, in this bidimensional model).

The heating produced by the coil may menace the adjacent tissue. To evaluate this effect we use the bioheat model, Eq. (6.15). The heat source is inside the coil only,

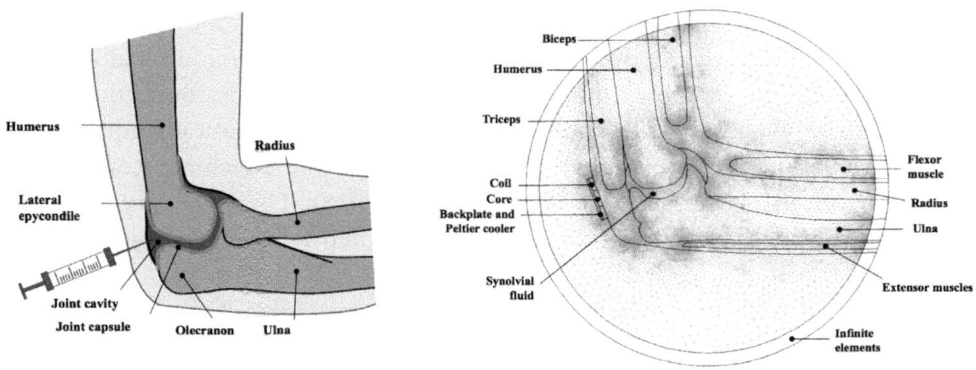

Figure 6.21 The magnetic field model in the MDT. The source is here an electromagnet (coreless coil). (A) Sagittal view through the elbow—a sketch. (B) The FEM mesh—the bordering layer is made of infinite elements, used to close the magnetic field.

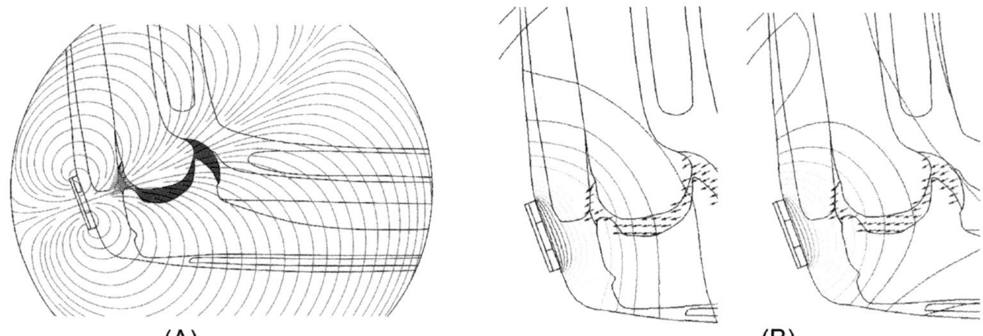

Figure 6.22 The magnetic field at the elbow level and the heating produced when a coil—A Péltier cooling system is used. (A) The magnetic flux density and forces. (B) Temperature after 10 min left, without PCS (T_{max} = 41.5°C), right with PCS (T_{max} = 37°C); h = 2 W/m^{2}K (convection heat transfer coefficient).

by Joule effect. The coil-tissue contact was found to reach over 41.5°C after just 60 s. To overcome this difficulty, the coil has to be cooled. For instance a Peltier element, placed on the opposite face of the coil, may provide for the right cooling effect. This method may keep the maximum temperature (at the skin-coil contact) within safe limits (37°C or less).

In *rheumatoid arthritis* the synovial membrane that delimits the joints is inflamed Synovitis, (2020) eventually leading to the joint dysfunction. The remedies range, according to the severity, from changes in the life style, local antiinflammatory therapy, and physical exercises to physiotherapy to direct infiltrations with glucocorticoids, nonsteroidal antiinflammatory drugs, analgesics, and hematopoietic and mesenchymal cell implant technologies [Stem Cell Clinic].

The magnetic field control in the MDT when applied to the knee joint is shown in Fig. 6.23. PMs are used here and the ROI (the knee joint) is motionless, which yields the mathematical model Eqs. (6.16), (6.17).

The magnetic forces are calculated using the virtual work at constant (magnetic) flux (Chapter 1: Physical, Mathematical, and Numerical Modeling). The numerical simulations have shown that, depending on B_{rem} (0.6−1.3 T) and the fraction of MNPs (0.1−0.5), the total magnetic force is 10−18 N/m (Manea et al., 2014).

A threedimensional model for this problem is presented in Fig. 6.24 (Manea et al., 2014; Manea, 2015). As this problem is of "open space" type too, a layer of "infinite" type elements is used to reduce the size of the computational domain.

As expected, the PM array is seen to attract the medication and it is suggested their positioning is such that they provide for the specific localization of the MD. Using the magnetic energy density map, Fig. 6.25 singles out the synovial capsule, and shows off

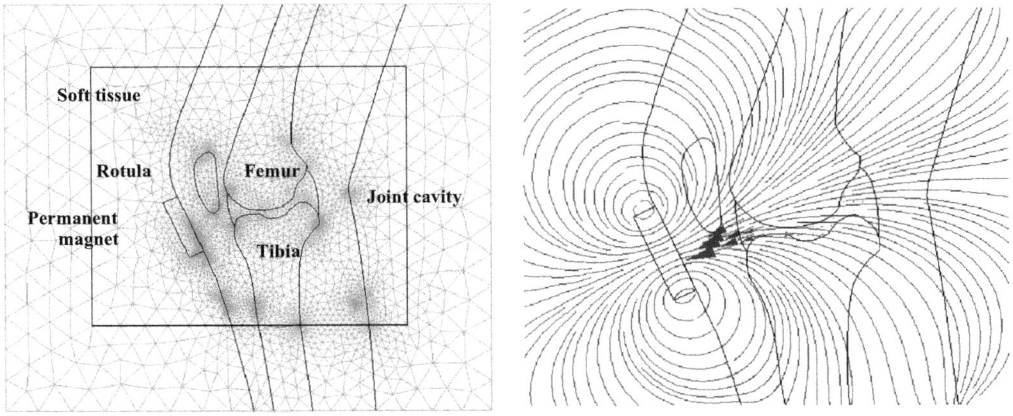

(A) (B)

Figure 6.23 A bidimensional model of the MDT therapy at the knee level. (A) The sagittal cross-section through the knee. (B) Magnetic flux density spectrum.

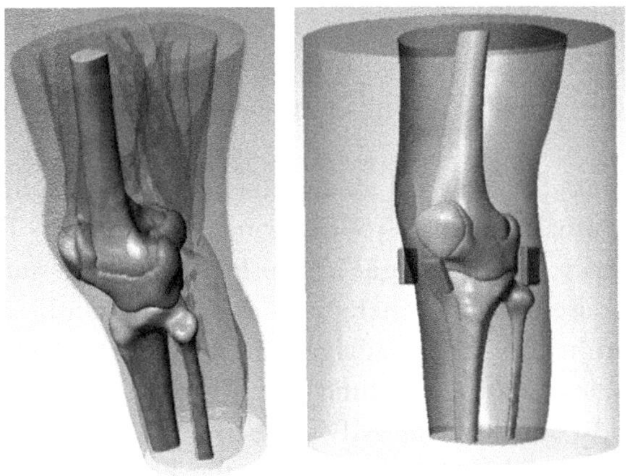

(A) (B)

Figure 6.24 A three dimensional model of the MDT therapy at the knee level. (A) The reconstruction of the knee. (B) The computational domain—a PM array is added.

the magnetic body forces. As expected, higher magnetic forces are associated with regions of higher magnetic energy gradients. Similar predictions for different arrangements PM arrays are detailed in Manea (2015).

Such information may help in optimizing the magnetic field used to control the MDT. To this same aim, computational domains produced through reconstruction techniques correlated with proper sizing and positioning of magnets, as this study advocates, may enhance the precision and efficiency of the treatment for personalized models (knee size, joint positioning, etc.).

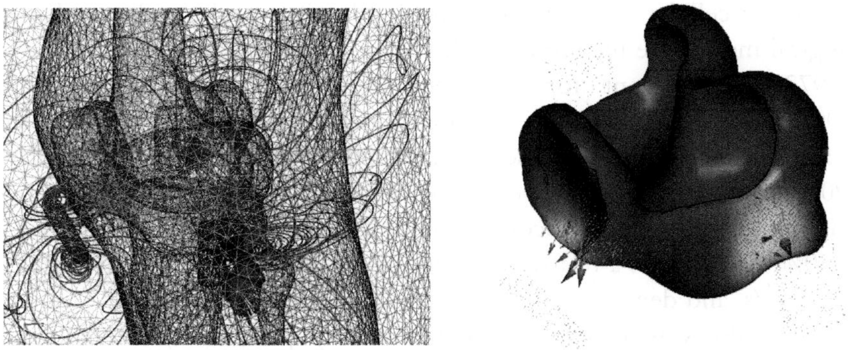

(A) (B)

Figure 6.25 The magnetic field produced by the array of magnets and the magnetic forces in the MDT therapy at the knee level. (A) The magnetic flux density and forces. (B) The magnetic energy density—surface map. The swarms indicate the permanent magnets positions. *MDT*, Magnetic drug targeting.

6.6 The magnetic drug transfer from the larger blood vessel to the region of interest

The magnetic drug transfer to the ROI occurs through two mechanisms: convection and diffusion. In the larger vessels the transfer is dominated by convection, whereas diffusion is prevalent for the vessels walls and in the adjacent microvascularized tissue. To these adds the magnetic field, which intervenes through magnetic body forces. Proportional to and led by the magnetic energy density gradient, its motive action adds as a control means in targeting the MD, directing it to higher energy density regions, that is, toward the magnetic source. Therefore qualitatively, the ROI has to be "on the path" from the delivery location to the higher magnetic energy region for this magnetic control to have the desired positive targeting effect.

Several transfer stages are discussed next: the MD transfer in larger vessels, where the blood flow is pulsatile; the transfer through smaller size vessels, where flow is stationary, and the transfer through the vessel wall and the adjacent region. The blood flow (larger and smaller blood vessels) is assumed laminar and incompressible.

Biorheological models in magnetic drug transfer

So far, in this chapter, the MAF was assimilated with a Newtonian fluid and its flow was assumed stationary. This line of study may be consistent with venous flows, and larger vessels. Arterial flows are pulsatile, and for small size vessels, the Newtonian rheological model is no longer representative. The low shear stress observed in the hemodynamic of microvessels with diameter less than 1300 μm shows off a blunted plug-flow type profile (Xu et al., 2005). Its constituents (red and white blood cells, and platelets suspended in the fluid) make the blood a complex fluid. Its mechanical behavior and its cellular components (red blood cells mainly) indicate that different biorheological models are recommendable [Shaw (2017)]: Casson models (Aroesty and Gross, 1972) are relevant for the hemodynamic od microvessel of diameter 130–1300 μm, and the Herschel–Bulkley model (Priyadharshini and Ponalagusamy, 2015) is adequate for the microvessel of radius 20–1000 μm (Misra et al., 1992; Pries et al., 2000). However, it is appealing to approximate blood as a homogeneous, Newtonian fluid, and its flow incompressible. The percentage of hematocrit (cells) is ~40%–50%, which gives a relative viscosity of 3–4, an equivalent kinematic viscosity $\nu = 0.032\ \mathrm{cm^2/s}$, and density $\rho = 1.05\ \mathrm{g/cm^3}$ are, for instance, used by Hunter (1972). For oscillatory flows with typically blunt velocity profiles the recommended the profile

$$u = \frac{\alpha}{2-\alpha} U\left[1 - \left(\frac{r}{R}\right)^{\frac{2-\alpha}{\alpha-1}}\right], U = \frac{1}{R^2}\int_0^R 2rudr, \alpha = \frac{1}{RU^2}\int_0^R 2ru^2 dr, \tag{6.18}$$

replaces the Hagen–Poiseuille (parabolic) profile, which is characteristic for swirl-free and axisymmetric Newtonian fluids. In Eq. (6.18) R is the radius of the circular vessel (the characteristic length recommended by Barnard et al., 1966), $u = u(r)$ the axial velocity, and U the mean axial velocity is a constant for the particular profile assumed. The value $\alpha = 1.1$ provides a velocity profile that matches satisfactorily empirical data. A similar profile was proposed by Smith et al. (2002)

$$u = U\left[1 - \left(\frac{r}{R}\right)^{\xi}\right], \tag{6.19}$$

to overcome the difficulty of actually solving the transport equations for a complex, non-Newtonian fluid, for example, $\xi = 9$ (Nacev et al., 2011; Köstler, 2016). However, this model is about the streamwise flow, essentially one dimensional. In MDT, though, the action of the magnetic field may be orthogonal to it, and this contributes to enhance (accelerate) or oppose (decelerate) it. Indirectly this streamwise effect may result in recirculation cells hence some flow orthogonal to it. If the MD is to cross the epithelial membrane, then the magnetic field force orthogonal to the streamwise flow is of interest.

In our study we use a Carreau-type rheological model (Morega et al., 2013a,b,c; Akbar and Nadeem, 2014)

$$\eta = \eta_\infty + \left(\eta_0 - \eta_\infty\right)\left[1 + (\lambda\gamma')^2\right]^{(n-1)/2}, \tag{6.20}$$

that provides for the blunt velocity profile specific to microcirculation. Here $\eta_\infty = 0.0032$ Pa s is the infinite shear rate viscosity, $\eta_0 = 45.6$ Pa s is the zero shear rate viscosity, γ' is the shear rate tensor, $\lambda = 10$ s, $n = 0.344$ are model parameters.

Magnetic drug transfer thorough larger vessels

The bidimensional study that follows is about the first stage of MDT: the MNPs arrive with the blood flow in the region where the PM acts. Progressively the vessel becomes a magnetizable medium, and magnetic body forces come into action. It is assumed that the magnetic field produced by PM is not perturbed throughout the dynamic change in the magnetic susceptivity of the vessel content. Eventually after 6–7 periods of the pulsating flow (e.g., at 60 bpm), the entire content becomes a magnetizable fluid. The magnetostatic field, the flow and the mass transfer model are

$$\nabla \times \left(\mu_0^{-1}\mu_r^{-1}\nabla \times \mathbf{A} - \mathbf{B}_{rem}\right) = 0, \tag{6.21}$$

$$\rho\left[\frac{\partial \mathbf{u}}{\partial t} + (\mathbf{u}\cdot\nabla)\mathbf{u}\right] = -\nabla\left[-p\mathbf{I} + \eta\left(\nabla\mathbf{u} + (\nabla\mathbf{u})^T\right)\right], \nabla\cdot\mathbf{u} = 0, \tag{6.22}$$

$$\frac{\partial c}{\partial t} + (\mathbf{u}\cdot\nabla)c = D\nabla^2 c. \tag{6.23}$$

The initial conditions for the flow and mass transfer are homogeneous. The MAF is the Carreau-type fluid Eq. (6.20). The boundary conditions for the flow are: zero velocity for the walls and constant pressure for the exit. Fig. 6.26 depicts the inlet velocity profile for several periods.

The inlet boundary condition for species is $C_b = 0.33$ mg/cm^3 (Soltani and Chen, 2013), and convection for outlet. The walls are impermeable, assuming that mass

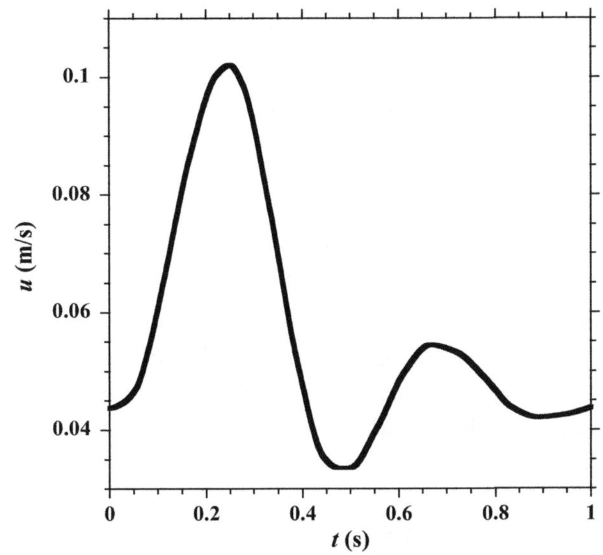

Figure 6.26 Inlet velocity profile for the flow in the MD transfer through vessels of larger size.

transfer through them is negligibly small during the time span on the study. In the absence of the magnetic field by the diffusive convection model

$$\frac{\partial c}{\partial t} + (\mathbf{u} \cdot \nabla)c = D\nabla^2 c, \tag{6.24}$$

where c is the species (MNPs), and D is the diffusivity of MNPs in the blood. The suggested order of magnitude balance is

$$\frac{1}{t_{MNP}}, \frac{U}{L} \sim \frac{D}{L^2}, \tag{6.25}$$

where t_{MNP} is the time scale of the mass transfer process, U is the velocity scale, and L is the streamwise length scale of capillary. Apparently two time scales are competing to lead the mass transfer: (1) the velocity time scale, $t_{U,MNP} \sim U/L$ and (2) the diffusion time scale, $t_{D,MNP} \sim L^2/D$.

In vitro studies (Lücker, 2018a,b) and experiments conducted on phantoms aimed at investigating Doppler ultrasound procedure for noninvasively measuring pulsatile capillary speed velocities for human and artificial blood (Law et al., 1989; Ting et al., 1992; Li et al., 1993; Razavi et al., 2018), and other studies indicate velocities of the order $U \sim 1$ mm/s. Although the no-slip assumption is arguable (Zeeshan et al., 2018), this reference is the leading transport quantity. For $L \sim 100$ μm (Almaça et al., 2018, Lücker, 2018a) and $D \sim 10^{-9}$ m^2/s (Li et al., 2008), it yields $t_{U,MNP} \sim 10^3$ s and $t_{D,MNP} \sim 10^3$ s, meaning that the two mechanisms (transport and diffusion) have the same time scale.

The contribution to the MD of the transport magnetic body forces produced by the magnetic field may be assessed use the stationary form of the momentum equation

$$\rho[(\mathbf{u} \cdot \nabla)\mathbf{u}] = -\nabla p + \mu\nabla^2\mathbf{u} + \mathbf{f}_{mg}, \tag{6.26}$$

which indicates the order of magnitude balance

$$\frac{\rho U_0^2}{d} \sim \frac{P_0}{d}, \frac{\mu U_0}{d^2}, F_{mg,0}, \tag{6.27}$$

where d is the vessel diameter (space scale), U_0 is the velocity scale, $P_0 = \rho U_0^2$ the pressure scale, $W_{mg,0} = B_0^2\mu_{mg}/d$ the magnetic energy density scale, B_0 the magnetic flux density scale, μ_{mg} the magnetic susceptibility, and $F_{mg,0} = B_0^2\mu_{mg}/d\mu_{mg} = W_{mg,0}/d$ the magnetic body forces scale. It should be noted that B_0 and d are not B_{rem}. Relation Eq. (6.27) then yields

$$1 \sim 1, \frac{1}{\mathrm{Re}_d}, \frac{W_{mg,0}}{\rho U_0^2}, \tag{6.28}$$

where $\text{Re}_d = (U_0 d/\mu/\rho) = (U_0 d/\nu)$ is the Reynolds group based on d, and a new group emerges, $W_{mg,0}/\rho U_0^2$, which is the ratio of (potential) magnetic energy density to kinetic energy density group. The usage of d as scale here is relevant because the gradient of the magnetic energy density (magnetic force) orthogonal to the flow is important—the magnetic field acts to extract the MD from the vessel into the ROI.

In fact, this energy group relates the contribution of the magnetic field to the MD transport with respect to the kinematic energy of the flow (provided by the "pump"-ing device) as leading term. Furthermore, the magnetic field may accelerate (upstream the magnet region) or decelerate (downstream it) the flow, but it always attracts the MNPs towards the magnet. As the magnetic body forces are the gradient of the magnetic energy density, this group may show off either

$$\frac{W_{mg,0}/d}{\rho U_0^2/d} = \frac{F_{mg,0,y}}{F_{mech}}, \quad \text{or} \quad \frac{W_{mg,0}/L}{\rho U_0^2/d}\frac{L}{d} = \frac{F_{mg,0,x}}{F_{mech}}\text{AR} \tag{6.29}$$

where $F_{mg,0,x}$ and $F_{mg,0,y}$ are the scales for the streamwise and orthogonal to the stream, respectively, magnetic body forces, F_{mech} is the scale for the streamwise mechanical body force, and AR is the geometric aspect ratio of the vessel.

The energy group, $\frac{W_{mg,0}}{\rho U_0^2}$, is essential to the MDT design, as it allows sizing the magnetic field source (its energy), either PM or electromagnet, for a specific situation. It becomes then a problem of design to position and optimize the magnet (Section 6.5) with respect to the ROI to achieve the required magnetomotive force. The magnetic to mechanical forces group Eq. (6.29) may be derived from the energy group to outline the order of magnitude of either the streamwise or the orthogonal to the flow forces (in the above U is the scale for the streamwise velocity, but the scale for the orthogonal velocity may be used), and AR is the conversion factor.

Two limiting cases may be noticed: (1) when the flow concurs with the magnetic field lines the magnetic has no "extraction" effect and (2) if the magnetic field lines are orthogonal to the flow, then the magnetic field has an drawing out effect. The streamwise action of the magnetic field is discussed next.

Initially the fluid inside the vessel segment in the model has no magnetic properties, but it eventually changes as the magnetizable species (MD) fills the vessel. This change in the magnetic susceptivity depends on mass concentration, and it is presented here as $\chi(t) = c(t)/C_b$. This form creates couplings between the magnetic field, flow, and species transfer such that Eqs. (6.21)–(6.23) have to be solved simultaneously. A segregated dynamic solver is recommended.

Fig. 6.27 shows the concentration profiles for several moments, and the flow field at $t = 16$ s. The computational domain is similar to that in Fig. 6.10, the PM has

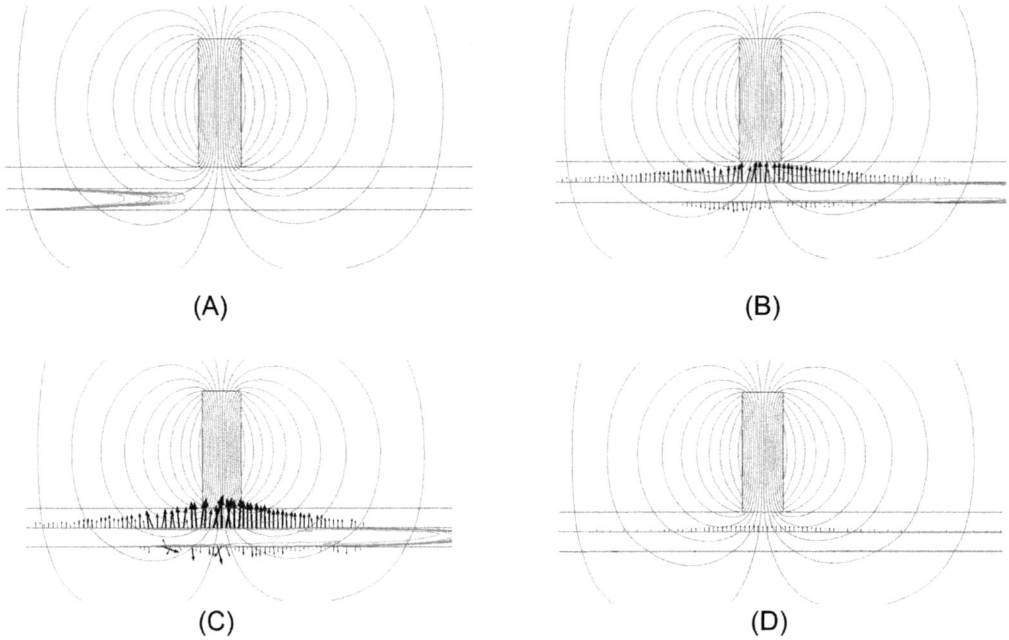

Figure 6.27 The magnetic flux density and magnetic body forces, the species (A–C) and the flow (D). (A) $t = 1$ s. (B) $t = 3$ s. (C) $t = 4$ s. (D) $t = 16$ s.

$B_{rem} = 1.3$ T, the blood vessel is 1 mm wide and 16 mm long. The MD (SPIONs) diffusivity in blood is $D = 1 \times 10^{-9}$ m^2/s (Sillerud, 2018).

The MD progressively fills in the vessels, and the magnetic body forces, although directed towards the region of high magnetic energy density, depart from the symmetric profile in Fig. 6.15. This is for the reason that the susceptivity of the vessel content changes in time, with the flow. And the flow here is unsteady.

Magnetic drug transfer through the membrane and tissue

The next step in MDT concerns the MNPs transfer thorough the membrane and the adjacent tissue. It is assumed that the content of the larger blood vessel is saturated with MD, of uniform magnetic susceptivity, and its walls behave as a species source, at constant concentration level, C_b, for the surrounding membrane and the surrounding tissue. The time scale of the vessel uniform filling with superparamagnetic species is much shorter than the time scale of the membrane transfer. The magnetic field model, in magnetic vector potential, is Eqs. (6.17) and (6.21).

Assuming that, whatever direction it may have, blood velocity in the microvasculature of the membrane and the tissue is small when compared with the MNPs velocity due to the magnetomotive effect, hence the magnetic field is intense, MD transfer through the endothelial membrane and tissue are described by, respectively,

$$\frac{\partial}{\partial t}c_M + \left(\mathbf{v}_{mg}\cdot\nabla\right)c_M = \nabla\cdot(D_M\nabla c_M), \tag{6.30}$$

$$\frac{\partial}{\partial t}c_T + \left(\mathbf{v}_{mg}\cdot\nabla\right)c_T = \nabla\cdot(D_T\nabla c_T), \tag{6.31}$$

where c_M and c_T denote the species (MNPs) in the membrane and tissue, respectively, D_M and D_T are the diffusivities of MNPs through the membrane and tissue, respectively. The recommended magnetic velocity, $\mathbf{v}_{mg}$, is (Grief and Richardson, 2005; Nacev et al., 2011)

$$\mathbf{v}_r = k\nabla\mathbf{H}^2, \quad k = \frac{a^2\mu_0\chi}{9\eta\left(1+\frac{\chi}{3}\right)}, \tag{6.32}$$

where k is the magnetic drift coefficient, which depends on: a—MNP radius, χ—MNPs susceptibility, and η the dynamic viscosity of blood.

In fact, the suggested approach to calculate the magnetic velocity, $\mathbf{v}_r$, that is used as a macroscopic quantity, is a homogenization technique that raises the physical model scale from the MNP nanometric level to a macroscopic, continuous, magnetizable medium, superparamagnetic like, which is characterized by an apparent susceptibility, χ_{app}, that depends on the (constant) viscosity of a Newtonian fluid—the coefficient k in Eq. (6.32), which is proportional with the susceptivity of the MNP and the dynamic viscosity of the solute fluid. In view of Eq. (6.32), for constant χ (same material for the MNP), $\eta \rightarrow 0$ yields $k \rightarrow \infty$, consequently less viscous the fluid is the higher the magnetic velocity is.

The bidimensional model in Fig. 6.28 refers to the membrane of a blood vessel and the surrounding tissue: a longitudinal section through a cylindrical blood vessel (R1) with a diameter $d = 5$ mm. The vessel is bounded by an endothelial layer (R2) of thickness $g = 0.5$ mm, and embedded into the target tissue (R5), 50 mm thick. The PM (R4, NdFeB, $B_{rem} = 1$ T, of size 10 mm $\times$ 30 mm) is outside the target tissue, and its magnetization is set to favor the transendothelial transfer of MNPs. MNPs may cross through the wall with a diffusion coefficient D_M. The magnetic field problem is closed by free space (R3). The membrane face inside the vessel has $C_b = 4$ mol/m^3, and all other parts of the species transfer boundary are impermeable. The properties are $D_M = 1.5 \times 10^{-12}$ m^2/s, $D_T = 1.2 \times 10^{-14}$ m^2/s, $a = 2.5 \times 10^{-3}$ m^2, $k = 5 \times 10^{-17}$ m^4/(A^2 s), and $\chi = 0.003$ (Swabb et al., 1974; Nacev et al., 2011; Soltani and Chen, 2013; Winner et al., 2016; Zhan et al., 2019).

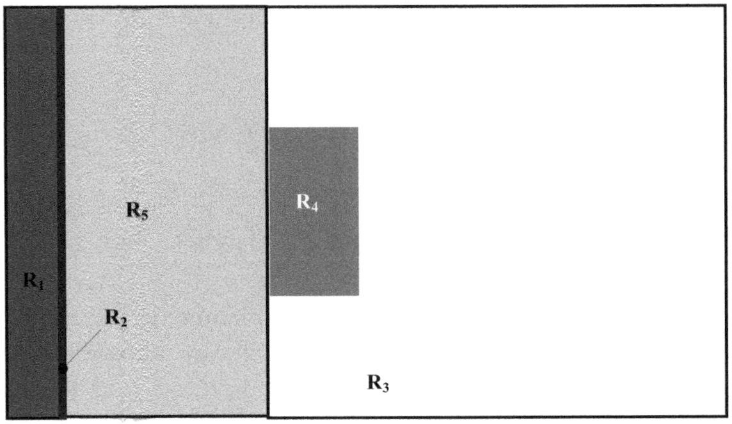

Figure 6.28 The bidimensional computational domain used to analyze the MD transfer through the epithelial membrane and the adjacent tissue. R1 - the blood vessel; R2 - the endothelial layer; R3 - free space, R4 - the permanent magnet.

The magnetic body forces that drive the MNPs are calculated as gradients of the magnetic energy density. They depend on magnetic susceptibility, permeability and concentration of the MD. The first assessment aims to find the most efficient PM configuration for MDT. Here, the horizontal component, F_x, needs to be maximized for the most efficient transfer of MNPs through the endothelial wall. Both magnetization directions commonly chosen for the PM were tested and the results are displayed in Fig. 6.29. This goal is best touched in the case of magnetization following the y direction.

Fig. 6.30 represents the MNPs concentration at the end of the diffusion process inside the target tissue, highlighted by color map and level contours (thin lines spectrum); both pictures show the same distribution of MD concentration. The red thick lines stand for the mass flux of MNPs, when the magnetic force is in action (upper picture) compared with free diffusion in the absence of the magnetic field (lower picture).

As one could observe, the presence of the magnetic field in this case favors the mass transfer at a nondecisive level. The optimization of the magnet (size, shape, position, and magnetization) might lead, however, to more efficient transport of the medication. Starting from this simplified layout, it is possible to simulate more accurate the whole process. Proper values for the physical properties, like the diffusion coefficients, or more accuracy for describing the anatomy structure would enhance the precision. Experimental results are useful for the calibration and validation of the numerical models, which could further become reliable research tools for the preinterventional procedures.

Keeping the same data, we focus next on the transient regime of the MD diffusion thorough the membrane and the adjacent tissue. However, the sizes of several domains

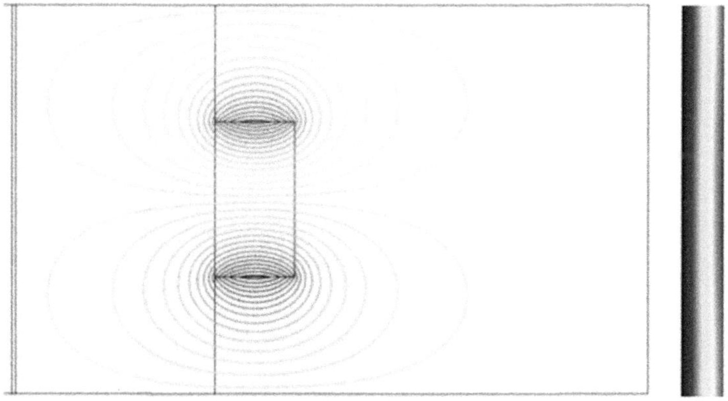
(A)

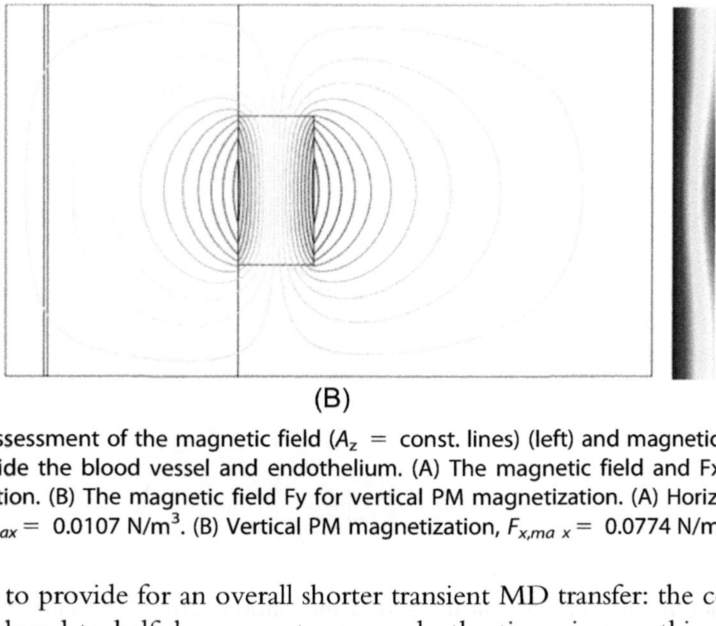
(B)

Figure 6.29 Assessment of the magnetic field (A_z = const. lines) (left) and magnetic force distribution (right) inside the blood vessel and endothelium. (A) The magnetic field and Fx for horizontal PM magnetization. (B) The magnetic field Fy for vertical PM magnetization. (A) Horizontal PM magnetization, $F_{x,max}$ = 0.0107 N/m^3. (B) Vertical PM magnetization, $F_{x,ma\ x}$ = 0.0774 N/m^3.

are modified to provide for an overall shorter transient MD transfer: the computational domain is reduced to half, by symmetry grounds, the tissue is now thinner (7.8 mm), the tumor is reshaped and closer to the blood vessel, and the PM (now B_{rem} = 1.3 T, horizontal) is sized to cover the tumor. Symmetry boundary conditions are set for the symmetry plane. The diffusivity of the MD (SPIONs) in the tumor, D_{Tu}, is 4×10^{-5} m^2/s (Sillerud, 2018).

Several instants of the MNPs migration from the blood vessel interior wall to the targeted region are shown in Fig. 6.31.

If for the vessel wall to become isoconcentration it takes 6–7 s (at 60 bpm), for the MD to start deliver to the ROI is lasts much longer, here $O(10^3)$ s.

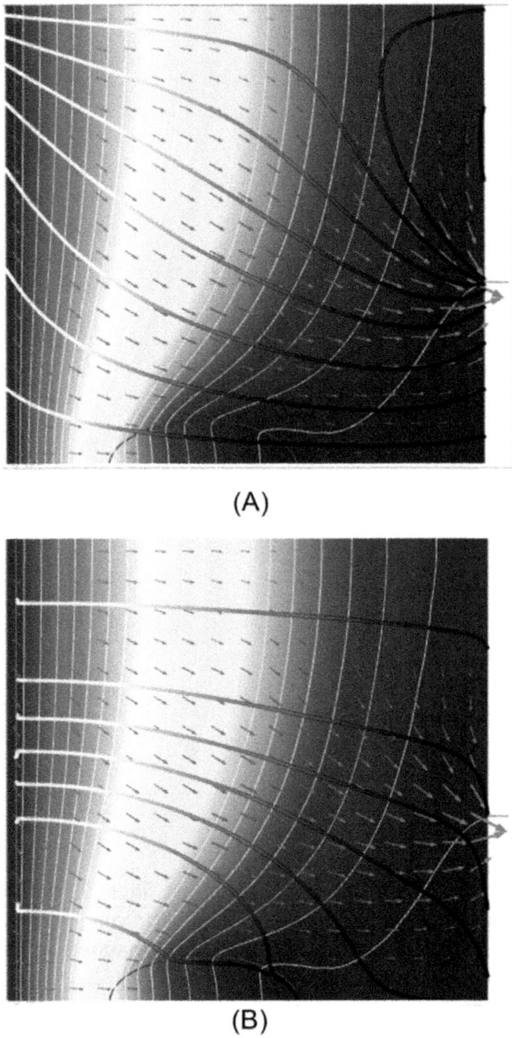

Figure 6.30 Concentration of MNPs at the end of the diffusion process through the target tissue (color map and level contours, $c_{max} = 0.33\ mg/cm^3$) and mass flux of MNPs (thick red lines). (A) In the presence of the magnetic field (diffusion under the action of the magnetic forces). (B) Without magnetic field (free mass diffusion). *MNP*, magnetic nanoparticles.

The progress of the MNPs to the target ROI is accompanied by the change in the susceptivity of the membrane and tissue. Consequently the magnetic field distribution is changing, hence the magnetic forces. It is also important to notice that gradient of the magnetic energy density, which is in fact a motive force, is not everywhere directed towards the magnet.

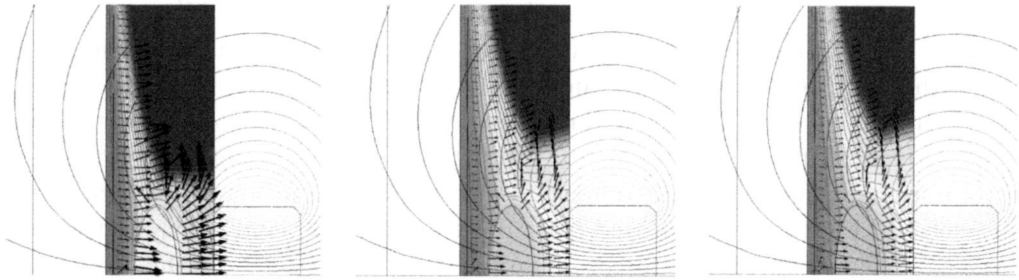

Figure 6.31 The dynamics of the MPNs transfer from the larger vessel to the target region ($c_{max} = c_b$)—species (contours), magnetic body forces. (A) $t =$ 500 s. (B) $t =$ 1000 s. (C) $t =$ 1170 s.

Eventually the tumor ROI is reached, due to the higher diffusivity of the MNPs here, it is covered with almost uniform concentration of MD, increasing in time.

It may be inferred that the MD has to produce its curative effect faster than the carrying MNPs leave to ROI to, eventually reach the magnet, because of the continuing action of the magnetic motive force. And, again, the PM should be placed such that the ROI is in between it and the larger vessel that advects the MD.

Eventually the MD at the vessel wall vanishes, and this entrains the extinction of the magnetic field controlled delivery.

On closing, the synthesis of MNPs covers a wide range of compositions and tunable sizes, fabrication, and surface engineering involve complex chemical, physical, and physicochemical multiple interactions. The promise of MNPs resides in natural properties of their superparmagnetic core combined with the drug loading capability and the biochemical properties that they can be enodwed with by means of a suitable coating. According to the current experience the polymer acts actually as the principal material in MNPs design for drug delivery.

The MDT is using MNPs (SIONs) as representative entities medical or biological purposes and, in numerical modeling their spatial distribution is usually consistent with the continuous media assumptions. The magnetic field sources in MDT are either permanent or coils. Time and space scales, couplings, which are typical to multiphysics problems, are important in the multiscaling analysis, with the aim to produce solvable numerical models consistent with the physics that they represent.

In larger vessels that transfer the MD, depending on the magnetic field spectrum, the magnetomotive forces may either enhance (accelerate) or oppose (decelerate) the flow, and they may indirectly induce recirculation. If the MD is to cross the epithelial membrane, the component of the MF force orthogonal to the streamwise flow is then of interest.

The energy group introduced here for the first time, $W_{mg,0}/\rho U_0^2$, may be used in sizing the magnetic field source (its energy), for a specific situation. This group relates the contribution of the magnetic field to the MD transport with respect to the kinematic energy of the flow (provided by the "pump"-ing device) as leading term. It is an important factor in the MDT design, positioning, and constructal optimization the magnet with respect to the ROI. The magnetic to mechanical forces group Eq. (6.29) is obtained from the energy group, and it may outline the order of magnitude of the forces streamwise or orthogonal to the flow. The streamwise velocity scale U_0 is used in general, but the scale for the orthogonal velocity may be used too, and aspect ratio of the vessel is the conversion factor.

The magnetic velocity, $\mathbf{v}_r$, that is used then as a macroscopic quantity to model th MNPs transfer in endothelial membrane and tissue is seen as a homogenization technique that maps the MNP nanometric physical process onto a macroscopic, continuous, magnetizable medium, superparamagnetic like, which is characterized by an apparent susceptibility, χ_{app}, that depends on the (constant) viscosity of a Newtonian fluid—the coefficient k in Eq. (6.32), which is proportional with the susceptivity of the MNP and the dynamic viscosity of the solute fluid.

References

Abd Elrahman, A.A., Mansour, F.R., 2019. Targeted magnetic iron oxide nanoparticles: Preparation, functionalization and biomedical application. J. Drug Deliv. Sci. Tech. 52, 702–712.

Akbar, N.S., Nadeem, S., 2014. Carreau fluid model for blood flow through a tapered artery with a stenosis. Ain Shams Eng. J. 5, 1307–1316.

Alexiou, C., Christoph, A., Arnold, W., Klein, R.J., Parak, F.G., Hulin, P., et al., 2000. Locoregional cancer treatment with magnetic drug targeting. Cancer Res. 60, 6641–6648.

Alexiou, C., Jurgons, R., Schmid, R.J., Bergemann, C., Henke, J., Erhardt, W., et al., 2003. Magnetic drug targeting-biodistribution of the magnetic carrier and the chemotherapeutic agent mitoxantrone after locoregional cancer treatment. J. Drug Target 11 (3), 139–149.

Alexiou, C., Diehl, D., Henninger, P., Iro, H., Röckelein, R., Schmidt, W., et al., 2006a. A high field gradient magnet for magnetic drug targeting. IEEE Trans Appl. Supercond. 16 (2), 1527–1530.

Alexiou, C., Schmid, R.J., Jurgons, R., Kremer, M., Wanner, G., Bergemann, C., et al., 2006b. Targeting cancer cells: magnetic nanoparticles as drug carriers. Eur. Biophys. J. 35 (5), 446–450.

Ali, A., Zafar, H., Zia, M., ul Haq, I., Phull, A.R., Ali, J.S., et al., 2016. Synthesis, characterization, applications, and challenges of iron oxide nanoparticles. Nanotechnol. Sci. Appl. 9, 49–67 (Review).

Almaça, J., Weitz, J., Rodriguez-Diaz, R., Pereira, E., Caicedo, A., 2018. The pericyte of the pancreatic islet regulates capillary diameter and local blood flow. Cell Met. 27, 630–644.

An, J., Zhanga, X., Guo, Q., Zhao, Y., Wu, Z., Li, C., 2015. Glycopolymer modified magnetic mesoporous silica nanoparticles for MR imaging and targeted drug delivery. Colloids Surf. A: Physicochem. Eng. Aspects 482, 98–108.

Aroesty, J., Gross, J.F., 1972. The mathematics of pulsatile flow in small vessels I. Casson theory. Microvas. Res. 4, 1–12.

Aslibeiki, B., Kameli, P., Manouchehri, I., Salamati, H., 2012. Strongly interacting superspins in Fe3O4 nanoparticles. Curr. Appl. Phys. 12, 812–816.

Avilés, M.O., Ebner, A.D., Chenb, H., Rosengartb, A.J., Kaminskic, M.D., Ritter, J.A., 2005. Theoretical analysis of a transdermal ferromagnetic implant for retention of magnetic drug carrier particles. J. Magn. Magn. Mater. 293, 605–615.

Avilés, M.O., Ebner, A.D., Ritter, J.A., 2008. Implant assisted-magnetic drug targeting: comparison of in vitro experiments with theory. J. Magn. Magn. Mater. 320, 2704–2713.

Babincova, M., Babinec, P., 2009. Magnetic drug delivery and targeting: principles and applications. Biomed. Pap., Med. Fac. Univ. Palacky Olomouc, Czech Repub. 153 (4), 243–250.

Bai, J., Wang, J.T.-W., Mei, K.-C., Al-Jamal, W.T., Al-Jama, K.T., 2016. Real-time monitoring of magnetic drug targeting using fibered confocal fluorescence microscopy. J. Control. Release 244, 240–246.

Barnard, A.C.L., Hunt, W.A., Timlake, W.P., Varley, E., 1966. A theory of fluid flow in compliant tubes. Biophys. J. 6, 717–724.

Barnsley, L.C., Carugo, D., Stride, E., 2016. Optimized shapes of magnetic arrays for drug targeting applications. J. Phys. D: Appl. Phys. 49 (225501), 17 pp.

Berry, C.C., 2009. Progress in functionalization of magnetic nanoparticles for applications in biomedicine. J. Phys. D: Appl. Phys. 42, 224003, 9 pp., (Topical Review).

Brown T., 2020. Transcription, translation and replication. In: *Nucleic Acid Structure, ATDBio*, <https://www.atdbio.com/content/14/Transcription-Translation-an,d-Replication> (accessed 03.20).

Butoescu, N., Seemayer, C.A., Palmer, G., Guerne P.-A., Gabay, C., Doelker, E., et al., 2009. Magnetically retainable microparticles for drug delivery to the joint: efficacy studies in an antigen-induced arthritis model in mice. Arthritis Res. Ther. 11, R72, 10 pp. Available From: https://doi.org/10.1186/ar2701

Chegini, S.P., Varshosaz, J., Taymouri, S., 2018. Recent approaches for targeted drug delivery in rheumatoid arthritis diagnosis and treatment. Artif. Cells Nanomed. Biotechnol. 46 (S2), 502–514.

Chomoucka, J., Drbohlavova, J., Huska, D., Adam, V., Kizek, R., Hubalek, J., 2010. Magnetic nanoparticles and targeted drug delivering. Pharmacol. Res. 62, 144–149 (Review).

Chuzawa, M., Mishima, F., Akiyama, Y., Nishijima, S., 2013. Precise control of the drug kinetics by means of non-invasive magnetic drug delivery system. Physica C 484, 120–124.

Cregg, P.J., Murphy, K., Mardinoglu, A., Prin-Mello, A., 2010. Many particle magnetic dipole–dipole and hydrodynamic interactions in magnetizable stent assisted magnetic drug targeting. J. Magn. Magn. Mater. 322, 2087–2094.

Cregg, P.J., Murphy, K., Mardinoglu, A., 2012. Inclusion of interactions in mathematical modelling of implant assisted magnetic drug targeting. Appl. Math. Model. 36, 1–34.

David, A.E., Cole, A.J., Chertok, B., Park, Y.S., Yang, V.C., 2011. A combined theoretical and in vitro modeling approach for predicting the magnetic capture and retention of magnetic nanoparticles in vivo. J. Control. Release 152, 67–75.

Dobre, A.A., 2012. Investigation Methods for the Analysis of Coupled Phenomena Specific to the Medical Engineering Field (Doctoral Thesis). Faculty of Electrical Engineering, University Politehnica of Bucharest.

Dobre, A.A., Morega, A.M., 2010. Numerical simulation in magnetic drug targeting. In: *The 12th Mediteranean Conference on Medical and Biological Engineering and Computing, MEDICON 2010*, Porto Carras, Chalkidiki, Greece, May 27–30, 2010, *IFMBE Proceedings*, vol. 29, issue 4, pp. 651–654.

Drochon, A., Robin, V., Fokapu, O., Rodriguez, D.A., 2016. Stationary flow of blood in a rigid vessel in the presence of an external magnetic field: considerations about the forces and wall shear stresses. Appl. Math. 7, 130–136.

Eckert, M.A., Vu, P.Q., Zhang, K., Kang, D., Ali, M.M., Xu, C., et al., 2013. Novel molecular and nanosensors for *in vivo* sensing. Theranostics 3 (8), 583–594.

Faraji, A.H., Wipf, P., 2009. Nanoparticles in cellular drug delivery. Bioorg. Med. Chem. 17, 2950–2962 (Review).

Feijóo, R.A., 2000. Computational methods in biology. In: *2nd Summer School LNCC/MCT*, Petrópolis, January 2000.

Furlani, E.J., Furlani, E.P., 2007. A model for predicting magnetic targeting of multifunctional particles in the microvasculature. J. Magn. Magn. Mater. 312, 187–193.

Gleich, B., Hellwig, N., Bridell, H., Jurgons, R., Seliger, C., Alexiou, C., et al., 2007. Design and evaluation of magnetic fields for nanoparticle drug targeting in cancer. IEEE Trans. Nanotechnol. 6 (2), 164–170.

Grief, A.D., Richardson, G., 2005. Mathematical modelling of magnetically targeted drug delivery. J. Magn. Magn. Mater. 293, 455–463.

Gupta, P.K., Hung, C.T., 1989. Magnetically controlled targeted micro-carrier systems. Life Sci. 44, 175–186.

Hawk's *Perch Technical Writing*, LLC. <https://www.understandingnano.com/medicine.html> (accessed.03.20).

Heidarshenas, B., Wei, H., Moghimi, Z.A., Hussain, G., Baniasadi, F., Naghieh, G., 2019. Nanowires in magnetic drug targeting. Mater. Sci. Eng. 3 (1), 3–9.

Hobbs, S.K., Monsky, W.L., Yuan, F., Roberts, W., Griffith, L., Torchilin, V.P., et al., 1998. Regulation of transport pathways in tumor vessels: role of tumor type and microenvironment. Proc. Natl. Acad. Sci. USA 95, 4607–4612. April.

Hoke, I., Dahmani, C., Weyh, T., 2008. Design of a high field gradient electromagnet for magnetic drug delivery to a mouse brain. In: *Proceedings of the COMSOL Conference*, Hannover.

Hoskins, P.R., Loupas, T., McDicken, W.N., 1990. A comparison of the Doppler spectra from human blood and artificial blood used in a flow phantom. Ultrasound Med. Biol. 16 (2), 141–147.

Hunter, P.J., 1972. Numerical Simulation of Arterial Blood Flow (MS Thesis). School of Engineering, University of Auckland.

Iacob Gh, R.O., Chiriac, H., 2004. A possibility for local targeting of magnetic carriers. J. Optoelectron. Adv. Mater. 6 (2), 713–717.

Jones, T.B., 1995. Electromechanics of Particles. Cambridge Univ. Press.

Kaminski, M.D., Nuñez, L., Ghebremeskel, A.N., Kasza, K.E., Fischer, P.F., Chang, F.-C., et al., 2003. Magnetically Responsive Microparticles for Targeted Drug and Radionuclide Delivery. A Review of Recent Progress and Future Challenges. Argonne National Laboratory, Argonne, IL.

Kheirkhah, P., Denyer, S., Bhimani, A.D., Arnone, G.D., Darian, R., Esfahani, D.R., et al., 2008. Magnetic drug targeting: a novel treatment for intramedullary spinal cord tumors. Sci. Rep. 8, 11417, 9 pp.

Koch, I., Josephson, L., 2009. Magnetic nanoparticle sensors. Sensors 9, 8130–8145.

Köstler, H., 2016. Magnetic Nanoparticle Simulation in Blood Flow (MS Thesis). Department of Computer Science, Friedrich-Alexander University Erlangen-Nürnberg.

Kudr, J., Haddad, Y., Richtera, L., Heger, Z., Cernak, M., Adam, V., 2017. Magnetic nanoparticles: from design and synthesis to real world applications. Nanomaterials 7 (9), 243.

Law, Y.F., Johnston, K.W., Routh, H.F., Cobbold, R.S.C., 1989. On the design and evaluation of a steady flow model for Doppler ultrasound studies. Ultrasound Med. Biol 15 (5), 505–516.

LexInnova, 2020. Nanoparticles smart drug delivery system for cancer. Available from: <http://www.wipo.int/edocs/plrdocs/en/lexinnova_nanoparticles_smart_delivery_system_for_tumors.pdf>.

Li, D., Ren, Y., 2018. High-gradient magnetic field for magnetic nanoparticles drug delivery system. IEEE Trans. Appl. Supercond. 28, 6, 4420107, 7 pp. (Technical Note).

Li, S., Hoskins, P.R., Anderson, T., McDicken, W.N., 1993. Measurement of mean velocity during pulsatile flow using time-averaged maximum frequency of Doppler ultrasound waveforms. Ultrasound Med. Biol. 19 (2), 105–113.

Li, X.L., Yao, K.L., Liu, Z.L., 2008. CFD study on the magnetic fluid delivering in the vessel in high-gradient magnetic field. J. Magn. Magn. Mater. 320, 1753–1758.

Lima, B.K., Tigheb, E.C., Kong, S.D., 2019. The use of magnetic targeting for drug delivery into cardiac myocytes. J. Magn. Magn. Mater. 473, 21–25.

Lin, P.A., Kumar, A., Mohan Sankaran, R., 2012. New insights into plasma-assisted dissociation of organometallic vapors for gas-phase synthesis of metal nanoparticles. Plasma Process Polym 9 (11-12), 1184–1193.

Liu, T.-L., 2019. A review of magnet systems for targeted drug delivery. J. Control. Release 302, 90–104.

Liu, C., Wu, S., Yan, Y., Dong, Y., Shen, X., Huang, C., 2019. Application of magnetic particles in forensic science. Trends Anal. Chem. 121 (115674), 10 pp.

Lübbe, A.S., Bergemann, C., Riess, H., Schriever, F., Reichardt, P., Possinger, K., et al., 1996. Clinical experiences with magnetic drug targeting: A phase I study with 4'-epidoxorubicin in 14 patients with advanced solid tumors. J. Cancer Res. 56, 4686–4693.

Lücker, A., Secomb, T.W., Barrett, M.J.P., Weber, B., Jenny, P., 2018a. The relation between capillary transit times and hemoglobin saturation heterogeneity. Part 2: capillary networks. Frontiers Physiol. 9 (1296), 10 pp.

Lücker, A., Secomb, T.W., Weber, B., Jenny, P., 2018b. The relation between capillary transit times and hemoglobin saturation heterogeneity. Part 1: theoretical models. Frontiers Physiol. 9 (420), 15 pp.

Mahmoudi, M., Sant, S., Wang, B., Laurent, S., Sen, T., 2011. Superparamagnetic iron oxide nanoparticles (SPIONs): development, surface modification and applications in chemotherapy. Adv. Drug Deliv. Rev. 63, 24–46.

Manea, L., 2015. Magnetic Drug Localization at Knee Level (Diploma thesis). Faculty of Medical Engineering, University Politehnica of Bucharest, Romania (in Romanian).

Manea, L., Morega, A.M., Morega, M., 2014. Numerical simulation of magnetic drug localization in the knee joint medication. In: *IEEE Proc. of the 2014 Intl. Symp. on Fundamentals of Electrical Engineering*, Bucharest, 28–29 November 2014.

Marinin A., 2012. Synthesis and Characterization of Superparamagnetic Iron Oxide Nanoparticles Coated with Silica (MS thesis). Division of Functional Materials, School of Information and Communication Technology Royal Institute of Technology, Stockholm.

Markides, H., Rotherham, M., El Haj, A.J., 2012. Biocompatibility and toxicity of magnetic nanoparticles in regenerative medicine. Hindawi Pub. Corp. J. Nanomater. 2012 (614094), 11 pp.

Martel, S., Mathieu, J.-B., Felfoul, O., Chanu, A., Aboussouan, E., Tamaz, S., et al., 2007. Medical and technical protocol for automatic navigation of a wireless device in the carotid artery of a living swine using a standard clinical MRI system Part I, LNCS 4791 In: Ayache, N., Ourselin, S., Maeder, A. (Eds.), *The Medical Image Computing and Computer Assisted Intervention Society* (MICCAI) 2007. Springer-Verlag, Berlin, Heidelberg, pp. 144–152.

Martel, S., Mohammadi, M., Felfoul, O., Lu, Z., Pouponneau, P., 2009. Flagellated magnetotactic bacteria as controlled MRI-trackable propulsion and steering systems for medical nanorobots operating in the human microvasculature. Int. J. Rob. Res. 28 (4), 571–582.

McBride, A.A., Price, D.N., Lamoureux, L.R., Elmaoued, A.A., Vargas, J.M., Adolphi, N.L., et al., 2013. Preparation and characterization of novel magnetic nano-in-microparticles for site-specific pulmonary drug delivery. Mol. Pharm. 10, 3574–3581.

Misra, J.C., Patra, M.K., Sahu, B.K., 1992. Unsteady flow of blood through narrow blood vessels – a mathematical analysis. Comput. Math. Appl. 24 (10), 19–31.

Mocanu, C.I, 1981. *The Theory of the Electromagnetic Field*, Ed. Didactică şi Pedagocică, Bucharest, Romania (in Romanian).

Morega A.M., Dobre A., Morega M., Mocanu D., 2009. Computational modeling of arterial blood flow. In: *MediTech Conference*, 23–26 September 2009, Cluj-Napoca, Romania.

Morega A.M., Dobre A.A., Morega M., 2010. Numerical simulation of magnetic drug targeting with flow–structural interaction in an arterial branching region of interest. In: *European COMSOL Conference*, Palais des Congrès de Versailles, 17–19 November 2010.

Morega, A.M., Dobre, A.A., Morega, M., 2011. Magnetic field – flow interactions in drug delivery through an arterial system. Rev. Roum. Sci. Techn. Électrotechn. et Énerg 56 (2), 199–208.

Morega, A.M., Dobre, A.A., Morega, M., 2012. Numerical simulation in electrical cardiometry. In: *The 13th International Conference on Optimization of Electrical and Electronic Equipment, OPTIM*, 24–26 May 2012, Brasov, Romania.

Morega, A.M., Savastru, C., Morega, M., 2013a. Numerical simulation of flow dynamics in the brahial-unar-radial system. In: *IEEE International Conference on e-Health and Bioengineering* (EHB), 21–23 November. Available from: <https://10.1109/EHB.2013.6707352>.

Morega, A.M., Savastru, C., Morega, M., 2013b. Numerical simulation of flow dynamics in the brachial-ulnar-radial arterial system. In: *The 4th IEEE International Conference on E-Health and Bioengineering – EHB*, Iasi, Romania.

Morega, A.M., Savastru, C., Morega, M., 2013c. An adaptive magnetic field source for magnetic drug fixation. Int. J. Monit. Surv. Tech. Res. 1 (4), 14 pp. Available from: https://doi.org/10.4018/ijmstr.2013100103.

Morega, A.M., Savastru, C., Morega, M., 2015. Numerical simulation of an adaptive magnetic field source concept in magnetic drug targeted transport. In: *Proc. of the 8th International Symposium on Advanced Topics in Electrical Engineering, IEEE-ATEE*, 23–24 May 2015, Bucharest, Romania.

Morega, A.M., Dobre, A.A., Morega, M., Sandoiu, A., 2018. Constructal optimization of magnetic field source in magnetic drug targeting therapy. In: *Proc. of the Romanian Academy, Series A-Mathematics Physics Technical Sciences Information Science*, vol. 19, pp. 123–128, Special Issue: SI.

Musielaka, M., Piotrowskia, I., Suchorsk, W.M., 2009. Superparamagnetic iron oxide nanoparticles (SPIONs) as a multifunctional tool in various cancer therapies. Rep. Pract. Oncol. Radiother. 2 (4), 307–314.

Nacev, A.N., 2013. Magnetic Drug Targeting: Developing the Basics (Doctoral thesis). The Faculty of the Graduate School of the University of Maryland, College Park.

Nacev, A.N., Beni, C., Bruno, O., Shapiro, B., 2011. The behaviors of ferromagnetic nano-particles in and around blood vessels under applied magnetic fields. J. Magn. Magn. Mater. 323 (6), 651–668.

Nedelcu, A., 2016. Electromagnetic and Heat Transfer Phenomena in Superconducting Devices (Doctoral thesis). Doctoral School of Electric Engineering, University Politehnica of Bucharest, Romania (in Romanian).

Neuberger, T., Schöpf, B., Hofmann, H., Hofmann, M., von Rechenberg, B., 2005. Superparamagnetic nanoparticles for biomedical applications: Possibilities and limitations of a new drug delivery system. J. Magn. Magn. Mater 293, 483–496.

Ningbo, 2019. Tongchuang Strong Magnet Material Co., Ltd. Available from: <https://tcmagnet.en.made-in-china.com/>.

Orekhova, N.M., Akchurin, R.S., Belyaev, A.A., Smirnov, M.D., Ragimov, S.E., Orekhov, A.N., 1990. Local prevention of trombosis in animal arteries by means of magnetic targeting of asp1 rin-loaded red cells. Thrombosis Res. 57, 611–616.

Osher, S., Sethian, J.A., 1988. Fronts propagating with curvature dependent speed: algorithms based on Hamilton-Jacobi formulations. J. Comput. Phys. 79, 12–49.

Pankhurst, Q.A., Connolly, J., Jones, S.K., Dobson, J., 2003. Applications of magnetic nanoparticles in biomedicine. J. Phys. D: Appl. Phys 36, R167–R181 (Topical Review).

Papisov, M.I., Savelyev, V.Y., Sergienko, V.B., Torchilin, V.P., 1987. Magnetic drug targeting. 1. In vivo kinetics of radiolabelled magnetic drug carriers. Int. J. Pharm. 40 (3), 201–206.

Papisov, M.I., Torchilin, V.P., 1987. Magnetic drug targeting. II. Targeted drug transport by magnetic microparticles: factors influencing therapeutic effect. Int. J. Pharm. 40, 207–214.

Plavins, J., Lauva, M., 1993. Study of colloidal magnetite binding erythrocytes: prospects for cell separation. J. Magn. Magn. Mater 122, 349–353.

Preis, K., Biro, O., Friedrich, M., Gottvald, A., Magele, C., 1991. Comparison of different optimization strategies in the design of electromagnetic devices. IEEE Trans. Magn. 27 (5), 4154–4157.

Price, D.N., Stromberg, L.R., Kunda, N.K., Muttil, P., 2017. *In vivo* pulmonary delivery and magnetic-targeting of dry powder nano-in-microparticles. Mol. Pharm. 14, 4741–4750.

Pries, A.R., Secomb, T.W., Gaehtgens, 2000. The endothelial surface layer. Pflügers Arch. Eur. J. Physiol. 440, 653–666.

Priyadharshini, S., Ponalagusamy, R., 2015. Biorheological model on flow of Herschel-Bulkley fluid through a tapered arterial stenosis with dilatation. Appl. Bionics Biomech. 2015 (406195), 12 pp.

Pyrhönen, J., Jokinen, T., Hrabovcová, V., 2009. Design of Rotating Electrical Machines. Wileycited by. Available from: https://en.wikipedia.org/wiki/Neodymium_magnet.

Quanyu, W., Xiaojie, L., Lingjiao, P., Weige, T., Chunqi, Q., 2017. Simulation analysis of blood flow arteries of the human arm. Biomed. Eng.: Appl. Basis Commun. 29 (4), 1750031.

Razavi, M.S., Shirani, E., Kassab, G.S., 2018. Scaling laws of flow rate, vessel blood volume, lengths, and transit times with number of capillaries. Frontiers Physiol. 9 (581), 15 pp.

Riegler, J., Lau, K.D., Garcia-Prieto, A., Price, A.N., Richards, T., Pankhurst, Q.A., et al., 2011. Magnetic cell delivery for peripheral arterial disease: A theoretical framework. Med Phys. 38 (7), 3932–3943.

Roots-Weiß, A., Papadimitriou, C., Serve, H., Hoppe, B., Koenigsmann, M., Reufi, B., et al., 1997. The efficiency of tumor cell purging using immunomagnetic CD341 cell separation systems. Bone Marrow Transpl. 19, 1239–1246.

Rosensweig, R.E., 1997. Ferrohydrodynamics. Dover Publications, New York.

Saiyed, Z.M., Telang, S.D., Ramchand, C.N., 2003. Application of magnetic techniques in the field of drug discovery and biomedicine. Biomagn. Res. Technol. 1, 8 pp. Available from: http://www.biomagres.com/content/1/1/2.

Salazar-Alvarez, G., Muhammed, M., Zagorodni, A.A., 2006. Novel flowinjection synthesis of iron oxide nanoparticles with narrowsize distribution. Chem. Eng. Sci. 61, 4625–4633.

Săndoiu, A.M., 2019. Electromagnetic Methods for Medical Therapies and Medication Management (Doctoral thesis). The Doctoral School of Electrical Engineering, University Politehnica of Bucharest, Romania.

Săndoiu, A.M., Morega, A.M., Morega, M., 2019. On the constructal optimal shape of magnetic field source used in magnetic drug targeting. UPB Sci. Bull. Series C Electr. Eng. Comput. Sci. 81 (1), 171–180.

Săvăstru, C., 2016. Electromagnetic Field, Heat Transfer and Structural Interactions in Biomedical Applications (Doctoral thesis). The Doctoral School of Electrical Engineering, University Politehnica of Bucharest, Romania (in Romanian).

Schenck, J.F., 2000. Safety of strong, static magnetic fields. J. Magn. Reson. Imaging 12, 2–19 (Review, Invited).

Schulze, K., Koch, A., Schopf, B., Petri, A., Steitz, B., Chastellainc, M., et al., 2005. Intra-articular application of superparamagnetic nanoparticles and their uptake by synovial membrane—an experimental study in sheep. J. Magn. Magn. Mater. 239 (1), 419–432.

Schütt, W., Grüttner, C., Häfeli, U., Zborowski, M., Teller, J., Putzar, H., et al., 1997. Applications of magnetic targeting in diagnosis and therapy—possibilities and limitations: a mini-review. Hybridoma 16 (1), 109–117.

Sethian, J.A., 1996. A fast marching level set method for monotonically advancing fronts. Proc. Natl. Acad. Sci. Appl. Math. 93, 1591–1595.

Sethian, J.A., Wiegmann, A., 2000. Structural boundary design via level set and immersed interface methods. J. of Comput. Phys. 163 (2), 489–528.

Shamsia, M., Sedaghatkish, A., Dejam, M., Saghafian, M., Mohammadia, M., Sanati-Nezha, A., 2018. Magnetically assisted intraperitoneal drug delivery for cancer chemotherapy. Drug Deliv. 25 (1), 846–861.

Shamloo, A., Amani, A., Forouzandehmehr, M., Ghoytasi, I., 2019. *In silico* study of patient-specific magnetic drug targeting for a coronary LAD atherosclerotic plaque. Int. J. Pharmaceutics 559, 113–129.

Shapiro, B., 2009. Towards dynamic control of magnetic fields to focus magnetic carriers to targets deep inside the body. J. Magn. Magn. Mater. **321**, 10, 1594, 13 pp.

Shapiro, B., Kulkarni, S., Nacev, A., Muro, S., Stepanov, P.Y., Weinberg, I.N., 2014. Open challenges in magnetic drug targeting. WIREs Nanomed. Nanobiotechnol. Available from: https://doi.org/10.1002/wnan.1311.

Sharma, S., Vkatiyar, K., Singh, U., 2015. Mathematical modeling for trajectories of magnetic nanoparticles in a blood vessel under magnetic field. J. Magn. Magn. Mater. 379, 102–107.

Shaw S., Mathematical model on magnetic drug targeting in microvessel. In: Magnetism and Magnetic Materials, IntechOpen, Chapter 5, pp. 83–107. <https://doi.org/10.5772/intechopen.73678> (accessed 03.20).

Shaw, S., Sutradhar, A., Murthy, P.V.S.N., 2017. Permeability and stress-jump effects on magnetic drug targeting in a permeable microvessel using Darcy model. J. Magn. Magn. Mater. 429, 227–235.

Shi, D., Gu, H., 2008. Nanostructured materials for biomedical applications. J. Nanomater. Article ID 529890, 2 pp., (Editorial).

Sillerud, L.O., 2018. Quantitative [Fe]MRI determination of the dynamics of PSMA-targeted SPIONS discriminates among prostate tumor Xenografts based on their PSMA expression. J. Magn. Reson. Imaging. 48 (2), 469–481.

Singh, N., Jenkins, G.J.S., Asadi, R., Doak, S.H., 2010. Potential toxicity of superparamagnetic iron oxide nanoparticles (SPION). Nano Rev. 1 (5358), 15 pp. Available from: https://doi.org/10.3402/nano.v1i0.5358.

Singh, R., Lillard, J.W., 2009. Nanoparticle-based targeted drug delivery. Exp. Mol. Pathol. 86, 215–223 (Review).

Smith, N.P., Pullan, A.J., Hunter, P.J., 2002. An anatomically based model of transient coronary blood flow in the heart. SIAM J. Appl. Math. 62 (3), 990–1018.

Soltani, M., Chen, P., 2013. Numerical modeling of interstitial fluid flow coupled with blood flow through a remodeled solid tumor microvascular network. PLoS One 8 (6), e67025, pp. 18, June.

Stanciu, D., 2016. Magnetic Drug Localization at Elbow Level (Diploma thesis). Faculty of Medical Engineering, University Politehnica of Bucharest, Romania (in Romanian).

Stem Cell Clinic. <https://www.stemcellclinic.com/revmatoidnyj-artrit/?lang = en> (accesed 03.20).

Sun, C., Jerry Lee, S.H., Zhang, M., 2008. Magnetic nanoparticles in MR imaging and drug delivery. Adv. Drug Deliv. Rev. 60, 1252–1265.

Sussman, M., Fatemi, E., Smereka, P., Osher, S., 1998. An improved level set method for incompressible two-phase flows. Comput. Fluids 27 (5–6), 663–680.

Swabb, E.A., Wei, J., Gullino, P.M., 1974. Diffusion and convection in normal and neoplastic tissues. Cancer Res. 34, 2814–2822. October.

Synovitis <https://www.hss.edu/condition-list_synovitis.asp> (accessed 03.20).

Thomas, R., Park, I.–K., Jeong, Y.Y., 2013. Magnetic iron oxide nanoparticles for multimodal imaging and therapy of cancer. Int. J. Mol. Sci. 14 (8), 15910–15930.

Ting, T.-H., Newhouse, V.L., Li, Y., 1992. Doppler ultrasound technique for measuring capillary-speed flow velocities with strong stationary echoes. Ultrasonics 30 (4), 225–231.

Veiseh, O., Gunn, J.W., Zhang, M., 2010. Design and fabrication of magnetic nanoparticles for targeted drug delivery and imaging. Adv. Drug Deliv. Rev. 62, 284–304.

Viota, J.L., Carazo, A., Munoz-Gamez, J.A., Rudzka, K., Gómez-Sotomayor, R., Ruiz-Extremera, A., et al., 2013. Functionalized magnetic nanoparticles as vehicles for the delivery of the antitumor drug gemcitabine to tumor cells. Physicochemical in vitro evaluation. Mater. Sci. Eng. C 33, 1183–1192.

Voltairas, P.A., Fotiadis, D.I., Michalis, L.K., 2002. Hydrodynamics of magnetic drug targeting. J. Biomech. 35, 813–821.

Wang, Y.X.J., 2015. Current status of superparamagnetic iron oxide contrast agents for liver magnetic resonance imaging. World J. Gastroenterol. 21 (47), 13400–13402.

Winner, K.R.K., Steinkamp, M.P., Lee, R.J., Swat, M., Muller, C.Y., Moses, M.E., et al., 2016. Spatial modeling of drug delivery routes for treatment of disseminated ovarian cancer. Cancer Res. 76 (6), 15. March 15.

Xu, C., Sun, S., 2013. New forms of superparamagnetic nanoparticles for biomedical applications. Adv. Drug Deliv. Rev. 65, 732–743.

Xu, H., Song, T., Bao, X., Lili Hu, L., 2005. Site–directed research of magnetic nanoparticles in magnetic drug targeting. J. Magn. Magn. Mater 293, 514–519.

Zeeshan, A., Fatima, A., Khalid, F., Bhatti, M.M., 2018. Interaction between blood and solid particles propagating through a capillary with slip effects. Microvas. Res. 119, 38–46.

Zhan, W., Rodriguez, F., Dini, B., Dini, D., 2019. Effect of tissue permeability and drug diffusion anisotropy on convection-enhanced delivery. Drug Deliv. 26 (1), 773–781.

Zheng, X., Silber-Li, Z., 2009. The influence of Saffman lift force on nanoparticle concentration distribution near a wall. Appl. Phys. Lett. 95, 124105, 3 pp.

CHAPTER 7

Magnetic stimulation and therapy

7.1 Introduction

From the arborizations of cerebellar Purkinje dendritic cells to motor neurons that transmit the control signals from the central nervous system to the distal muscles, the complexity and spatial extent of neurons vary significantly (Niebur, 2008). Arborescent networks allow neurons to accept and transmit a vast number of synaptic inputs that may interact in complex ways to enable dendritic computations.

The difficulties encountered in measuring the distal parts of the cell limit the knowledge of the integration of the neurons' distal synaptic input. The nervous fiber has a complex structure (Villegas and Villegas, 1965; Cole, 1968; Plonsey and Barr, 1988). During the biological evolution, its structure, size, and shape morphed into an optimal construct that is consistently described through the cable theory (Plonsey and Barr, 1988; Malmivuo and Plonsey, 1995; Morega, 1999; Timotin, 2004). The cable theory presents how electrical signals from different synapses merge and circulate through the system of branches that forms the dendritic tree of a cell. In this approach, the electrical signal (voltage) varies mainly along the axis of the neural process than orthogonal to it, and *neurites* (also called "neural processes") are used to model this part of a dendrite or axon—these limitations are the consequence of reducing the morphological complexity of a neuron by neglecting the small radial variations and only considering the variation along its axis. This assumption, which is based on scale analysis has important consequences: it reduces the model from three dimensions to just one, paving the path to an equivalent electric circuit and the pending cable model. The first section in this chapter is devoted to the *cable theory and model* as needed for stimulation analysis and analytic solutions are proposed to this end.

Magnetic stimulation (MS) represents, in fact, the activation of excitable tissues (nerve, muscle, or sensitive cells), which is produced noninvasively by applying an electric stimulus near the cell membrane, by electromagnetic induction. Electric current impulses are released through a stimulation applicator (coil), which is placed at the surface of the skin; the variable magnetic field induces eddy current impulses inside the body. MS began to be scientifically studied and tested for clinical use by a group of researchers from the University of Sheffield, UK, in the early 1980s. From the stimulation of peripheral nerves [peripheral magnetic stimulation (PMS)], the procedure has evolved to apply to the central nervous system through transcranial magnetic stimulation (TMS). It is more and more used in neurology, both for therapeutic purposes

Computational Modeling in Biomedical Engineering and Medical Physics
DOI: https://doi.org/10.1016/B978-0-12-817897-3.00007-5

and in diagnosis and may help identify and repair damaged transmission pathways in the central and peripheral nervous system or can be used to map the motor cortex Davey et al. (1994).

In medical practice, MS is accepted for its noninvasive and painless nature. It produces minimal discomfort to the subject because the pain sensors (sensitive cells) in the skin do not react to a magnetic field, while the induced electric field is too low to be perceived as pain. MS also has the advantage of the accessibility of the equipment, which is not sophisticated in principle. Lately, efficient technical solutions have been designed for applicator coils and control electronics. Although the physics are straightforward, stimulation by an induced electric field is a medically sensitive task, as it requires precision to locate the target inside the body, good focus and adequate intensity of the stimulus, while minimizing side effects. Mathematical simulation of the distribution of stimuli inside the body correlated with the optimization of technical characteristics of magnetic applicators helps the progress of this medical procedure, by enhancing its precision.

MS is characterized by a vast palette of medical and clinical applications, starting from noninvasive nerve stimulation, TMS, motor-evoked potentials to neuropsychiatric applications, drug-resistant depression, obsessive–compulsive disorders, and many others. One of the first applications of MS, widely spread nowadays, is the TMS, a procedure based on a high-frequency magnetic field generated using magnetic coils positioned in the proximity of the cranium, able to induce stimulating electrical currents (Chokroverty, 1990; Hallet, 2007; Rossini and Rossi, 2007; Babbs, 2014). This technique easily became an alternative to the painful *transcranial electrical stimulation* (TES) (Berényi et al., 2012; Paulus, 2011).

TMS is currently used to cure neuropsychiatric diseases, depression, migraine, and to understand different neural processes, for example, Baxendale (2009). It can act either localized or at reasonably distant regions of interest, influencing the neuronal activity as modulated by the stimulating magnetic field frequency (Wasserman and Linsaby, 2001). Also TMS is used in brain mapping (cognitive functions, sensorial processes, motor cortex mapping, cranial nerve muscles mapping, etc.) by activating or inhibiting brain regions.

The increasing interest in TMS, TES, and their associated clinical applications led to the development of numerical models, especially targeted at the magnetic field source analysis and optimization. One of the TES approaches is presented in a finite element method (FEM) model based on the computational domain of the head segmented out of an MRI image set (Datta et al., 2013). The numerical simulation results in presenting voltage maps of the scalp, generated by the applied stimulation electrical pulses and using different electrode setups.

Repetitive TMS (rTMS) proves to be highly efficient in the treatment of depression. For patients suffering from epilepsy, for whom the temporal lobectomy (Novelly et al., 1984; Meyer et al., 1986; Sperling et al., 1996) is meant to improve the quality of life, the "Wada test" (Elliott et al., 2013; Woermann et al., 2003; Baxendale, 2009) is conducted to determine what side of the subject's brain is used for language processing: sodium amytal is injected through the carotid arteries until the hemispheric function is suppressed, and the speech arrest occurs on the side of the brain that needs to be detected for lobectomy. rTMS can be considered a Wada test alternative when positioning the excitation coil above the left inferior frontal region and of the patient and using frequencies between 8 and 25 Hz for 10 s (Hallet, 2000).

Besides rTMS, another technique is the paired-pulse TMS (Kobayashi and Pascual-Leone, 2003; Thickbroom et al., 2006; Peinemann et al., 2001), which is used for the evaluation of facilitatory and inhibitory intracortical mechanisms in the motor cortex. This method combines a subthreshold conditioning stimulus with a suprathreshold test stimulus applied at short time intervals (1−20 ms) provided by the same magnetic coil. Paired-pulse TMS is also used when stimulating two different brain regions using the same magnetic stimulus for evaluating interhemispheric and intracortical connections in motor control or movement disorders.

In neurosurgery, TMS is an important tool for surgical planning and intraoperative monitoring (Bestmann et al., 2006; Hartwigsen et al., 2010). Along with functional MRI, TMS can provide information regarding the causality between brain activity and behavior, thus estimating if damaging a certain brain region is likely to become a postsurgical failure.

Pain is defined as "an unpleasant sensory and emotional experience associated with actual or potential tissue damage or described in terms of such damage" (Merskey and Bogduk, 1994; Dowalti, 2017). In particular, back pain is one of the most widespread cause of people's use of absence at work, and it is a primary reason for disability worldwide (Mayo, 2020). Chronic back pain persists for more than 12 weeks and is caused by degenerative or traumatic vertebral processes. It is also the most expensive benign health condition and one of the main causes of activity restrictions for persons less than 45 years of age.

The spine is a complicated structure that provides essential functions such as support, movement, and protection. The spinal cord harm is a disturbing condition that can produce prejudiced or complete loss of control of movements. It could distress the patient's corporeal, psychosomatic, and social well-being. The *lumbar magnetic stimulation* (LMS) is applied to restore motor functions, it is used to map the motor cortex and may decrease the spasticity of the lower limbs, it may help recuperate the bladder control while diminishing the risks of recurrent catheter usage (ScienceDaily, 2019), and it may be used to diagnose lumbosacral motor radiculopathy, or lumbar spondylosis (Davey et al., 1994).

FDA approves MS procedure in bladder rehabilitation but its role is experimental, for the time being. LMS is used also to stimulate the expiratory muscles (Lin et al., 2001) and to control the neurogenic bowel dysfunction in patients with spinal cord injury

(Tsai et al., 2009), chronic back pain, motor function and posture in Parkinson's disease (Hofstoetter et al., 2014; Yang et al., 2018) and to diagnose lumbar spondylosis or lumbosacral motor radiculopathy (Krause et al., 2004). Moreover, LMS is used in neurology for therapeutic purposes and in diagnosis, as it has been proven to help to identify and repairing injured transmission ways in the central and peripheral nervous system, in stimulating the expiratory muscles, in controlling neurogenic bowel dysfunctions, and posture in Parkinson's disease.

In this chapter we present some of our FEM approaches to MS procedures based on 3D, image reconstructed, morphologically accurate computational domains combined with CAD-generated models (applicator, coil geometries). These numerical tools are cost-efficient, noninvasive, bioethics compliant, and easy-to-use for understanding the MS techniques and optimize the associated magnetic field source to create patient-specific solutions.

Magnetotherapy (MT) is a basic physiotherapy procedure (Bednarčík, 2019). In its static form, using a permanent magnet (e.g., Aydin and Bezer, 2011), it was used since time immemorial as one of the natural healing sources. The interest in using it was reinvigorated by the low-frequency *pulsed magnet therapy* (PMT) (e.g., Shupack, 2003; Assiotis et al., 2012), whose effects are up to 100 times more effective than the application of a stationary magnetic field, which elicits PMT as one of the most common physiotherapy methods at present. In chronic pains in degenerative articular diseases, PMT has proven successful as therapy with long-term remedial effect even when other therapy methods were less successful. MT may be recommended for usage in combination with pharmacotherapy, whose effects are, in general, supported by MT. These benefits features advocate the usage of MT in case of a comprehensive approach to treatment, rather than monotherapy. The physiological response of the body to the EMF implies the analgesic, antiedematous, antiphlogistic, trophic, myorelaxant and spasmolytic, vasodilatation effects. A glimpse in the EMF distribution used in MT and the associated heat transfer process is touched through numerical simulation in this chapter too.

7.2 Magnetic stimulation of long cell fibers, a reduced mathematical model

Cable theory and the activating function

Cable theory for cells similar to long cylindrical fibers (Plonsey and Barr, 1988; Malmivuo and Plonsey, 1995; Morega, 1999) as the axons of peripheral nerves, provides a mathematical representation for the equivalent electrical circuit of the nerve and the stimulus, with its active component, the so-called "activating function" (AF), both for electrical stimulation and for MS. It is further intended to highlight the mathematical expressions of AF and to present an analytical calculation method for estimates of AF in a simple concept of PMS, useful for quickly assessing the behavior of different applicators and their effectiveness.

Fig. 7.1A shows the electrical circuit with distributed elements (*lower image*) for an equivalent section of the *cable model* (*upper image*), corresponding to the element of length Δx; the stimulation current i_s, is considered as local injection of charge on the external surface of the membrane.

The inner and external cell currents, i_i and i_e, flow along the x-direction, through the corresponding resistances $r_i\ \Delta x$ and $r_e\ \Delta x$ drive the corresponding voltages $V_i(x+\Delta x) - V_i(x)$ and $V_e(x+\Delta x) - V_e(x)$ at the level of the elementary cable length, while the transmembrane voltage, defined as $U_m = V_i - V_e$, could be determined for each section. The central box, marked with a dashed line and crossed by the current $i_m\ \Delta x$, represents the equivalent circuit for the membrane; it could be shown *at rest* (electrically polarized) using the Goldman–Hodgkin–Katz membrane model, or *active* (depolarized) with a scheme derived from the Hodgkin–Huxley model, as Fig. 7.1B shows in its upper and, respectively, lower diagrams. Membrane equivalent resting conductance $g_m\ \Delta x$, ionic channels conductances, $g_{\mathrm{Na}}\ \Delta x$, $g_{\mathrm{K}}\ \Delta x$, $g_L\ \Delta x$, membrane capacitance, $c_m\ \Delta x$, refer to the elementary length Δx of the long cylindrical fiber, while the voltage generators refer to the resting membrane voltage U_{m0} and to the Nernst electrochemical potentials U_{K}, U_{Na}, and U_L, respectively (where Na and K are the symbols for sodium and potassium ions and L comes from the less significant *Leakage* ions).

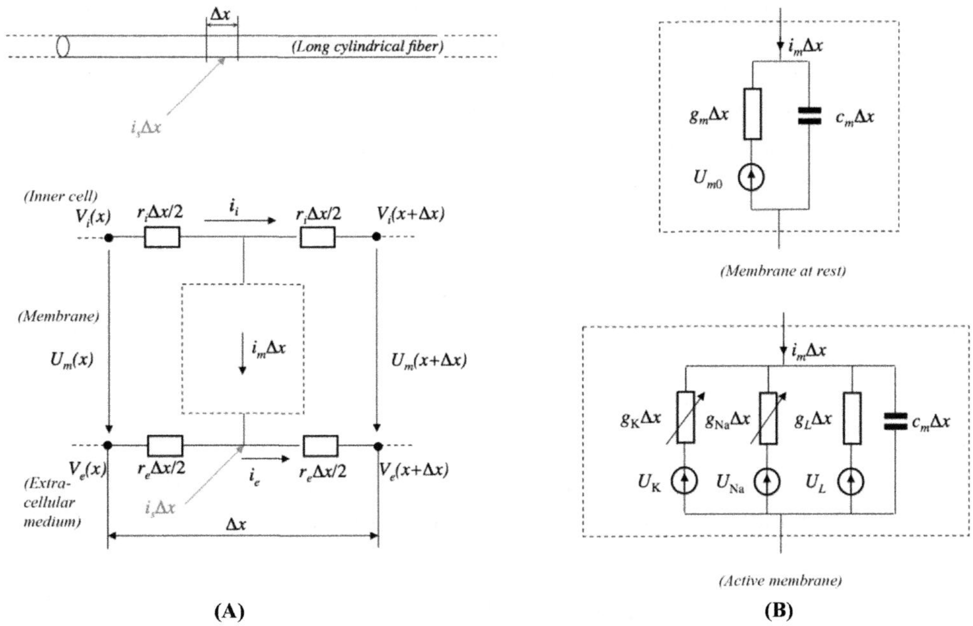

Figure 7.1 The equivalent electrical circuit for the cylindrical long fiber in the cable theory (Morega, 1999). (A) The equivalent electrical circuit of the long cylindrical fiber. (B) The equivalent circuit for the membrane.

Combined Ohm's law and Kirchhoff's theorems applied to the equivalent circuit with distributed elements presented here lead to the spatial and temporal distribution of the transmembrane voltage, which is illustrated by the second-order partial differential equation:

$$\frac{\partial^2 U_m}{\partial x^2} = r_e i_S + (r_i + r_e) i_m, \tag{7.1}$$

and where the expression of the transmembrane current, i_m, shows the difference between the resting and active states of the membrane, as in the following equations:

$$i_m(U_m, t) = (U_m - U_{m0}) g_m + c_m \frac{\partial U_m}{\partial t}, \tag{7.2a}$$

for the resting (polarized) membrane, and

$$i_m(U_m, t) = (U_m - U_{Na}) g_{Na} + (U_m - U_K) g_K + (U_m - U_L) g_L + c_m \frac{\partial U_m}{\partial t}, \tag{7.2b}$$

for the active (depolarized) membrane; Na and K channels conductances are not constant, they depend on time and transmembrane voltage, as the Hodgkin–Huxley model states (Chapter 4: Electrical Activity of The Heart).

When the stimulus is applied, the transmembrane voltage, U_m, rises from the resting value, U_{m0} (specific to the polarized state), toward a threshold value, U_p, which marks the limit between polarized and depolarized states; this process is characterized by Eq. (7.1) with i_m given by Eq. (7.2a). For a sufficiently strong stimulus, the threshold is exceeded and an *action potential* is generated, a process described by Eqs. (7.1) and (7.2b), which is a mathematical model much more complicated than the previous one, solvable only by numerical methods. Practically if only the effectiveness of the stimulus is evaluated, no complexity is needed, and the first path allows the study of the membrane electrical behavior from the rest state to reach the threshold. Equations (7.1) and (7.2a) yield the expression of the polarized membrane behavior under the action of an electric stimulus

$$\frac{\partial^2 U_m}{\partial x^2} = r_e i_s + (r_i + r_e)\left[(U_m - U_{m0}) g_m + c_m \frac{\partial U_m}{\partial t}\right], \tag{7.3a}$$

where, after a few processing, the canonical form of a second-order PDE (Chapter 1: Physical, Mathematical, and Numerical Modeling) is obtained

$$\lambda^2 \frac{\partial^2 u_m}{\partial x^2} - \tau \frac{\partial}{\partial t} u_m - u_m = r_e \lambda^2 i_s. \tag{7.3b}$$

The transmembrane true voltage U_m was replaced considering its variation relative to the resting state, so that the new voltage variable is $u_m = U_m - U_{m0}$, while space and time constants of the membrane $\lambda = \sqrt{1/(g_m(r_i + r_e))} = \sqrt{r_m/(r_i + r_e)}$ and $\tau = c_m/g_m = r_m c_m$ are also identified in the equation; the parameters λ and τ depend on cell dimension (cylinder radius) and dielectric properties.

The right term in Eq. (7.3b) evidence the rise of the transmembrane voltage, from rest to the depolarization threshold, $\Delta u_m = r_e \lambda^2 i_s$, where $AF = r_e i_s$ is evidenced as the AF for electrical stimulation.

Experiments performed by Sudhansu (1990) proved that $\Delta u_m \approx 10$ mV for a successful depolarization of axons. Further processing of AF and in connection with the cable model (Fig. 7.1) one could reach the specific expression of AF for MS.

$$\mathrm{AF} = r_e i_s = (i_s \Delta x)(r_e \Delta x)\frac{1}{(\Delta x)^2} = \frac{r_e \Delta x}{(\Delta x)^2}[i_e(x+\Delta x) - i_e(x) - i_m \Delta x], \tag{7.4a}$$

where $i_m \Delta x < < i_s \Delta x$ looks like a reasonable approximation for the membrane under the depolarization threshold (Morega, 1999); under this assumption $i_m \Delta x$ is neglected, yielding

$$\begin{aligned}\mathrm{AF} &= \frac{r_e \Delta x}{(\Delta x)^2}[i_e(x+\Delta x) - i_e(x)] = \frac{r_e \Delta x}{(\Delta x)^2}\left[\frac{V_e\left(x+\frac{\Delta x}{2}\right) - V_e(x+\Delta x)}{\frac{r_e \Delta x}{2}} - \frac{V_e(x) - V_e\left(x+\frac{\Delta x}{2}\right)}{\frac{r_e \Delta x}{2}}\right] \\ &= -2\frac{V_e(x+\Delta x) - 2V_e\left(x+\frac{\Delta x}{2}\right) + V_e(x)}{(\Delta x)^2} = 2\frac{\partial}{\partial x}\left(-\frac{\partial V_e}{\partial x}\right) = 2\frac{\partial E_x}{\partial x},\end{aligned} \tag{7.4b}$$

$\mathrm{AF} = 2(\partial E_x/\partial x)$ shows that in the MS of a cable-like fiber, the active physical quantity is the spatial derivative of the induced electric field along the fiber (in fact, the tangential component of the electric field strength relative to the fiber direction is derived concerning the same space coordinate as its direction). In these circumstances, a computational model for magnetic stimulation should focus not only on the assessment of the induced electric field but also especially on the derivative of its strength along the fiber.

A computational model for the induced electric field and the activating function

Finding the distribution of the electric field inside a semiinfinite dispersive space (which could be assimilated to an anatomical structure), as the result of electromagnetic induction from an external current-carrying coil is the objective of this subsection. The computational model is based on analytical methods applied in electromagnetism and its importance lies in highlighting some physical aspects that are fundamental for medical procedures based on the generation of the electric field by electromagnetic induction. It was introduced and discussed in Esselle and Stuchly (1992) for applications in PMS and further used for the assessment of the electric field and AF distribution and optimization of applicators relative to the quality of the AF (Stuchly and Esselle, 1992; Morega, 1999; Morega, 2000). The same model could be applied to transcutaneous energy transfer or mild heating of implanted metallic devices for medical purposes.

The computational domain is shown in Fig. 7.2, relative to a Cartesian coordinate system. The biological medium is the homogeneous and isotropic conductive half-space for $z<0$, while $z>0$ corresponds to the other half-space, of air; the magnetic field applicator is represented here by a current-carrying coil with the contour Γ, where $i(t)$ is the variable current and dl is the elemental length of the cable, at a certain position of coordinates (x_0, y_0, z_0). The induced electric field is computed inside the conductive half-space at the location $P(x, y, z)$, for example along a nervous trunk, as is the case in PMS. The Euclidian measure of the distance between the elemental current, idl and point $P(x, y, z)$ is $R=\sqrt{(x-x_0)^2+(y-y_0)^2+(z-z_0)^2}$, and its projection on the air–tissue interface (xOy, $z = 0$ plane) is $\rho=\sqrt{(x-x_0)^2+(y-y_0)^2}$.

Another assumption of the model is the quasisteady regime; the frequency of the excitation current ($f<10$ kHz) and the average tissue conductivity (common values for body tissues at low frequency electric stress, like $\sigma \approx 0.5$ S/m) lead to a penetration depth much larger than the characteristic dimensions of anatomical domains, which in turn are larger than the dimensions of the coils currently used in MS.

The relative balance between the sizes of the coils, the depth of the target for the induced electric field, the operating frequency, the dielectric properties of biological tissues support the simplified assumptions taken here into consideration.

First, the strength of the elemental electric field induced in the conductive half-space, d**E**, due to the time-variable elemental electric current through the coil, idl, is expressed as the sum of the contributions of two elemental components: the incident field, d$\mathbf{E}_1$, and the reaction field, d$\mathbf{E}_2$, each, in turn, formed by the contributions of a solenoidal component and a potential component. At the process scale, the potential part of the incident field and the solenoidal part of the reaction field are negligible, as shown by

$$d\mathbf{E} = d\mathbf{E}_1 + d\mathbf{E}_2 = d\left(-\frac{\partial \mathbf{A}_1}{\partial t} - grad\,V_1\right) + d\left(-\frac{\partial \mathbf{A}_2}{\partial t} - grad\,V_2\right) \cong d\left(-\frac{\partial \mathbf{A}_1}{\partial t} - grad\,V_2\right), \quad (7.5)$$

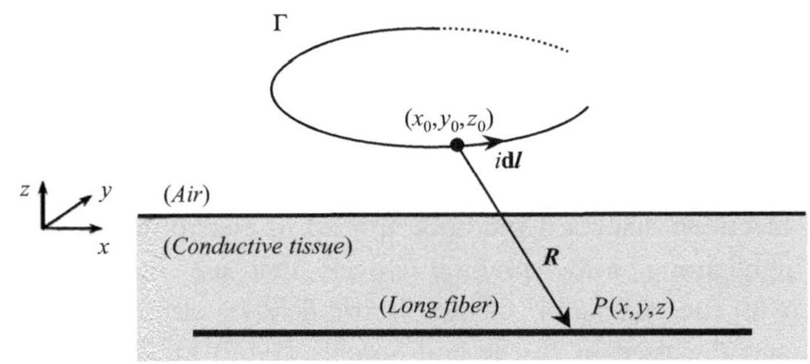

Figure 7.2 The geometry of the computational model for the induced electric field (Morega, 2000).

where $\mathbf{A}_1$ is the magnetic vector potential of the incident magnetic field and V_2 is the electric potential produced by charge separation at the air–tissue interface; the indices are further removed for simplification and Eq. (7.5) becomes

$$\mathrm{d}\mathbf{E} = \mathrm{d}E_x\mathbf{i} + \mathrm{d}E_y\mathbf{j} + \mathrm{d}E_z\mathbf{k} = -\frac{\mu_0}{4\pi R}\frac{\mathrm{d}i}{\mathrm{d}t}\mathrm{d}\mathbf{l} - \mathrm{d}(grad\,V). \tag{7.6}$$

The scalar electric potential could be derived as the solution of the Laplace equation in the conductive half-space by the separation of variables, as presented in Esselle and Stuchly (1992); the complete solution is expressed with Bessel functions. The term $\mathrm{d}(grad\,V)$ needed for the complete computation of Eq. (7.6) is fully derived in Morega (1999) and the result is as follows

$$\mathrm{d}\mathbf{E} = -\frac{\mu_0}{4\pi}\frac{\mathrm{d}i}{\mathrm{d}t}\left\{\left[\frac{\mathrm{d}l_x}{R} + \frac{(x-x_0)\mathrm{d}l_z}{\rho^2}\left(1+\frac{z-z_0}{R}\right)\right]\mathbf{i} + \left[\frac{\mathrm{d}l_y}{R} + \frac{(y-y_0)\mathrm{d}l_z}{\rho^2}\left(1+\frac{z-z_0}{R}\right)\right]\mathbf{j}\right\}. \tag{7.7}$$

As one could observe, the electric field strength has a null component along z-axis direction, normal to the skin surface. In Fig. 7.2, the long cylindrical fiber is shown in the direction of the x-axis, which implies that only the x-component in Eq. (7.7) is important for the evaluation of the AF, that is, the expression of the field derivative $\partial E_x/\partial x$ is only needed and AF results by the integration of its expression along the coil contour, $\mathrm{AF} = \oint_\Gamma \mathrm{d}(\partial E_x/\partial x)$, where

$$\mathrm{d}\left(\frac{\partial E_x}{\partial x}\right) = \frac{\mu_0}{4\pi}\frac{\mathrm{d}i}{\mathrm{d}t}\left\{\frac{(x-x_0)}{R^3}\mathrm{d}l_x + \left[\frac{(x-x_0)^2-(y-y_0)^2}{\rho^4}\left(1+\frac{z-z_0}{R}\right) + \frac{(x-x_0)^2(z-z_0)}{\rho^2R^3}\right]\mathrm{d}l_z\right\}. \tag{7.8}$$

Equation (7.8) is built as the differential of the electric field component, derived from Eq. (7.7) on the same x-direction.

The activating function produced by circular coils

Coils commonly used in PMS are wound with circular turns, either concentrated or distributed in various modes, such as double coil (in the shape of figure eight) or quadruple coil (like a flower) with the turns in a parallel or inclined positions relative to the body surface, like a butterfly or o slinky coil (e.g., Ren et al., 1995). To compute AF as shown here, Eq. (7.8) must be integrated on the contour of the stimulation coil (all ampere-turns) and that respective contour must be described by a suitable geometry capable of providing a convenient expression for the analytical calculation, or the integration must be approached numerically. In Morega (2000) change of variables is applied for different geometries and positions of coils with circular turns. According to Fig. 7.3, the elementary turn is circular, of radius r; its basic position, shown in Fig. 7.3A, is coplanar with the reference plane (xOy, $z = 0$), tangential to the axes. The turn can be rotated at an angle α, keeping point A fixed, as shown in Fig. 7.3B; "d" is a reference axis attached to the moving turn.

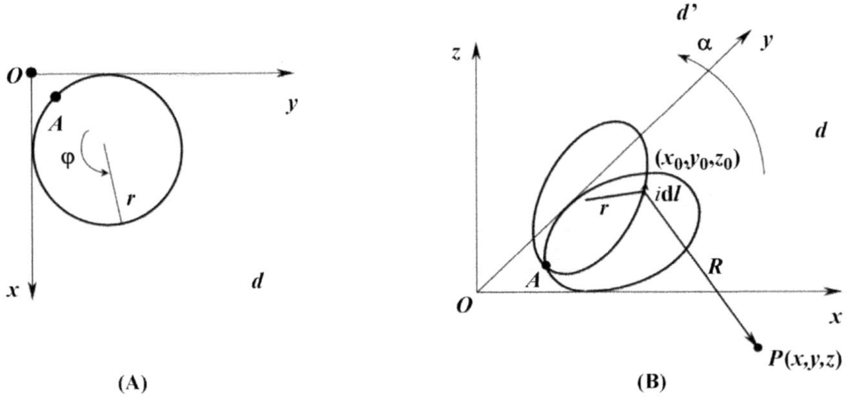

Figure 7.3 Geometric characteristics of the elementary circular turn of a stimulating coil (Morega, 2000). (A) $\alpha = 0$, the turn is in the (xOy, $z = 0$) plane. (B) The turn can rotate by $0 \le \alpha \le \pi/2$, with A as a fix point.

From the geometry shown in Fig. 7.3, the equations connecting the variables of the geometry described in Fig. 7.2 to the new variables (angles α and φ) are given as follows:

$$OA = r\left(\sqrt{2} - 1\right), \tag{7.9a}$$

$$x_0 = \frac{r}{\sqrt{2}}\left(\sqrt{2} - 1 + \cos\alpha + \sin\varphi - \cos\alpha\cos\varphi\right), \tag{7.9b}$$

$$y_0 = \frac{r}{\sqrt{2}}\left(\sqrt{2} - 1 + \cos\alpha - \sin\varphi - \cos\alpha\cos\varphi\right), \tag{7.9c}$$

$$z_0 = r\sin\alpha(1 - \cos\varphi). \tag{7.9d}$$

$$\begin{aligned} \mathrm{d}\mathbf{l} &= \mathrm{d}l_x\mathbf{i} + \mathrm{d}l_y\mathbf{j} + \mathrm{d}l_z\mathbf{k} = \mathrm{d}x_0\mathbf{i} + \mathrm{d}y_0\mathbf{j} + \mathrm{d}z_0\mathbf{k} \\ &= \left[\frac{r}{\sqrt{2}}(\cos\varphi + \cos\alpha\sin\varphi)\mathbf{i} + \frac{r}{\sqrt{2}}(-\cos\varphi + \cos\alpha\sin\varphi)\mathbf{j} + r\sin\alpha\sin\varphi\,\mathbf{k}\right]\mathrm{d}\varphi. \end{aligned} \tag{7.9e}$$

With the new variables, Eq. (7.8) becomes

$$\begin{aligned} \mathrm{d}\left(\frac{\partial E_x}{\partial x}\right) &= \frac{\mu_0}{4\pi}\frac{\mathrm{d}i}{\mathrm{d}t}r\left\{\frac{x - x_0}{R^3}\frac{\cos\varphi + \cos\alpha\sin\varphi}{\sqrt{2}}\right. \\ &\left. + \left[\frac{(x-x_0)^2 - (y-y_0)^2}{\rho^4}\left(1 + \frac{z - z_0}{R}\right) + \frac{(x-x_0)^2(z - z_0)}{\rho^2R^3}\right]\sin\alpha\sin\varphi\right\}\mathrm{d}\varphi, \end{aligned} \tag{7.10}$$

with $R = \sqrt{(x-x_0)^2 + (y-y_0)^2 + (z-z_0)^2}$ and $\rho = \sqrt{(x-x_0)^2 + (y-y_0)^2}$ as defined earlier. The total contribution of the coil ampere-turns to AF results in any point $P(x,y,z)$ by the integration of Eq. (7.10), only after the variable represented by the φ angle, on a complete circumference, from 0 to 2π, and following all the turns of the coil magnetic field applicator.

The spatial distribution of AF depends on the geometry of the magnetic field applicator. Computational models like the one presented here are helpful during the design process of the applicator and the electric circuit, by the control of shape, size, the position of the coils and the equivalent circuit elements (inductance and resistance). An optimization process could be easily conducted on criteria as follows: (1) maximization of stimulation efficiency through the control of some parameters: AF peak directed to the target zone for stimulation and precisely focused, while minimizing side effects (other peaks with lower amplitude or with opposite polarity); (2) handling the applicator easily and without risk, by highlighting functional correlations between design aspects and physical characteristics of the AF distribution inside the body; and (3) the design of the electric circuit that includes the applicator for optimal morphology of the AF waveform, lowest energy consumption, lowest heating, and adequate time constants of the therapeutical process.

Example of activation function distribution inside the body

Illustrative results obtained with the assessment method presented here were published by Esselle and Stuchly (1992) and Stuchly and Esselle (1992) for applicators with circular and rectangular coils, either parallel with the air-body demarcation surface or perpendicular to it, purposing the maximization of the stimulating main peak, while minimizing the occurrence or effect of other nontherapeutical peaks of the AF spatial distribution. In a study by Morega (2000) an analysis of some applicators with circular coils was performed, aiming to compare the efficiency of operational performances. Two indices are defined: (1) *focalization* (*F*) that quantifies the spatial concentration of AF and is equivalent to the target tissue area; *F* is determined by the projection of the negative AF peak on the plane that includes the embedded nerve—it should be minimized for best targeting the action on the intended nerve region and (2) *efficiency ratio* (ER) defined by the absolute value of the ratio between the negative and positive peaks of AF in neighboring areas—it should be maximized to avoid hyperpolarization or chaotic stimulation in the proximity. Three types of applicators were compared (as shown in Fig. 7.4): (1) the concentrated single-coil, (2) the double coil (figure eight), and (3) the quadruple coil (like a flower). Besides the geometry of the coils (left column), Fig. 7.4 presents their performances: corresponding AF spatial distribution in the plane where the long cylindrical fiber is embedded (xOy, $z = -0.01$ m) (central column) and the distribution along the fiber (x-axis direction, at $y = 0$, $z = -0.01$ m) where the highest negative peak occurs on the nerve (right column).

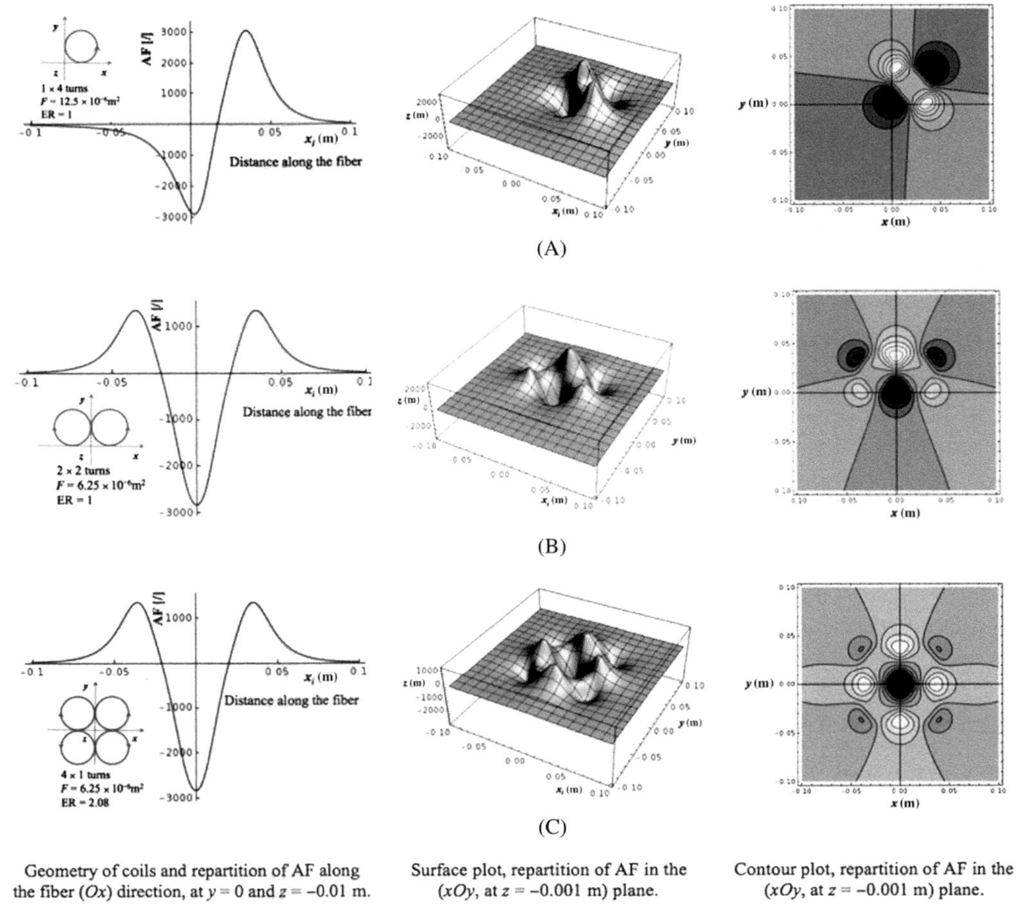

Figure 7.4 Assessment of operational characteristics for three magnetic applicator coils (Morega, 2000). (A) Concentrated coil. (B) Double coil (figure of eight shape). (C) Quadruple coil (flower-like shape). *AF*, Activating function.

The superiority of the quadruple coil is evident, for the following criteria: the best focalization $F = 6.25 \times 10^{-6}\ \text{m}^2$, the same as for the double coil but half compared with the single coil, and the highest efficiency, with ER = 2.08, compared with 1 for the other two configurations. The negative peak of AF is located at the cross-section of its symmetry axes ($x = 0$, $y = 0$ in the considered coordinate system) that ensures more precision and safety to the procedure. It is expected that quadruple coils with an eccentric position of turns could better concentrate AF at the intersection of their axes of symmetry, as Zhang and Edrich (1996) illustrates with eccentric, yet concentrated coils. They also suggest a different problem that occurs in MS: the uneven distribution of AF in the depth of the tissue (z-axis direction). For horizontal turns, the peak values of AF are maxim at the skin surface and decrease (approximately exponentially) in depth. That could cause unpleasant, even painful reactions at the skin surface, a region rich in sensitive nerves.

A possibility to reduce AF values at the skin surface and increase them in depth is the use of combined coils, made of horizontal and inclined turns. This is the reason for considering in the model the possibility to rotate the turns (Fig. 7.3) with the alpha angle to the skin surface. Fig. 7.5 presents the distribution of AF in the depth of the tissue (z-axis direction) for three cases: case (a) the quadruple coil introduced earlier (it has eight turns, two per section); case (b) a combination of a similar quadruple coil with three turns for each section and a quadruple coil with one turn for each section, with the turns inclined ($\alpha = \pi/3$) and with opposite current polarity; and case (c) the same combination as (b), but with double magnitude for the radius of turns.

As Fig. 7.5 shows, the rate of AF values at $z = 0$ and at $z = -0.01$ m (the considered target area) is $\sim$2.5 in case (a), 1.8 in case (b), and 1.15 in case (c). The inclination and radius of turns could provide good control for the localization of the AF values in the depth of the tissue.

The quadruple coils produce the best concentration of AF at the target stimulating area, minimizing the inhibition effects on surrounding regions. The combination of quadruple coils with horizontal and inclined turns could result into the control of the

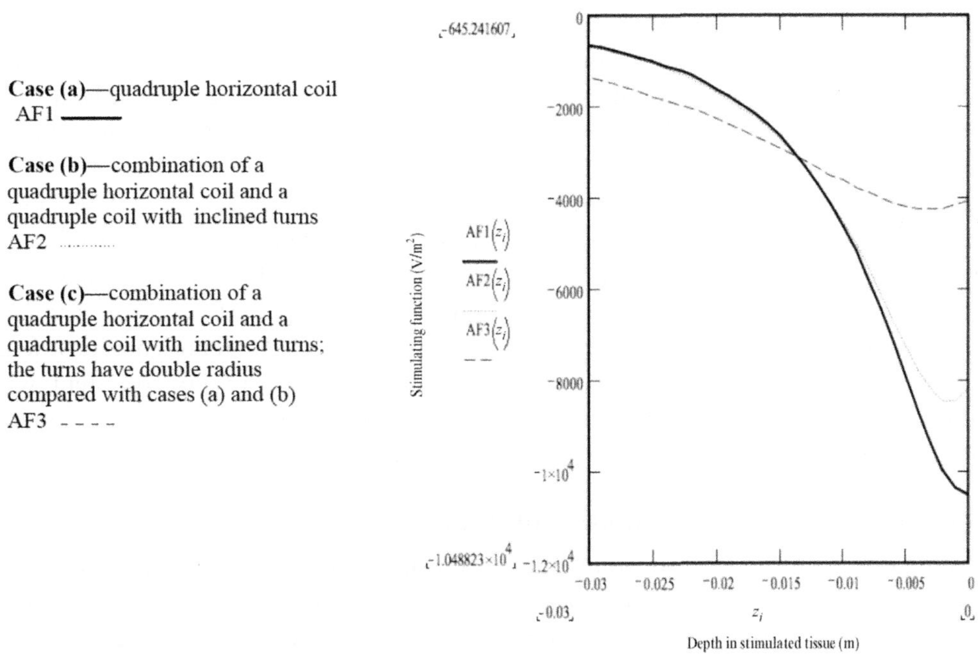

Figure 7.5 Distribution of *AF* in the depth of tissue for three forms of the stimulating coils (Morega, 2000).

magnitude of AF in the depth of stimulated tissue and could concentrate the peak values at the fiber depth level, rather than at the skin surface. It is possible that the forms of the coils could be refined, considering quadruple coils with eccentric turns. This geometry requires a slight modification of Eqs. (7.5)−(7.7), for circular turns that are not tangential to the coordinate axes.

The analytic solutions introduced in Section 7.2 provide valuable insights into the MS aiming the activation of excitable tissues (nerve, muscle, or sensitive cells), produced by applying an electric stimulus near the cell membrane, which is produced noninvasively, by electromagnetic induction. The main features of the electromagnetic field and the electric current impulses released through a stimulation coil, which is placed at the surface of the skin are the key results. However, numerical modeling may be required to analyze the MS of the spinal cord, when computational domains are to approach more realistic. Either CAD or image reconstruction of the lumbar spine is needed to predict the distribution of the eddy current impulses. The sequel provides a glimpse into such MS studies.

7.3 Magnetic stimulation of the spinal cord

Numerical modeling may be used as an assertive noninvasive evaluation of effects, pre-operational stage in designing the LMS. To provide consequential insights, the assessment has to be patient-related and to this aim customized computational domains are valuable. This subsection concerns the numerical simulation of LMS when the computational domain is patient-related, constructed using medical imaging techniques. The magnetic field and the electric field are calculated to guide the optimal positioning and adjustment of the external excitation coils that are used to excite the lower spinal cord. Attention is devoted to the numerical modeling of LMS using patient-related anatomic information, to add predictive valences to the available analytic models presented before. The computational domain used for the lumbar spine is obtained by imaging reconstruction of the spine, fused with CAD constructs for the coils (Baerov et al., 2019). Different studies rely on CAD representations (e.g., Darabant et al., 2013; Soltoianu, 2018). Although producing results consistent with known understanding and knowledge, CAD models are less relevant for patient-related LMS evaluations.

Modeling the lumbar magnetic stimulation

Turning our attention to the computational domain, DICOM computed tomography (CT) images in *axial*, *coronal*, and *sagittal* planes (e.g., WholeScan, 2017) are used as input data. Fig. 7.6 presents the five vertebrae of interest and their associated intervertebral discs, the spinal cord, and the lumbar spinal nerves of concern, for one (Fig. 7.6, left) and two (Fig. 7.6, middle) MS coils. When two coils are used (Fig. 7.6, middle)

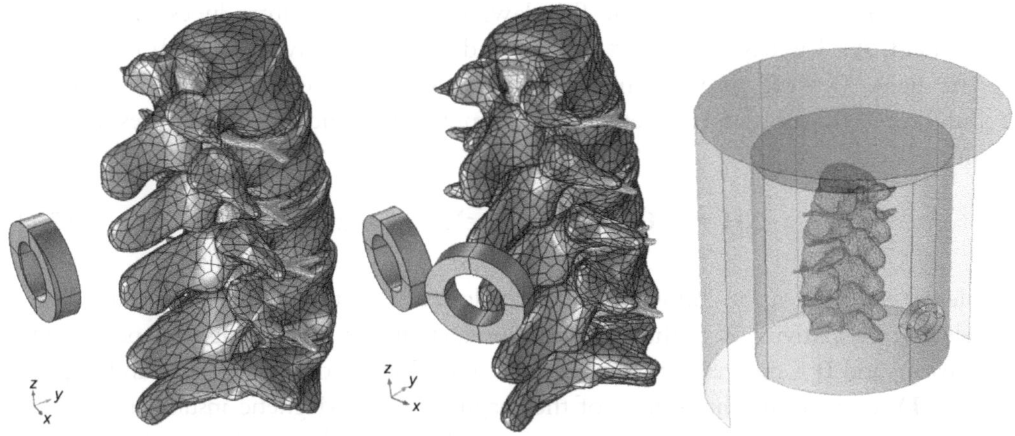

Figure 7.6 Lumbar spine models, for one (left) and two (middle) LMS coils, and the FEM computational domain (Baerov et al., 2019). The FEM mesh is made of ~3,000,000 tetrahedral quadratic elements for the geometry. *FEM*, Finite element method; *LMS*, lumbar magnetic stimulation.

these are oriented such that their axes focus on the nerve outline. The construction process is detailed in Chapter 3, Computational Domains.

The complexity of the model created in the first place using a medical planning model builder (e.g., 3D Slicer, 2020) was reduced using the "Quadric Edge Collapse Decimation" and "Screened Poisson Surface Reconstruction" of MeshLab (2020) that creates "watertight" surfaces and smoothes out the model. Then a CAD tool (e.g., Autocad, 2020) is used to convert these constructs to solid entities, assemble the entire lumbar spine model, add the MS coil(s), and present it in FEM compatible format to the FEM solver (e.g., Comsol, 2020).

The most frequently used stimulation coil is the circular one. It is used to stimulate a larger area without the need for precise positioning but with a limited focus. The "eccentric figure-eight coil," which uses two (noncoplanar, "butterfly") circular coils, provides for more intense stimulation in the region where the axes of the two coils come across (Maccabee et al., 1991; Sekino et al., 2020). Other MS methods are available or under investigation—for instance, implanted pulse generators used for spinal cord stimulation (Lempka and Patil, 2018; Risson et al., 2018; Mayfield, 2020).

The size and complexity of the spine construct may exceed regular, available computational resources therefore the containing torso segment is here replaced with an equivalent cylindrical enclosure, as is the containing volume of the magnetic field, Fig. 7.6, right.

LMS may use pulsed, PWM (pulse width modulated), or harmonic stimuli. Here the excitation is presented as harmonic (ac) at $f = 100$ Hz—a sinusoidal signal per se or the fundamental harmonic of a PWM pulses train.

The harmonic quasistationary EMF model (Chapter 1: Physical, Mathematical, and Numerical Modeling) is described through the magnetic circuit law, $\nabla \times \underline{\mathbf{H}} = \underline{\mathbf{J}}$, where $\underline{\mathbf{J}} = \sigma\underline{\mathbf{E}} + j\omega\underline{\mathbf{D}}$, the magnetic flux law, $\underline{\mathbf{B}} = \nabla \times \underline{\mathbf{A}}$ (it introduces the magnetic vector field, **A**), Faraday's law, $\underline{\mathbf{E}} = -j\omega\underline{\mathbf{A}}$, and the constitutive (material) laws for the electric field, $\underline{\mathbf{D}} = \varepsilon\underline{\mathbf{E}}$, and the magnetic field, $\underline{\mathbf{B}} = \mu\underline{\mathbf{H}}$, Mocanu (1982) yields the Helmholtz PDE for the magnetic field strength, **H** (underscore denotes complex quantity)

$$\Delta\underline{\mathbf{H}} = \gamma^2\underline{\mathbf{H}}, \quad \gamma^2 = \omega^2\mu\varepsilon^*, \quad \varepsilon^* = \varepsilon + \frac{\sigma}{j\omega}. \tag{7.11}$$

Here $j = \sqrt{-1}$, $\omega = 2\pi f$ is the angular velocity, ε^* is the complex permittivity, ε is the electric permittivity, σ is the electrical conductivity, μ is the magnetic permeability (free space), and $\underline{\mathbf{B}}$ is the magnetic flux density. The electric conduction, $\sigma\underline{\mathbf{E}}$, and displacement, $j\omega\underline{\mathbf{D}}$, currents are the sources of the magnetic field. Magnetic insulation boundary condition closes the model ($\mathbf{n}\cdot\underline{\mathbf{B}} = 0$, where **n** is the outward normal to the boundary).

The electromagnetic field produced by the coils diffuses through the body and causes the excitation of neurons (Ugawa et al., 1989; Maccabee et al., 1991). The induced electric field depends on the coil type (shape), its position, and, of course, on the electric properties of the tissues. The working frequency is $f = 100$ Hz, and Table 7.1 lists the properties that are used (Gabriel, 1996; Andreuccetti et al., 2020).

Due to the limitations in the hardware resources available at the time (Baerov et al., 2019), the thorax is presented as an ensemble of equivalent, homogenized subdomain. Its electrical permittivity and conductivity have average, volume-weighted values over the anatomic subdomains. The spinal cord and spinal nerves (Liu et al., 2015) are homogenized too, their properties have average, volume-weighted values of the nerve, white matter, gray matter, and cerebrospinal fluid, respectively (Table 7.2).

Table 7.1 Volume-weighted average electrical properties for different anatomical regions at $f = 100$ Hz.

Tissue	Relative permittivity (–)	Electrical conductivity (S/m)
Vertebrae	5.85×10^3	2.01×10^{-2}
Intervertebral disc	6.1×10^1	8.30×10^{-1}
Spinal cord and spinal nerves	1.51×10^6	5.43×10^{-1}
Thorax, average	4×10^3	3.3×10^{-1}

Table 7.2 Electrical properties for tissues used in numerical simulations.

Tissue	Relative permittivity (–)	Relative permittivity (–)
Nerve	4.66×10^5	2.8×10^{-2}
White matter	1.67×10^6	5.81×10^{-2}
Gray matter	3.91×10^6	8.9×10^{-2}
Cerebrospinal fluid	1.09×10^2	2

The applicator coils have 20 turns each, with 4 mm^2 overall cross-sections areas. The calculated inductance of such a coil is 10.89 μH, consistent with values reported in the literature (e.g., Darabant et al., 2013). They are powered by ac voltage sources, at 5 V (effective value), which keeps the electric current density for the coil conductor (Copper) within safe limits.

The PDE Eq. (7.11) was integrated numerically using FEM, Comsol. Linear elements for the magnetic field (vector potential) were used. The algebraic system was solved using a Flexible Generalized Minimal Residual solver Morikuni et al. (2012), without preconditioning.

Numerical simulation results

Fig. 7.7 singles out the lumbar spine and shows a slice distribution of the magnetic flux density (left). When a single coil is used, $B_{max} \sim 3.5$ mT, whereas for two, focused coils, $B_{max} \sim 4.7$ mT. As the mathematical model Eq. (7.11) is linear and the electric and magnetic properties are constant, the stimulation levels for other excitations are readily available, by scaling the solution with the scaling factor of the field source (ampere-turns, or "total" current of the applicator). The cut-line in Fig. 7.7, middle is used to approximately draw the magnetic (and electric) field along the nerve principal direction (its real trajectory is much more complex but traceable though).

The magnetic flux density (Fig. 7.7, right) shows off a bell-shaped profile, with maxima consistent with reported results in similar cases (e.g., Darabant et al., 2013).

The electric field strength (its effective value) is rendered in Fig. 7.8 using slices selected to showcase the distribution in a vertical plane that passes through the nerve's central branch, and in horizontal planes elected to intercept the horizontal prolongations (xOy plane) of the nerve.

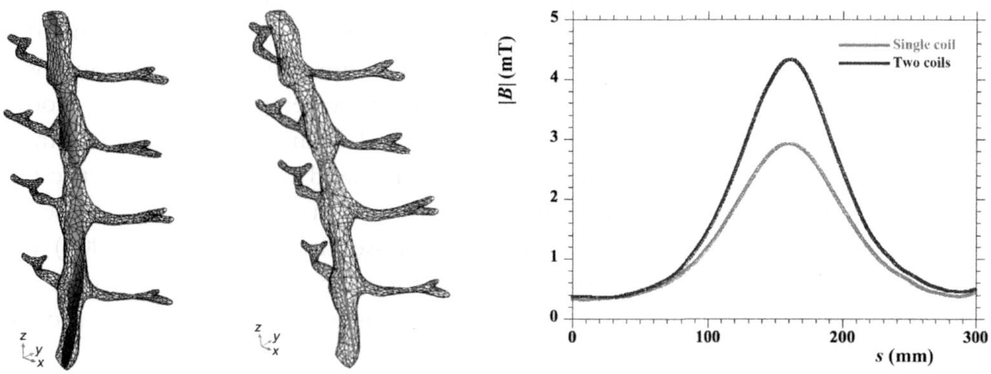

Figure 7.7 A slice distribution of the magnetic flux density through the spine (left) and the magnetic flux density (right) along a line traced through the spine (middle).

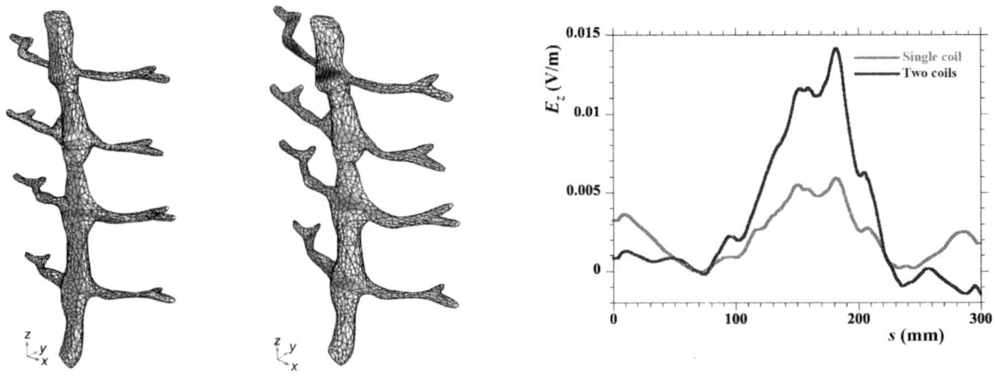

(A) (B) (C)

Figure 7.8 The electric field strength in LMS obtained by numerical simulation. (A) One coil, $E_{z,max}$ ~7 mV/m. (A) Two coils, $E_{x,max}$ ~16 mV/m. (C) E_z component, along the cut-line (Fig. 7.7, middle). *LMS*, Lumbar magnetic stimulation.

The vertical (Oz) and the horizontal (Ox) components of the electric field strength, respectively, are represented through color map slices in Fig. 7.8A and B. Local maxima, where the higher current density is expected to occur, are identifiable. Moreover, Fig. 7.8C presents the E_z component of the electric field strength profile (effective value), which is the stimulus delivered by the LMS coil(s), along the cut-line depicted in Fig. 7.7, middle. The "shape-8" coil generates a stronger, and of a higher gradient, electric field, and stimulation current.

7.4 Transcranial magnetic stimulation

The increasing interest in TMS motivates the development of numerical simulations aiming to analyze and optimize the magnetic field source for different applicators and setups. To apprehend the outlining processes, 3D models using simplified geometries may be valuable in analyzing the induced current within the gray matter volume during the TMS (e.g., Wagner et al., 2004). However, as for the LMS, when purposing to contemplate the procedure in a preinterventional stage, patient-specific models, numerical "twins," ought to be used. Numerical models and computational domains of the head segmented out of an MRI image sets are used to unveil the electric field strength and electric current density maps that show off the EMF profile used in TMS. For instance, interest in this area is also expressed by Salinas et al. (2009) who evaluates the induced electric field (the source of the induced stimulating electric currents) produced by TMS, using the boundary element method (BEM) instead of the more popular FEM.

The study shows that the computational domain morphology is crucial: an anatomically accurate geometry outlines the presence of an Oz (vertical) component of the induced electrical field, while simplified spherically symmetrical models of the head concluded the absence of this. The same observation is highlighted by Nummenmaa et al. (2013), who also contributes with more information on this topic by specifying that the volume conductor models (Chapter 4: Electrical Activity of The Heart) influence the numerical evaluations of the electric field produced by TMS. BEM is a good compromise between numerical accuracy and computational cost and the anatomically accurate geometries are better for TMS navigation, especially when considering prefrontal regions of interest, usually targeted in medical therapies.

Further on, Yang et al. (2006, 2007, 2010) and Xu et al. (2005) developed numerical models for the analysis and geometrical optimization of the magnetic field source used in TMS. Three different coil configurations were studied, starting from a single circular coil placed above the scalp, then moving on to coil arrays based on two and seven circular coils placed above the head. The best behavior was achieved with the array made of seven coils, which generated the most intense stimulation currents and the lowest magnetic field attenuation in-depth when compared with other simulated coil setups.

Other studies are focused on more detailed aspects of the TMS magnetic field source. For example, Chen and Mogul (2009) embeds a highly detailed 3D geometry of the brain in a FEM numerical study for TMS, created using image-based segmentation and optimized meshing. The level of anatomical accuracy goes deep down to the *cerebral gyri* and *sulci* all reached by a skillful combination between CT and MRI image sources. Regarding the TMS magnetic source field, an in-depth study of the magnetic coil numerical modeling, published by Petrov et al. (2017), presents the effects of idealized coil geometries upon the generated stimulating magnetic field. Thus after analyzing three coil geometries (a simple circular coil, a coil with in-plane spiral winding turns, and one with stacked spiral windings) the FEM results showed that notable differences occurred. The numerical results were empirically validated.

There are different TMS protocols—applicator, position, duration, etc. For instance, in rTMS the applicator is positioned above the left inferior frontal region and uses frequencies in the range 8–25 Hz for 10 s (Hallet, 2000). The "continuous theta-burst stimulation" consists of trains of uninterrupted TBS (e.g., 20 s) with bursts of three pulses at 50 Hz, repeated every 200 ms (i.e., 5 Hz), for a total number of pulses (e.g., 300 pulses; Noh et al., 2015). The numerical simulation presented next is about a continuous harmonic stimulation, at 10 Hz, using a planar, circular coil applicator.

Modeling the transcranial magnetic stimulation

The computational domain presented next, a numerical phantom, was created using imaging-based segmentation techniques applied on a high-resolution CT dataset, Fig. 7.9, the upper row (Chapter 3: Computational Domains). Three types of tissue are targeted: the brain, the cranium bones, and the surrounding tissue (a homogenous muscle, fat, and skin compound). The masks produced, after conveniently setting the threshold filter parameters, are presented in Fig. 7.9, middle row. The limited level of detail concerning the complexity and functionality of the real case is due to the limited accuracy of the images set and computational resources available at the time. This simplified geometry brings out a good balance between the computational costs and the excessive computational domain idealization, which can provide for misleading results and false conclusions, as already stated in other papers (Wagner et al., 2004; Salinas et al., 2009; Nummenmaa et al., 2013).

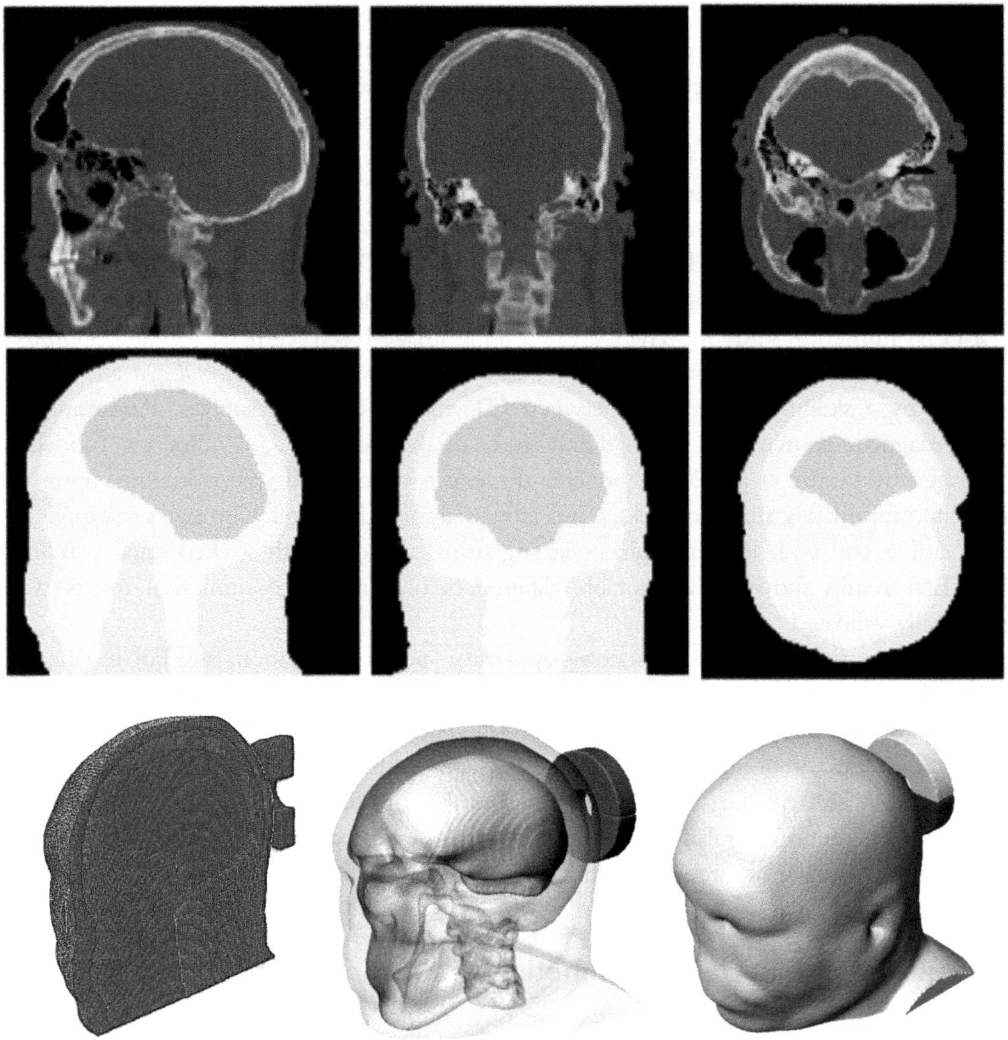

Figure 7.9 The head and the applicator used in TMS numerical simulation. *TMS*, Transcranial magnetic stimulation. Source images (upper row) and label maps (middle row). The bottom row: FEM mesh (left), computational domain (middle and right).

Table 7.3 Properties for tissues used in the numerical simulation of the TMS.

Tissue	Electric permittivity (F/m)	Electric conductivity (S/m)
Brain	1475	0.105
Bone	200	0.006
Muscle	3000	0.04

TMS, Transcranial magnetic stimulation.

The applicator is a circular coil placed in the occipital region close to the scalp, for TMS therapy aimed at improving the tactile sensations of visually impaired or blind subjects (Beckers and Hömberg, 1991; Ptito et al., 2008; Mullin and Steeves, 2011). The coil is CAD-generated and manipulated until the desired region is achieved. Thus a hybrid computational model is obtained (Fig. 7.9, the bottom row), comprised of an image-based segmented, anatomically accurate morphology of the head's ROI, and a CAD coil model.

The computational domain is enclosed by a containing volume, which represents the surrounding space. The hybrid 3D solid model was then FEM-discretized with a mesh of ~216,000 tetrahedral elements (Fig. 7.9, bottom row, right) and imported in the FEM solver.

The EMF model is time-harmonic, presented for the magnetic field strength (in complex, simplified form), $\underline{\mathbf{H}}$, described by the Helmholtz Eq. (7.11). The simplified geometry is associated with homogenization techniques used to model the different tissues as domains with equivalent properties. Hence, anatomic domains are homogeneous media with uniform, equivalent properties. Their electrical permittivity and electrical conductivity are averaged, volume-weighted values of the constituent anatomical regions: cortical bone, trabecular bone, and bone marrow for bone, and skin, fat, muscle, and blood for soft tissue, calculated at $f = 10$ Hz, Table 7.3, using ITIS (2020).

The boundary condition that closes the problem is magnetic insulation ($\mathbf{n} \cdot \underline{\mathbf{B}} = 0$). The inductor electric current density is set 5×10^5 A/m^2, such that the applicator provides for 2600 kA-turns. This excitation produces a maxim magnetic flux density of ~50 mT, at 10 Hz.

It should be noted that the EMF problem inside the head is here of interest and not the EMF, heat transfer, and mechanical stability of the applicator itself. Moreover, because the PDE Eq. (7.11) presents a linear problem, its solution (the magnetic field strength and the magnetic flux density) is proportional to the excitation. Therefore if other levels of stimulation are acquired, say 1−1.5 T, then the inductor ampere-turns have to be scaled correspondingly. Then, the results shown next need just to be scaled—the EMF problem is not to be solved again.

Numerical simulation results

The mathematical model was integrated using FEM, in Galerkin formulation (Comsol, 2020). Quadratic and cubic vector elements were used to compensate for

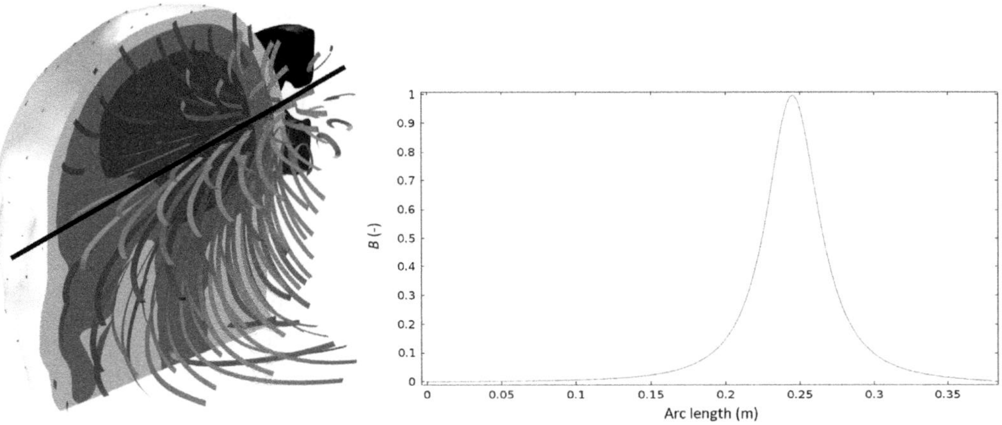

Figure 7.10 Magnetic flux density (right) along the symmetry axis (left) of the applicator—nondimensional values.

the limited resolution of the numerical grid. The magnetic flux density, scaled by its maximum, along the axis of the applicator is seen in Fig. 7.10. The peak value is recorded at the applicator level.

The magnetic flux density, the electric field strength, and the electric current density produced by TMS have the same orientation as the inductor current in the coil, Fig. 7.11. The local electric properties (permittivity and conductivity) "morph" the electric field strength tube lines into the tube lines of the electric current density larger noted beyond the skull, inside the head, in the occipital lobe region.

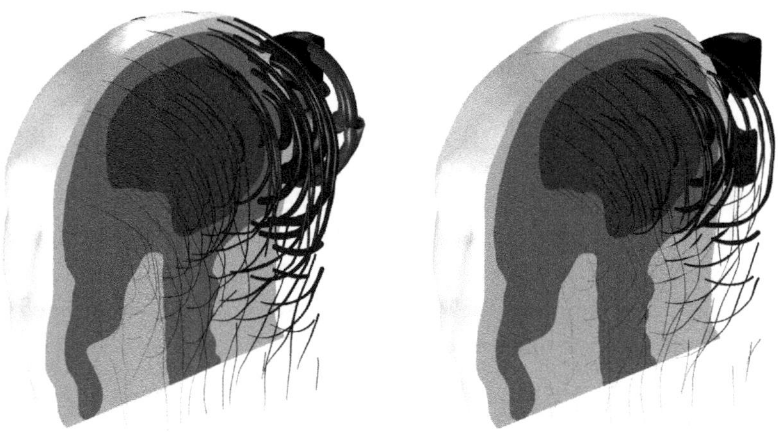

Figure 7.11 The electric field produced through TMS: the electric field strength (left) and the electric current density (right). *TMS*, Transcranial magnetic stimulation.

These results outline processes that could hardly (if possible) be perceived using other means and support the merit of the numerical simulation as an assistive, valuable aid in the preparing the TMS and evaluating the main, stimulating effect as well as the side effects—thermal, mechanical, etc.

7.5 Magnetic therapy

Magnetic field therapy (MFT) is using EMF exposure at 10–100 Hz (up to 170 Hz) such that the magnetic field component prevails the electric field component. However, in case of metallic implants, especially those that are fixed in soft tissues and are not made of antimagnetic materials, the magnetic component may be significant. In this frequency range, biological entities are diamagnetic and their molecules orient to minimize the field energy. In biological tissues, such actions are against the bonds between atoms, molecules, and ions, which accordingly influence the cellular processes (Monzel et al., 2017; Fei et al., 2019; Peng et al., 2019).

The physiological response of the body to the application of MT comprises multiple effects: analgesic, antiedematous, trophic (acceleration of healing by growth), vasodilation and muscle relaxation, reduced pain, costs, and duration of treatment (Markov, 2007; Chalidis et al., 2011; BTL, 2020).

Time-variable magnetic fields are significantly more effective than static magnetic fields for various therapies, and they can be combined with other physiotherapy procedures or with pharmacotherapy. The vascular tree (including lymph and blood), the peripheral nerves, the central neural system and its path—the moving conductor—and, not the least, the individual ions and charges on cellular membranes—the traveling charges—are electrically conductive constituents, such that in variable magnetic field physiotherapy, the electrodynamics effects are important.

Different types of magnetic field excitations are used: stationary, for increased bleeding conditions, acute states, postoperative conditions; alternating, for nerves or muscle distortion, triangular, for cartilages or tendons dysfunctions, and pulse, for bone diseases. However, for pulse MF it is yet to clarify whether the electric field component is more intense, prevailing.

As for TMS and LMS, the MFT uses applicators (electric current-carrying coils) to provide focused exposure to magnetic field (Markov, 2007; Krawczy et al., 2017; Physiomed, 2020). The coils function may be enhanced using permanent magnets embedded within the disc-type applicators. The applicators are placed as close as possible to the patient's body, to limit the dispersive magnetic flux. The minimum exposure time is at least 10 min, and the total daily exposure should not exceed 40 min (BTL, 2020). For best results exposures should be repeated.

In the electrodynamics of conductive moving media and variable magnetic fields, the electric currents are accompanied by heating effects, and in this section, we are concerned with this possible side effect of the MFT when applied, for example, for postsurgical healing of a fractured femur shaft. This kind of fracture is a severe handicap, the associated recovery lasts and it is accompanied by immobilization and discomfort. The specific treatment is, in general, surgical and purposed to stabilize the fractured bone with metallic structures. Plates and screws outside the bone or an external fixator are the common solutions (Femur, 2020; Uptodate, 2020) but, in severe cases, a rod is introduced in the center of the medullar canal and fastened with screws (Su et al., 2015). Intensive numerical modeling is devoted, in the presurgical phase, to predict the structural stability and dynamic response of such consolidated structures under various solicitations, mainly mechanical (e.g., Coquim et al., 2018).

The unavoidable immobilization rapidly leads to the depreciation of the muscle, the reduction of its volume and its physical fitness, and even musculoskeletal and neuronal disorder can occur. Physiotherapy may help speeding up restoring mobility, and one of the most popular and successful is MFT—used also in orthopedics and rheumatology or for the treatment of internal diseases. MFT has been proven to help relieve pain and accelerate the healing time. It has an important influence on trophic stimulation of collagen and bones by producing microcurrents that speed up osteogenesis (èada-Tondrya, 2019; Baerov et al., 2020; Efisioterapia.Net, 2020).

We touch also the possible amplification of the side effects of the MFT due to the electromagnetic heating of metallic implants that are the siege of induced currents. Numerical simulations are conducted using computational models that represent the anatomy of the upper leg and an orthopedic implanted device that is exposed to a harmonic magnetic field—first a sketchy 3D CAD construct and then a medical images-based reconstructed domain.

Modeling the magnetic field therapy

For a CAD rendering of the anatomy, the tissue is approximated through a cylinder that models the soft tissues, which hosts another cylinder that represents the femur, Fig. 7.12, left. Stainless steel and titanium plates are usually used (Sahoo et al., 1994) but, recently, there is a growing in using platinum in medical applications due to biocompatibility, inertness within the body, durability, electrical conductivity, and radiopacity (Cowley and Woodward, 2011), including orthopedics (spinal fixation; hip implants, knee implants—Biomet, Johnson & Johnson, Stryker, and Zimmer producers). Here, a platinum plate (parallelepiped), which mimics the fixator, is attached to the femur—screws and other ancillary parts are discarded. An outer cylindrical enclosure, capped (top and bottom) with infinite elements subdomains, contains the anatomic volumes and closes the MF.

Once the outlining MFT processes are evidenced using this CAD construct, a realistic model is essential to achieve consistent results for medical applications to add predictability and enhance the diagnosis accuracy. To this aim, reconstruction

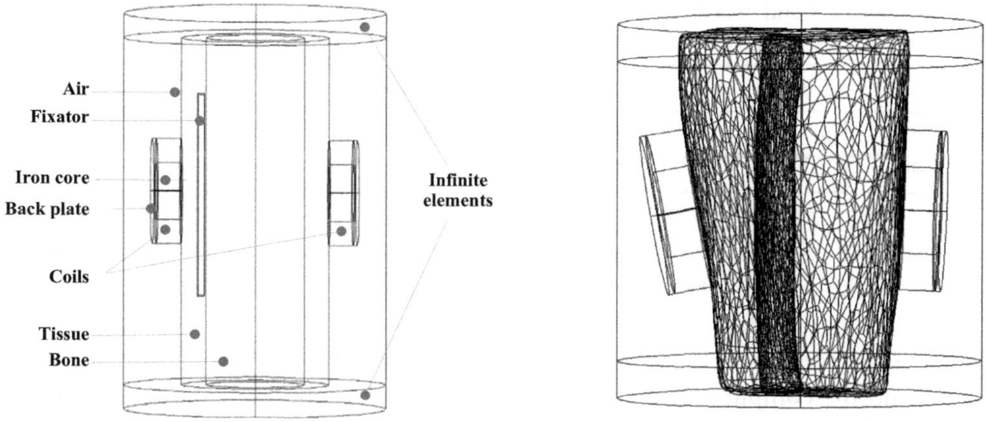

Figure 7.12 Simplified computational domain—CAD model (left) and reconstructed model (right). Infinite elements are used to close it within a conveniently limited volume (Baerov et al., 2020).

techniques (Chapter 3: Computational Domains) fused with CAD parts for the MF (coils, the implant, and the containing cylinder) are used to build the computational domain that represents the region of interest. Along this path, an imaging reconstruction tool is used first, for example, 3D Slicer (2020), to create an anatomically realistic 3D geometry out of CT images in *axial*, *coronal*, and *sagittal* planes (e.g., Embodi3D, 2020). The final model must contain the femur and the surrounding tissue.

The next step, important for higher resolution 3D models, is to eliminate unwanted or spurious details, reduce the complexity of construct, and save it in a CAD compatible format (e.g., Meshlab, 2020). The models are then converted to 3D SOLID entities and assembled and finally saved in a format compatible with the FEM solver (e.g., Comsol, 2020, Fig. 7.12, right).

The time-harmonic MF is described through the Helmholtz PDE and formulated for the magnetic field strength, $\underline{\mathbf{H}}$, Eq. (7.11). The boundary condition that closes the problem is magnetic insulation $(\mathbf{n}\cdot\underline{\mathbf{B}} = 0)$. The mathematical model is solved numerically.

The applicators are two circular copper coils, of 130 mm diameter, with 200 turns each. The electric current density is set within safe limits (e.g., 5×10^5 A/m^2). This excitation level provides a magnetic flux density of $\sim$60 mT along the axes of the coils (at 100 Hz). As for the TMS numerical simulation, because of the PDE Eq. (7.11) is a linear problem, its solution (the magnetic field strength and the magnetic flux density) is proportional to the excitation. So, if other levels of stimulation are needed, the excitation (inductor current) has to be scaled correspondingly. The results need just to be scaled—the EMF problem is not to be solved again.

The applicators and the ampere-turns are oriented such that their magnetic fields are additional, focused upon the targeted region. To strengthen the

Table 7.4 Electrical and heat transfer properties of the anatomical regions and implant (Gabriel, 1996; Andreuccetti, 2020; IT'IS Foundation, 2020).

Region	ε (F/m)	σ (S/m)	*Cp* (J/kg K)	ρ (kg/m^3)	*k* (W/m K)
Femur	8.7×10^{-7}	5.08×10^{-2}	2080	1273	0.27
Soft tissue	2.1×10^{-5}	0.25	3195	1040	0.40

magnetic and reduce its dispersion, the coils are provided with iron cores and metallic backplates, made of an alloy containing nickel, iron, and molybdenum, Fig. 7.12, left (e.g., BTL, 2020).

The intricacy of the anatomic systems, the limited precision of the input CT images, and the computational resources available at the time limit the resolution scale of the details accounted for. To circumvent these difficulties, homogenization techniques, consistent with the continuous media paradigm are used to model the different tissues as domains with equivalent properties.

The anatomic domains are homogeneous media with uniform, equivalent properties. Their electrical permittivity and electrical conductivity are mean, volume-weighted values of the constituent anatomical regions: cortical bone, trabecular bone, and bone marrow, for bone, and skin, fat, muscle, and blood, for soft tissue, calculated at $f = 100$ Hz, Table 7.4.

Numerical simulation results

The electric and magnetic fields spectra are seen in Fig. 7.13.

The presence of the magnetic field concentrators (iron core and nickel alloy backplate) is perceived through the effective values of the fields, Table 7.5. The CAD model is preferable here due to the numeric effectiveness it has while preserving a satisfactory consistency with the real situation.

At the working frequency (100 Hz) the concentrators indeed enhance the EMF but the fixator has a screening effect for the magnetic field, mainly due to the induced electric field (and currents) that damp the magnetic field, as the fixator material is a nonmagnetic but good electric conductor. In the absence of the fixator, the EMF diffuses in a larger area and penetrates deeply into the bone tissue. So, the plate can influence the MT procedure. This intuitive finding may suggest the repositioning of the applicators concerning the fixator and the adjustment of the excitation level such as to circumvent these effects and provide the expected MFT result.

Fig. 7.14 shows the EMF field when the imaging-based computational domain is used.

The applicators are provided with concentrators, and the coils are energized as before. The electric field induced within the bone tissue may lead a system of eddy currents that could contribute to enhancing the mass transfer process in the osseous healing process.

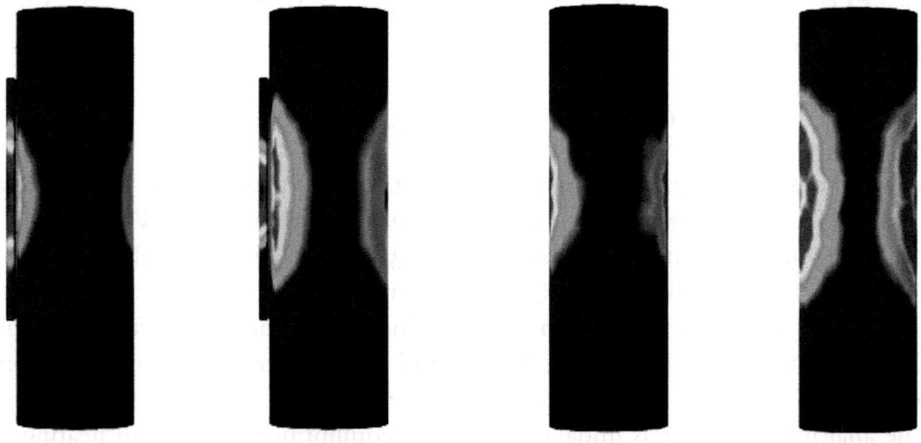

(A) (B) (C) (D)

Figure 7.13 Magnetic flux density (left) and electric field strength (right) in the femur with a fixation plate; magnetic field concentrators are attached to the coils (Baerov et al., 2020). (A) Magnetic flux density with fixator. (B) Electric field strength with fixator. (C) Magnetic flux density without fixator. (D) Electric field strength without fixator.

Table 7.5 Electric and magnetic fields maxima in the MFT simulation.

		Concentrator and fixator	Fixator	Concentrator
Fixator	B_{max} (mT)	40	20	–
Bone	B_{max} (mT)	20	10	30
	E_{max} (V/m)	0.44	0.24	0.28

MFT, Magnetic field therapy.

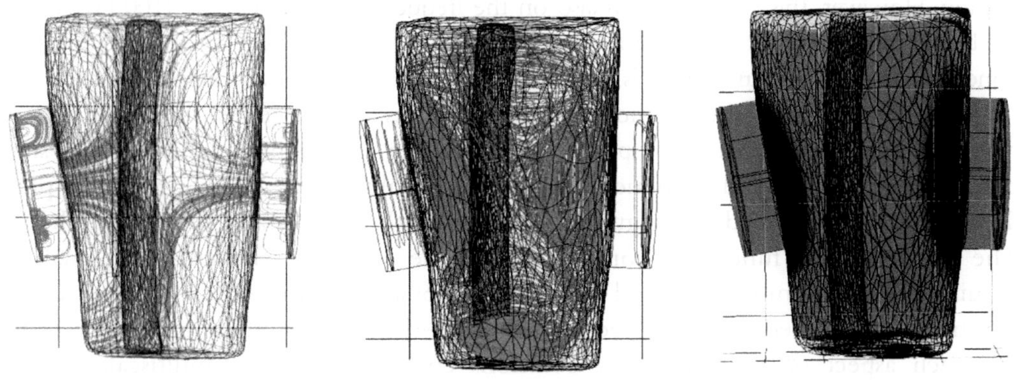

(A) (B) (C)

Figure 7.14 Numerical simulation result for MFT (Baerov et al., 2020). (A) Magnetic field density. (B) Electric field strength. (C) Temperature after 15 min ~37°C. *MFT*, Magnetic field therapy.

The EMF heating effect that seems to accompany the MFT, is described by the energy equation

$$\rho C_p \frac{\partial T}{\partial t} + \nabla \cdot \mathbf{q} = Q + Q_{perfusion} + Q_{met}, \tag{7.12}$$

where ρ is the mass density, C_p is the specific heat at constant pressure, T is the temperature, $\mathbf{q} = -k\nabla T$ [W/m^2] is the heat flux, k is the thermal conductivity, Q [W/m^3] is the heat source (EMF source, volumetric loss density), and Q_{met} [W/m^3] is the metabolic heat source (neglected here). The bioheat term, $Q_{perfusion}$ [W/m^3] (Chapter 1: Physical, Mathematical and Numerical Modeling) accounts for the heat in/outflow through perfusion. It will be concluded later that it may be ignored. The thermal properties of the constituent anatomical regions, listed in Table 7.4, are volume-weighted average values.

The anatomic structure is initially in thermal equilibrium (37°C) and heating starts when the applicators are energized. The boundary condition that closes the heat transfer problem is convection

$$q_0 = h\,(T_{ext} - T), \tag{7.13}$$

where q_0 [W/m^2] is the inward heat flux, h is the heat transfer coefficient (2 W/[m^2 K]), and T_{ext} is the ambient temperature (20°C).

Insignificant heating by MFT is seen after 15 min, when the temperature rise seems to stabilize, Fig. 7.14C. The procedure may take longer (60–80 min), but the situation is unlikely to change. The occurrence of the bioheat term could only strengthen this inference, because the perfusion driven convection would act to cool the regions down to the biological equilibrium temperature—unaffected by the MFT power level, here ~79.4 W/m^3 (Baerov et al., 2020).

The applicators themselves may warm up the tissue, which could produce mild local heating. However, this effect depends also on the frequency used but MFT (100 Hz) is not used for a thermal effect. It may then be inferred that MFT is a safe method for healing bone tissue even when an implant is present, a plate with screws, or an external fixation.

From the understanding of the underlying physics and processes, the optimization of the pending equipment and procedures, and the evaluation of the possible side effects, mathematical and numerical modeling are valuable means that may add predictive value to the medical diagnose associated with the MS and therapy methods. The structural and functional complexities of the anatomic entities under analysis require consistent models related to computational domains—from simpler CAD constructs to detailed anatomic reconstructions—that may produce relevant results. For each aspect of concern, some solutions compete with the multiscale nature (space and time) of the problems and the available computational resources. Therefore it is always to find the trade-off that satisfactorily accommodates all these constraints and objectives intractable computational endeavors.

References

Andreuccetti, D., Fossi, R., Petrucci, C., 1996. Available from: http://niremf.ifac.cnr.it/tissprop. Based on data published by C. Gabriel et al. in 1996 (accessed 06.20).

Assiotis, A., Sachinis, N.P., Chalidis, B.E., 2012. Pulsed electromagnetic fields for the treatment of tibial delayed unions and nonunions. A prospective clinical study and review of the literature. J. Orthop. Surg. Res. 7, 24.

Autocad. (2020) https://www.autodesk.com. (accessed 2020).

Aydin, N., Bezer, M., 2011. The effect of an intramedullary implant with a static magnetic field on the healing of the osteotomised rabbit femur. Int. Orthop. (SICOT) 35, 135–141.

Babbs, C.F., 2014. A compact theory of magnetic nerve stimulation: predicting how to aim. Biomed. Eng. Online 13 (53), 1–22.

Baerov, R., Morega, A.M., Veli, Y., Morega, M., 2019. Numerical simulation of magnetic stimulation of spinal cord—an assistive planning technology. In: The 11th Int. Symp. Adv. Topics El. Eng. (ATEE 2019), 28–30 March 2019, Bucharest, Romania.

Baerov, R.M., Morega, A.M., Morega, M., 2020. Analysis of magnetotherapy effects for post-traumatic recovery of limb fractures, Rev. Roum. Sci. Techn.– Électrotechn. et Énerg. 65 (1–2), 145–150.

Baxendale, S., 2009. The Wada test. Curr. Opin. Neurol. 22 (2), 185–189.

Beckers, G., Hömberg, V., 1991. Impairment of visual perception and visual short term memory scanning by transcranial magnetic stimulation of occipital cortex. Exp. Brain. Res. 87, 421–432.

Bednarčík, P., 2019. Principle of pulse electromagnetic field therapy (PEMF). https://www.biomag-medical.com/about-magnetotherapy/. (accessed 06.20).

Berényi, A., Mao, D., Buzsáki, G., 2012. Closed-loop control of epilepsy by transcranial electrical stimulation. Science 337 (6095), 735–737.

Bestmann, S., Oliviero, A., Voss, M., Dechent, P., Lopez-Dolado, E., Driver, J., et al., 2006. Cortical correlates of TMS-induced phantom hand movements revealed with concurrent TMS-fMRI. Neuropsychologia 44 (14), 2959–2971.

BTL. Magnetotherpay. https://www.btlnet.com/products-physiotherapy-magnetotherapy. (accessed 07.2020).

Chalidis, B., Sachinis, N., Assiotis, A., Maccauro, G., 2011. Stimulation of bone formation and fracture healing with pulsed electromagnetic fields: biologic responses and clinical implications. Int. J. Immunopath. Pharm. 24 (1, suppl. 2), 17–20.

Chen, M., Mogul, D.F., 2009. A structurally detailed finite element human head model for simulation of transcranial magnetic stimulation. J. Neurosci. Methods 179 (1), 111–120.

Chokroverty, S. (Ed.), 1990. Magnetic Stimulation in Clinical Neurophysiology. Butterworths Publishers, Reed Publishing Inc.

Cole, K.S., 1968. Membranes, Ions and Impulses. University of California Press, Berkeley.

Comsol Multiphysics, (2020) v.3.5a. . .5.3a, https://www.comsol.com/.

Coquim, J., Clemenzi, J., Salahi, M., Sherif, A., Avval, P.T., Shah, S., et al., 2018. Biomechanical analysis using FEA and experiments of metal plate and bone strut repair of a femur midshaft segmental defect. Biomed. Res. Int. 2018 (4650308), 11 pp.

Cowley, A., Woodward, B., 2011. A healthy future: platinum in medical applications. Platin. Met. Rev. 55 (2), 98–107.

Darabant, L., Cretu, M., Darabant, A., 2013. Magnetic stimulation of the spinal cord: experimental results and simulations. IEEE Trans. Mag. 49 (5), 1845–1848.

Davey, K., Luo, L., Ross, D.A., 1994. Toward functional magnetic stimulation (FMS). Theory and experiment. IEEE Trans. BME 41 (11), 1024–1030.

Datta, A., Zhou, X., Su, Y., Parra, L.C., Bikson, M., 2013. Validation of finite element model of transcranial electrical stimulation using scalp potentials: implications for clinical dose. J. Neural Eng. 10 (036018), 10 pp.

Dowalti, E., 2017. Spinal cord anatomy, pain, and spinal cord stimulation mechanisms. Semin. Spine Surg. 2 (9), 136–146.

Efisioterapia.Net. The value of the magnetic field in bone fractures. https://www.efisioterapia.net/en/the-value-of-the-magnetic-field-in-bone-fractures-t-7066.html. (accessed 06.20).

Elliott, R.E., Bollo, R.J., Berliner, J.L., Silverberg, A., Carlson, C., Geller, E.B., et al., 2013. Anterior temporal lobectomy with amygdalohippocampectomy for mesial temporal sclerosis: predictors of long-term seizure control. J. Neurosurg. 119, 261–272.

Embodi3D, 2020. The biomedical printing community. https://www.embodi3d.com/files/file/33822-femur/. (accessed 12.19).

Esselle, K., Stuchly, M., 1992. Neural stimulation with magnetic fields: analysis of induced electric fields. IEEE Trans. BME 39 (7), 693–700.

Fei, Y., Su, L., Lou, H., Zhao, C., Wang, Y., Chen, G., 2019. The effects of 50 Hz magnetic field–exposed cell culture medium on cellular functions in FL cells. J. Radiat. Res. 60 (4), 424–431.

Femur Fracture. Symptoms, treatment, recovery, complications. https://www.reginamaria.ro/dictionar-de-afectiuni/fractura-de-femur. (accessed 06.20) (in Romanian).

Gabriel, C., 1996. Compilation of the Dielectric Properties of Body Tissues at RF and Microwave Frequencies. Department of Physics, Kings College, London.

Hallet, M., 2000. Transcranial magnetic stimulation and the human brain. Nature 406, 147–150.

Hallet, M., 2007. Transcranial magnetic stimulation: a primer. Neuron 55 (2), 187–199.

Hartwigsen, G., Siebner, H.R., Stippich, C., 2010. Preoperative functional magnetic resonance imaging (fMRI) and transcranial magnetic stimulation (TMS). Curr. Med. Imag. 6 (4), 220–231.

IT'IS Foundation. Heat capacity, itis.swiss. https://itis.swiss/virtual-population/tissue-properties/database/heat-capacity. (accessed 06.20).

Hofstoetter, U.S., Danner, S.M., Minassian, K., 2014. Paraspinal magnetic and transcutaneous electrical stimulation. Encycl. Comp. Neurosci. 1–21.

Kobayashi, M., Pascual-Leone, A., 2003. Transcranial magnetic stimulation in neurology. Lancet Neurol. 2 (3), 145–156.

Krause, P., Edrich, T., Straube, A., 2004. Lumbar repetitive magnetic stimulation reduces spastic tone increase of the lower limbs. Spinal Cord. 42 (2), 67–72.

Krawczy, A., Murawski, P., Korzeniewska, E., 2017. Medical and technical analysis of magnetotherapeutical devices. In: IEEE 2017 Int. Conf. on Modern Elec. Energy Systems (MEES), 15–17 November, Kremenchuk, Ukraine.

èada-Tondrya, E., 2019. The impact of applicator size on distribution of electromagnetic field used in magnetotherapy. Prz. Elektrotech. 1 (7), 27–30.

Lempka, S.F., Patil, P.G., 2018. Innovations in spinal cord stimulation for pain. Curr. Opin. Biomed. Eng. 8, 51–60.

Lin, V., Hsiao, I., Zhu, E., Perkash, I., 2001. Functional magnetic stimulation for conditioning of expiratory muscles in patients with spinal cord injury. Arch. Phys. Med. Rehab. 82 (2), 162–166.

Liu, Y.T., Zhou, X.J., Ma, J., Ge, Y.B., Cao, X., 2015. The diameters and number of nerve fibers in spinal nerve roots. J. Spinal Cord. Med. 38 (4), 532–537.

Maccabee, P.J., Amassian, V.E., Eberle, L.P., Rudell, A.P., Cracco, R.Q., Lai, K.S., et al., 1991. Measurement of the electric field induced into inhomogeneous volume conductors by magnetic coils: application to human spinal neurogeometry. Electroencephalogr. Clin. Neurophysiol. 81 (3), 224–237.

Malmivuo, J., Plonsey, R., 1995. Bioelectromagnetism: Principles and Applications of Bioelectric and Biomagnetic Fields. Oxford University of Press.

Markov, M.S., 2007. Expanding use of pulsed electromagnetic field therapies. Electromagn. Biol. Med. 26 (3), 257–274.

Mayfield Brain & Spine. Spinal cord stimulation. p. 6. https://mayfieldclinic.com/pe-stim.htm. (accessed 06.20).

Mayo. https://www.mayoclinic.org/diseases-conditions/back-pain/symptoms-causes/syc-20369906 (accessed 06.20).

Merskey, H., Bogduk, N. (Eds.), 1994. Classification of Chronic Pain Descriptions of Chronic Pain Syndromes and Definitions of Pain Terms. second ed. IASP Press, Seattle.

Meshlab. www.meshlab.net. (accessed 06.20).

Meyer, F.B., Marsh, W.R., Laws, E.R., Sharbrough, F.W., 1986. Temporal lobectomy in children with epilepsy. J. Neurosurg. 64 (3).

Monzel, C., Vicario, C., Piehler, J., Coppeya, M., Dahan, M., 2017. Magnetic control of cellular processes using biofunctional nanoparticles. Chem. Sci. 8, 7330–7338.

Mocanu, C.I., 1982. Electromagnetic Field Theory. Ed.D.P., Bucharest (in Romanian).

Morega, M., 1999. Bioelectromagnetism. Ed. Matrix Rom, Bucharest (in Romanian).

Morega, M., 2000. Design of coils for magnetic neural stimulation. Efficiency criteria and technical solutions. ACTA Electroteh. 41 (1), 133–138.

Morikuni, K., Reichel, L., Hayami, K., 2012. Generalized minimal residual method, NII-2012-001E Technical report. National Institute of Informatics, Tokyo, Japan. Available from: http://www.nii.ac.jp.

Mullin, C.R., Steeves, J.K.E., 2011. TMS to the lateral occipital cortex disrupts object processing but facilitates scene processing. J. Cognit. Neurosci. 23 (12), 4174–4184.

Niebur, E., 2008. Electrical properties of cell membranes. Scholarpedia 3 (6), 7166. (accessed 08.20).

Noh, N.A., Fuggetta, G., Manganotti, P., 2015. Theta-burst transcranial magnetic stimulation alters the functional topography of the cortical motor network. Malays. J. Med. Sci., Spec. Issue Neurosci. 36–44.

Novelly, R.A., Novelly, R.A., Augustine, E.A., Mattson, R.H., Glaser, G.H., Williamson, P.D., et al., 1984. Selective memory improvement and impairment in temporal lobectomy for epilepsy. Ann. Neurol. 15 (1), 64–67.

Nummenmaa, A., Stenroos, M., Ilmoniemi, R.J., Okada, Y.C., Hämäläine, M.S., Raij, T., 2013. Comparison of spherical and realistically shaped boundary element head models for transcranial magnetic stimulation navigation. Clin. Neurophysiol. 124 (10), 1995–2007.

Paulus, W., 2011. Transcranial electrical stimulation (tES – tDCS; tRNS, tACS) methods. Neuropsychol. Rehab. 21 (5), 602–617.

Peinemann, A., Lehner, C., Conrad, B., Siebner, H.R., 2001. Age-related decrease in paired-pulse intracortical inhibition in the human primary motor cortex. Neurosci. Lett. 313 (1–2), 33–36.

Peng, J., Zhao, J., Long, Y., Xie, Y., Nie, J., Chen, L., 2019. Magnetic materials in promoting bone regeneration. Front. Mater. 6 (268), 14 pp.

Petrov, P.I., Mandija, S., Sommer, I.E.C., van den Berg, C.A.T., Neggers, S.F.W., 2017. How much detail is needed in modeling a transcranial magnetic stimulation figure-8 coil: measurements and brain simulations. PLoS One 20 pp.

Physiomed. https://www.physiomed.de/en/products/mag-expert-with-set-of-applicators/. (accessed 06.20).

Plonsey, R., Barr, R.C., 1988. Bioelectricity. A Quantitative Approach. Plenum Press, New York.

Ptito, M., Fumal, A., de Noordhout, A.M., Schoenen, J., Gjedde, A., Kupers, R., 2008. TMS of the occipital cortex induces tactile sensations in the fingers of blind Braille readers. Exp. Brain. Res. 184, 193–200.

Ren, C., Tarjan, P.P., Popović, D.B., 1995. A novel electric design for electromagnetic stimulation – the slinky coil. IEEE Trans. BME 42 (9), 918–925.

Risson, E.G., Serpa, A.P., Berger, J.J., Heil Koerbel, R.F., Koerbel, A., 2018. Spinal cord stimulation in the treatment of complex regional pain syndrome type 1: is trial truly required? Clin. Neurol. Neurosurg. 171, 156–162.

Rossini, P.M., Rossi, S., 2007. Transcranial magnetic stimulation: diagnostic, therapeutic, and research potential. Neurology 68 (7).

Sahoo, N., Anand, S.C., Bhardwaj, J.R., Sachdeva, V.P., Sapru, B.L., 1994. Bone response to stainless steel and titanium bone plates. Med. J. Armed. Forces India 50 (1), 10–14.

Salinas, F.S., Lancaster, J.L., Fox, P.T., 2009. 3D modeling of the total electric field induced by transcranial magnetic stimulation using the boundary element method. Phys. Med. Biol. 54 (12), 3631–3647.

ScienceDaily, 2019. Neuroscientists restore significant bladder control to 5 men with spinal cord injuries: magnetic stimulation of lower spine eliminates need of catheter for up to 4 weeks. https://www.sciencedaily.com/releases/2018/08/180822082648.htm.

Sekino, M., Ohsaki, H., Takiyama, Y., Yamamoto, K., Matsuzaki, T., Yasumuro, Y., et al. Eccentric figure-eight coils for transcranial magnetic stimulation. https://core.ac.uk/download/pdf/148778587.pdf. (accessed 06.20.

3DSlicer. https://www.slicer.org/. (accessed 06.20).

Shupack, N.M., 2003. Therapeutic uses of pulsed magnetic-field exposure: a review. Radio. Sci. Bul. 307, 9–32.

Soltoianu, C., 2018. Magnetic stimulation of spinal cord (BS thesis). Faculty of Medical Engineering, University Politehnica of Bucharest (in Romanian).

Sperling, M.R., O'Connor, M.J., Saykin, A.J., Plummer, C., 1996. Temporal lobectomy for refractory epilepsy. JAMA 276 (6), 470–475.

Stuchly, M., Esselle, K., 1992. Factors affecting neural stimulation with magnetic fields. Bioelectromagn. Suppl. 1, 191–204.

Su, X.-Y., Zhao, J.-X., Zhao, Z., Zhang, L.-C., Li, C., Li, J.-T., et al., 2015. Three-dimensional analysis of the characteristics of the femoral canal isthmus: an anatomical study. BioMed. Res. Int. 2015 (459612), 9 pp.

Sudhansu, C. (Ed.), 1990. Magnetic Stimulation in Clinical Neurophysiology. Butterworths Pub., Reed Publishing Inc.

Thickbroom, G.W., Byrnes, M.L., Edwards, D.J., Mastaglia, F.L., 2006. Repetitive paired-pulse TMS at I-wave periodicity markedly increases corticospinal excitability: a new technique for modulating synaptic plasticity. Clin. Neurophysiol. 117 (1), 61–66.

Timotin, A., 2004. La structure de la fibre nerveuse: un projet optimal. Proc. Romanian Acad. A 5 (1), 65–72.

Tsai, P., Wang, C., Chiu, F., Tsai, Y., Chang, Y., Chuang, T., 2009. Efficacy of functional magnetic stimulation in neurogenic bowel dysfunction after spinal cord injury. J. Rehab. Med. 41 (1), 41–47.

Ugawa, Y., Rothwell, J.C., Day, B.L., Thompson, P.D., Marsden, C.D., 1989. Magnetic stimulation over the spinal enlargements. J. Neurol. Neurosurg. Psychiatry 52, 1025–1032.

Uptodate. Uptodate.Com. https://www.uptodate.com/contents/midshaft-femur-fractures-in-adults. (accessed 06.20).

Villegas, R., Villegas, G.M., 1965. Les couches superficielles de la fibre nerveuse du calmar. Actualités neurophysiologiques. Sixième série, Masson, Paris.

Wagner, T.A., Zahn, M., Grodzinsky, A.J., Pascual-Leone, A., 2004. Three-dimensional head model simulation of transcranial magnetic stimulation. IEEE Trans. Biomed. Eng. 51 (9), 1586–1598.

Wasserman, E.M., Linsaby, S.H., 2001. Therapeutic application of repetitive transcranial magnetic stimulation: a review. Clin. Neurophysiol. 112, 1367–1377.

Whole Spine (Cervical-Dorsal-Lumbar-Sacral). CT scan, embodi3D.com, 04 September 2017. https://www.embodi3d.com/files/file/11735-whole-spine-cervical-dorsal-lumbar-sacral-ct-scan/.

Woermann, F.G., Jokeit, H., Luerding, R., Freitag, H., Schulz, R., Guertler, S., et al., 2003. Language lateralization by Wada test and fMRI in 100 patients with epilepsy. Neurology 61 (5).

Xu, G., Chen, Y., Yang, S., Wang, M., Yan, W., 2005. The optimal design of magnetic coil in transcranial magnetic stimulation. In: Proc. IEEE Eng. Med. Biol. 27th Annual Conf., 1–4 September, Shanghai, China, pp. 6221–6224.

Yang, S., Xu, G., Wang, L., Chen, Y., Wu, H., Li, Y., et al., 2006. 3D realistic head model simulation based on transcranial magnetic stimulation. In: Proc. 28th IEEE Int. Conf. Engineering in Medicine and Biology Society, 30 August–3 September, New York.

Yang, S., Xu, G., Wang, L., Gen, D., Yang, Q., 2007. Electromagnetic field simulation of 3D realistic head model during transcranial magnetic stimulation. In: 1st Int. Conf. Bioinf. Biomed. Eng., Wuhan, China, pp. 656–659.

Yang, S., Xu, G., Wang, L., Geng, Y., Yu, H., Yang, Q., 2010. Circular coil array model for transcranial magnetic stimulation. IEEE Trans. Appl. Supercond. 20 (3), 829–833.

Yang, C., Guo, Z., Peng, H., Xing, G., Chen, H., McClure, M.A., et al., 2018. Repetitive transcranial magnetic stimulation therapy for motor recovery in Parkinson's disease: a meta–analysis. Brain Behav. 8 (11), e01132.

Zhang, T., Edrich, J., 1996. Eccentric coils for focused neuromagnetic stimulation and biomagnetic detection. In: Proc. of the 18th Annual Int. Conf. IEEE-EMBS, 31 October—3 November 1996, Amsterdam.

CHAPTER 8

Hyperthermia and ablation

8.1 Thermotherapy methods

Thermal therapy medical procedures rely on either raising or lowering the local temperature in tissues with healing aims. In the hyperthermia group, the emerging radiofrequency (RF) and microwave (MW) hyperthermic procedures are used in breast cancer detection and the angioplasty, in the ablation of the liver and heart, benign prostate hypertrophy (Westermark, 1989; Rosen and Rosen, 1995; Rosen et al., 2002), and the enumeration may continue (Dahl et al., 1995; Habash et al., 2006). Current researches on the thermal therapy-related biological effects (Michaelson and Lin, 1987) are concerned with the viability of the tumor cells decreases when exposed to even lukewarm elevated temperatures, which yields them sensitive to radiation and chemotherapy (Horsman and Overgaard, 2007; Poulou et al., 2015; Ryu et al., 2004). External applicators are positioned either near or around the region of interest (ROI), inserted through the skin or just below its surface, and power is focused on the tumor to increase or decrease its temperature, and either heating or cooling of the targeted area, such as a tumor, is produced using various methods. Different types of power sources may be used, of which electromagnetic field (EMF) sources, ultrasound sources (US), and cryogenic sinks (CS) are most common. Medical thermal applications based on EMF power sources (Rosen and Rosen, 1995; Strezer, 2002; Vander Vorst et al., 2006) concern a wide range of therapeutic effects, with some existing variability in thermal therapy methodology (Stauffer and Goldberg, 2004). The dosage of the parameters (exposure time, intensity) distinguishes between hyperthermia therapy and ablation (surgery). The heating of tissue depends on the properties of the EMF source, its frequency, the heat transfer properties of the tissues, and their vascularization.

Hyperthermia

General hyperthermia methods are nonfocused and thus have limited ability to deliver a uniform, sufficient thermal dose to metastases, which limits its clinical efficacy. However, they are particularly successful as an antineoplastic agent used in the treatment for intraperitoneal metastases from ovarian tumors, increasing the efficacy of certain chemotherapeutics (Ryu et al., 2004; Wust et al., 2003). According to the National Cancer Institute classification (NIH), in hyperthermia (or thermal therapy or thermotherapy) body tissue is exposed to temperatures high enough (up to 45°C or 113°F) to damage the proteins and the

DOI: https://doi.org/10.1016/B978-0-12-817897-3.00008-7

structures within into killing cancer cells, in general with minimal damage to normal tissues (Hildebrandt et al., 2002; Van der Zee, 2002). The precise mechanisms by which cells are destroyed by heating are yet to be elucidated (Lepock, 2003), but it is suggested that thermal injury and the reduction in the cell growth rate may increase with temperature up to some critical threshold, above which growth is sharply inhibited, and necrosis may occur. It is thought that cellular necrosis is produced by the thermal denaturation of critical targets in the cell (Miles, 2006), and cytotoxicity, radiosensitization, and thermotolerance responses that occur in the hyperthermic region are most likely temperature-induced alterations in the molecular pathways (Lepock, 2005).

Hyperthermia methods may be either regional (RH) or local (LH). RH aims parts of the body (e.g., organ, limb, or body cavity), and it is usually combined with chemotherapy or radiation therapy. RH methods include: regional perfusion (RP) or isolation perfusion (IP), continuous hyperthermic peritoneal perfusion (CHPP), also called hyperthermic intraperitoneal chemotherapy (HIPEC), and deep tissue hyperthermia (DTH).

In RP (or IP) the blood in that part of the body is heated outside the body, and chemotherapy can be added in at the same time. For CHPP (or HIPEC), during surgery, heated anticancer drugs flow from a warming device through the peritoneal cavity, whose temperature raises from 41°C to 42°C. The DTH uses devices, which are positioned on the surface of the body cavity or organ to deliver RF or MW power to heat the ROI.

Local hyperthermia methods include *intraluminal* or *endocavitary* methods and *interstitial* methods that may solve the focalization problem, and it may be expected also that they could greatly enhance chemotherapy by decreasing the necessary dose and diminish normal tissue damage (Petryk et al., 2009). The intraluminal treatment of tumors situated nearby or within body cavities (e.g., rectum and esophagus) employs probes, which are positioned inside the cavity and inserted into the tumor for the direct heating of the tumor. The interstitial methods use probes or needles, which are inserted, under anesthesia, into the tumor to treat tumors deep inside the body, as for instance brain tumors. The tumor is thus heated to higher temperatures than the external techniques can do. Accompanying imaging techniques, for instance, ultrasound may help properly guiding the probe within the tumor.

Depending on the desired temperature and power levels, procedural duration, it may be distinguished between (1) *diathermia*—up to 41°C, used in physiotherapy (rheumatism and related diseases); (2) *hyperthermia*, from 41°C to 45°C, used in oncology to enhance the efficiency of other cancer treatments; (3) *thermal ablation* (TA), above 45°C, used to destroy localized tumor formations. Table 8.1 summarizes the thermal therapy methods that utilize EMF sources.

Hyperthermia in the range of 42°C–45°C for periods of 30–60 min causes irreversible cellular deterioration through protein denaturation (Dutreix et al., 1978) as shown in Habash et al. (2006), and the time to irreversible cellular harm decreases exponentially when the tissue temperature rises to 50°C.

Table 8.1 Electromagnetic field and thermal therapy methods.

Medical therapy	Hyperthermia and ablation		Laser ablation
Frequency range	10^2 kHz–10 MHz (RF)	10^2 MHz–10 GHz (MW)	10^2 THz–10 PHz (light)
EMF radiation	Nonionizing		

EMF, Electromagnetic field; *RF*, radio frequency.

The Joule heat injury on the tumor is determined by the local, total electric work, the tumor microenvironment, and the tumor biology (Nikfarjam et al., 2005a), as cited in Habash et al., (2006). Although the underlying thermosensitivity mechanisms imply complex cellular and subcellular interactions within tumor tissue, it is the cell membrane seems to be the most vulnerable to heat injury. *In vitro* (Dickson and Calderwood, 1980) and *in vivo* (Overgaard and Overgaard, 1972) [as cited in Habash et al. (2006)] studies evidence that tumor cells are killed at temperatures lower than for the normal cells.

The hyperthermia application triggers a progression in tissue damage through several factors: induced cancer cell death through apoptosis, microvascular destruction, and ischemia–reperfusion related injury (Brown et al., 1992), *Kupffer cells* activation (Tsutsui and Nishiguchi, 2014) and alteration of cytokine peptides expression (Welc et al., 2012), and modulation of the immune response (Baronzio et al., 2006). These factors depend on tissue temperatures produced by the applied total power, the heat exhaust rate, and the thermal properties of the tissue (Christensen and Durney, 1981; Dewhirst et al., 2003a; Haemmerich and Laeseke, 2005; Haemmerich et al., 2005; Nikfarjam et al., 2005b; Osepchuk, Petersen, 2001; Seegenschmiedt and Vernon, 1995; Vander Vorst et al., 2006). The underlying relationships between thermal exposure and the pending damage, the collateral effects endured by the normal tissue, and methods for converting one time-temperature protocol to time at standardized temperature are reviewed in Dewhirst et al. (2003b). Table 8.2 summarizes the temperature-related effects produced to biological tissues (Germer et al., 1998; Habash et al., 2006; Haemmerich et al., 2005; Lepock, 2005; Stauffer, 2005).

Table 8.2 Thermal effects, temperature range, duration.

Temperature range and duration	Effects
Below −50°C for more than 10 min	Complete cellular killing by freezing
30°C–39°C for no specific duration 40°C–46°C for 30–60 min	Hyperthermic cellular death
47°C–50°C for more than 10 min	Protein denaturation, necrosis
Over 50°C after ~2 min	Necrosis, cell death
60°C–140°C for seconds	Protein denaturation, membrane rupture, cell stricture, ablation
100°C–300°C for seconds	Vaporization of extracellular steam vacuoles
Over 300°C for fraction of a second	Carbonization

More recently Jain et al. (2005) and Zhang and Misra (2007) reported *in vitro* studies on ion nanoparticle (iron oxide-based and others) platforms, which may convey chemotherapeutic for the sustained and stimuli-triggered release (by heating, electric fields, US, UV light, and radiation). A promising avenue is the hyperthermia-enhanced radiotherapy (Sekhar et al., 2007). In this respect, hyperthermia-mediated magnetic nanoparticle (MNP)-delivered radionucleotides is prospectively attractive (Pankhurst et al., 2003), as cited by Giustini et al. (2010).

Ablation

TA is a relatively new treatment, which leads to tissue necrosis through coagulation. This minimally invasive procedure for cancer treatment is an efficient alternative to the classical surgical resection of small, less than 3 cm, tumors. TA is produced either by heating or by freezing the tumoral tissue either to raise its temperature [above 60°C, in TA (Brace, 2011)] or to lower it below the tissue living threshold [below −40°C, in cryoablation (Mayo Clinic; Lepock, 2005)] the tissue living threshold. Highly localizable inside the tumoral tissue, TA is used in the treatment of lung, bone, renal, and liver cancers. Radiofrequency ablation (RFA), which is thought of as a type of interstitial hyperthermia, uses RF waves to heat and destroy cancer cells.

Ablation through hyperthermia or cryogenic processes occurs when the work of the external source (EMF, US, and CS) produces excessive heat absorption or release by the tissue that results in its permanent damage when a critical (necrotic) temperature is reached. The amount of damaged tissue by ablation may be evaluated by computing the direct calculation (integration) of the change in the internal energy or by the duration of the exposure to the necrotizing temperature.

In the energy analysis, the rate of tissue injury is given by the Arrhenius equation (Xu et al., 2008) $k(T) = \frac{d\Omega}{dt} = Ae^{-\frac{E_a}{RT}}$, and it yields the amount of damaged tissue

$$\Omega = \int_0^t Ae^{-\frac{E_a}{RT}} dt, \tag{8.1}$$

where t is the time since the procedure starts, $R = 8.314$ J/mol K is the universal constant of gasses, A [1/s] is a frequency parameter (the number of times two molecules collide), a constant, tissue-dependent quantity, and E_{ava} [J/mol] is the activation energy needed to trigger the irreversible damage reaction, a tissue-dependent quantity (Bhowmick et al., 2004; Hasgall et al., 2015; Jacques et al., 1996; Pop et al., 2003; Rossmann and Haemmerich, 2016; Xu et al., 2008). Typical values for the liver are $A = 7.39 \times 10^{39} s^{-1}$ and $E_a = 2.577 \times 10^5$ J/mol (Jacques et al., 1996), and the fraction of destroyed tissue is $\Gamma = 1 - e^{-\Omega}$.

In the damage integral analysis (either TA or cryoablation) of the damaged tissue, Ω, two distinct cases are identified: (1) ablation occurs immediately when the temperature reaches the critical value (T_h respectively T_c), and (2) the temperature exceeds (hyperthermia) or decreases below (cryoablation) the critical value for a certain time (t_h, respectively t_c).

A *damage tissue indicator*, α_d, and a *necrosis time indicator*, α_t, are defined through

$$\alpha_d = \frac{1}{t_{d,h}}\int_0^t \delta_{d,h}\left(T_{d,h}\right)\mathrm{d}t + \frac{1}{t_{d,c}}\int_0^t \delta_{d,c}\left(T_{d,c}\right)\mathrm{d}t,\ \alpha_n = \frac{1}{t_{n,h}}\int_0^t \delta_{n,h}\left(T_{n,h}\right)\mathrm{d}t + \frac{1}{t_{n,c}}\int_0^t \delta_{n,c}\left(T_{n,c}\right)\mathrm{d}t, \tag{8.2}$$

where $\delta_h = 1$ for $T > T_h$, and 0 otherwise, and $\delta_c = 1$ for $T < T_c$ and 0 otherwise. The subscript "*d*" stands for damage, "*n*" for necrosis, "*h*" for hyperthermia, and "*c*" for cryoablation (Comsol, 2010–2019). A concise expression for the overall fraction of damaged tissue is

$$\Omega = \begin{cases} 1 & \text{if } \alpha_n > 0, \\ \min\left(\alpha_d,\ 1\right) & \text{otherwise.} \end{cases} \tag{8.3}$$

Effective thermal conductivity, $k_{eff} = \theta_d k_d + (1-\theta_d)k$, and effective heat capacity at constant pressure, $(\rho C_p)_{eff} = \theta_d \rho_d C_{p,d} + (1-\theta_d)\ \rho C_p$, are introduced to account for the tissue injury—here, ρ_d is the mass density, $C_{p,d}$, the heat capacity at constant pressure, k_d the thermal conductivity for the damaged tissue of the damaged tissue, and θ_d is a weight term.

As already discussed, numerical modeling may provide useful and unique information on the underlying heat transfer paths and mechanisms in thermal therapy that could assist the preoperational planning, improve the diagnosis, and could suggest therapeutic protocols. In what follows, this chapter presents several localized hyperthermia and ablation models with EMF and US power sources and emphasizes the effects of the local vascularization upon the heat transfer balance. The optimization of the applicators, the control of their location, and the adjustment of the heating protocols are currently possible, and numerical simulation may bring a significant help in improving the procedures based on the predictive analysis of the thermal behavior of tissues.

8.2 Radiofrequency thermotherapy

Radiofrequency localized hyperthermia ablation of tumors is a relatively new, promising, minimum invasive, and highly effective tumor extirpation procedure that may be used to eradicate bone, lung, liver, and renal smaller tumors (Ramanathan et al., 2010). The heat source is provided by the electrothermal (Joule) effect.

In RFA, high-frequency electrical currents flow through a needle electrode introduced into the tumor to some ground pads positioned on the body, to create a focal hotspot that kills the cancer cells situated in the electrode range (Radiology). The heat source is then the electrothermal (Joule) effect. Numerical modeling may complement other medical assertive elements in the preinterventional phase with the aim of precisely positioning the antenna and adjust its power level. An as simple and as accurate possible prediction of the RFA protocol is a desideratum, and to this aim difficulties related to the realistic representation of the tumor volume have to be surmounted.

Microwave ablation (MWA) is also a minimally invasive intervention used for the same indications as RFA to heat and kill the tumor. MWA provides low risk and a short hospital stay as an outpatient procedure, with overnight supervision in the hospital if general anesthesia is recommended. The electrodes for RFA and interstitial MWA, specialized needle-like probes (Mulier et al., 2005), are inserted using an image-guided technique: magnetic resonance imagery (MRI), ultrasound (US), or computed tomography (CT). Multiple tumors may be treated simultaneously, and the procedure can be repeated if remittance occurs. RFA electrodes for soft tissue (kidney and liver) are expanding rapidly (Goldberg et al., 2003). Recently the International Working Group on Image-Guided Tumor Ablation (IWGIGTA) proposed their classification. Many new electrodes, commercial and experimental, have been introduced since then, including "multiple electrode systems".

The heat source characterization and its numerical modeling are one significant part of the problem. The other part is the heat transfer in vascularized tissues. This "physics" is usually addressed by the continuous media bioheat equation (Pennes) (Chapter 1: Physical, Mathematical, and Numerical Modeling), which accounts for the blood vessels heat transfer. However, this homogenization approach may be prone to underestimation of the required power levels for a successful procedure when larger than the capillaries vessels are present too.

This chapter presents several models of localized (interstitial) thermal therapy (hyperthermia and ablation) with a focus on modeling the power sources (power level and duration) and their sizing when applied to tissues where heat is conveyed through hemodynamic flow in larger size vessels and the tissue, approximated to be a saturated porous medium, called here the general heat transfer (GHT) model. The results are compared to those obtained by using the bioheat (BHT) model. The exposure to EMF is limited in time and power level to provide for the success of the procedure and to avoid the thermally produced damage of the neighboring healthy tissue. Numerical simulations may add to the experimental work in solving this matter (Sands and Layton, 2000), as affordable, noninvasive and accurate methods used to solve thought experiments and studies focused on patient-specific data, electrode design (Koda et al., 2011), procedure duration, and power level. Numerical experiments are normally performed on validated and reasonably realistic models of utmost utility for exploring a wide spectrum of correlations between various interventional parameters and for the assessment of corresponding physical consequences. For instance the power level may be adjusted to provide for a local overheating of the tumor, while avoiding in as much as possible the damage of the neighboring healthy tissue.

Thermal ablation of a kidney tumor

EMF ablation, either RF or MW, is used to extirpate kidney cancer (Mayo Clinic), and its success may rely on a preoperative assessment, where numerical simulation is a useful, convenient, affordable, and accurate mean. On the other hand, the structure of the kidney is complex (Layton, 2013; TBQ Editors, 2013), and some of its functions are still not fully elucidated. At this time, a consistent physical model for it is still a desideratum, and some convenient, reasonable yet realistic assumptions are required. Along this line, the kidney and the tumor (tissue and capillaries) are assumed here to be a saturated porous media (Durlofsky and Brady, 1987; Truskey et al., 2004), hemodynamically connected to the local, embedded hepatic arterial and venous trees (Bachmann et al., 1965). A one-temperature model, meaning that the solid phase of the porous medium and the plasma are in local thermal equilibrium, would be a fair realistic choice. The actual multiscale, directional blood flow is replaced by an equivalent directional flow whose convective contribution to the heat transfer replaces the average perfusion-related source term in the energy equation as per the bioheat homogenization-based analysis (Chapter 1: Physical, Mathematical, and Numerical Modeling), which is a more common approach. The two methods lead to different predictions, and to evidence discrepancies, we consider the RF ablation (Knavel and Brace, 2013; Zagoria, 2004). An order of magnitude analysis of the time scales of the three concurring "physics" may suggest stationary forms for the EMF and the hemodynamic flow and transient form for the heat transfer (Dobre et al., 2017).

Mathematical modeling

Medical image-based reconstruction techniques fused with CAD constructs (Chapter 2: Shape and Structure Morphing of Systems with Internal Flows) might render (if possible) the complex morphology of the kidney, resulting in a computational domain. Here, to prove the concept and to provide a glimpse in the pending physics, a CAD computational domain [e.g., by using Dassault Systemes, 2016] that mimics as realistically as possible an adult kidney may be found satisfactory (Fig. 8.1).

Each kidney receives a renal artery but up to six arteries may exist (Bacmann). The renal artery here splits into two daughters to supply the kidney: the upper (UD), and lower (LD) divisions (Wacker et al., 2018), which in turn split into two smaller arteries each. These constructs mimic the natural morphological branching (bifurcations, lengths, and diameters of the "mother"/"daughters" segments), consistently with the allometric laws (Chapter 1: Physical, Mathematical, and Numerical Modeling and Chapter 2: Shape and structure morphing of systems with internal flows) (Bejan, 2000;

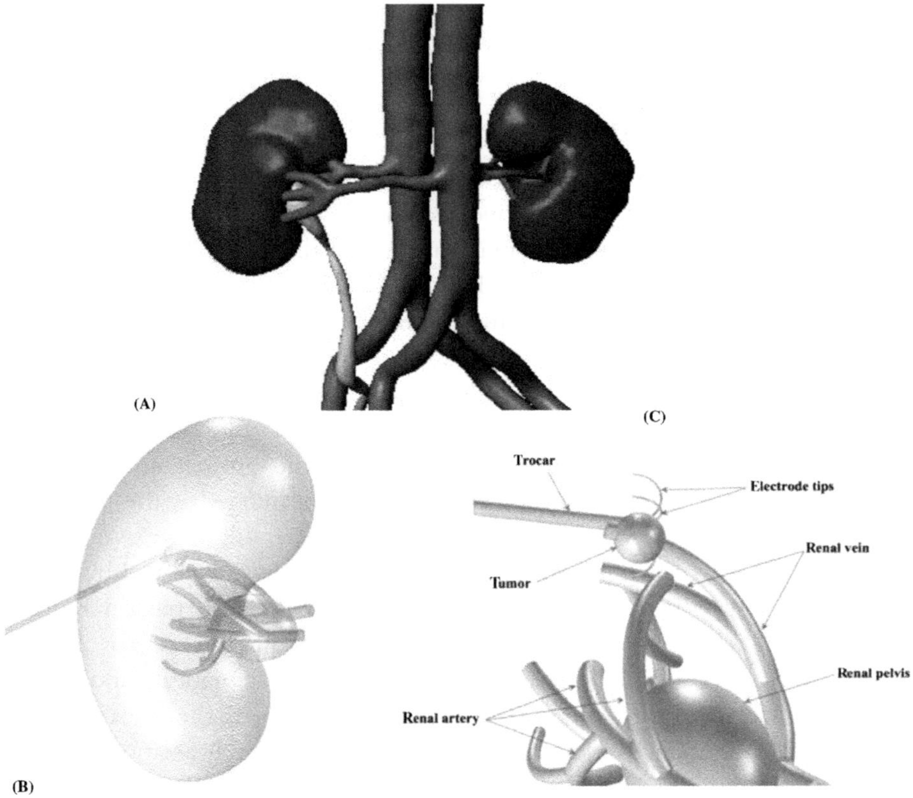

Figure 8.1 Qualitative view of the kidney, its vascularization, and the RF antenna: (A) medical scans reconstruction; (B) and (C) CAD construct. *RF*, Radio frequency.

Murray, 1926) and the constructal law predictions (Bejan, 2000; Bejan and Zane, 2012). The arterial segment belongs to the group of resistance vessels, Fig. 8.2.

Kidneys filter the blood and regulate the arterial pressure. Arterial blood flows through *nephrons* to *glomeruli* where *tubules* collect the toxins eventually drained through the urine (NIH, 2019). The venous system returns the filtered blood. In humans, depending on the resting heart rate, the whole content of the circulatory system ($\sim$5 L for an adult) is filtered every 1/2–1 hour. The *renal pelvis* of the *ureter*, comprising the *calyces* (cuplike extensions) that collect the urine before it flows on into the urinary bladder, is considered here too, Fig. 8.1. Considering that the EMF exposure lasts, perhaps, up to 10 min, on one hand, and for 1 L of filtered blood, $\sim$0.01 L of urine is produced, on the other hand, the fluid loss to urine generation may be neglected. Although it seems reasonable to assume that the volume fluid intake of the renal pelvis is stagnant, it may however absorb some RF-produced heat. Consistent with available experimental data, urine thermal properties are those of the water (ITIS, 2019; Putnam, 1971).

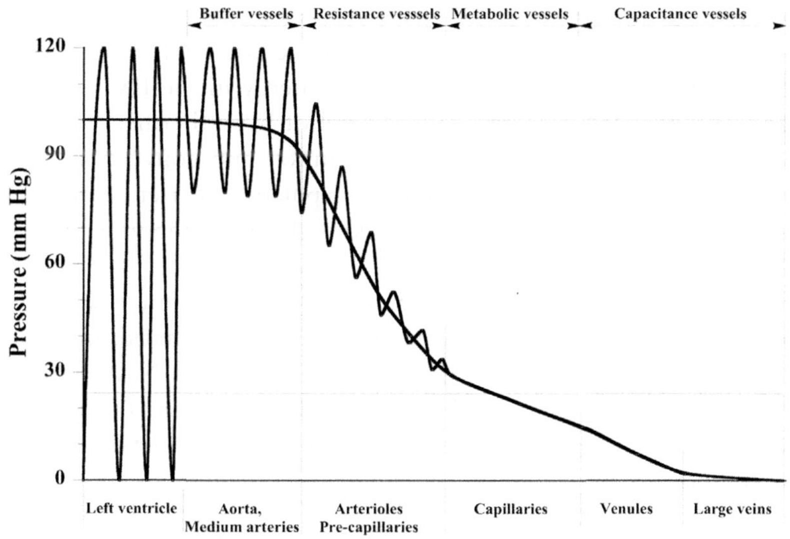

Figure 8.2 Blood vessels classification. After Feijóo, 2000. Computational methods in biology. In: Proceedings of the 2nd Summer School LNCC/MCT, Petrópolis, Brazil.

Using online imaging techniques (e.g., unenhanced CT) (Zagoria, 2004), a "LeVeen" array of electrodes is positioned close to the tumor using a trocar guideway (Fig. 8.1B and C) (Boston, 2013). The tumor volume, modeled by an ellipsoid here, is located in a region that is prone to its proliferation through genesis and angiogenesis (Maeshima and Makino, 2010; Osteaux and Jeanmart, 1979).

The time scales of the physics that concur suggest that from the first law perspective, the RF-EMF heat and electric work interactions are consistently modeled as electrokinetic (Morega et al., 2020)

$$\nabla\cdot(-\sigma\nabla V)=0, \tag{8.4}$$

where V is the electric potential and σ is the electrical conductivity. Dirichlet boundary conditions are set for the kidney surface (ground) and for the electrode tips ($V=22$ V), which provide the required power level for a successful ablation (Fig. 8.3, left). The inlet arterial velocity (uniform profile) is 0.1 m/s and a uniform pressure profile is set for the vein outlet.

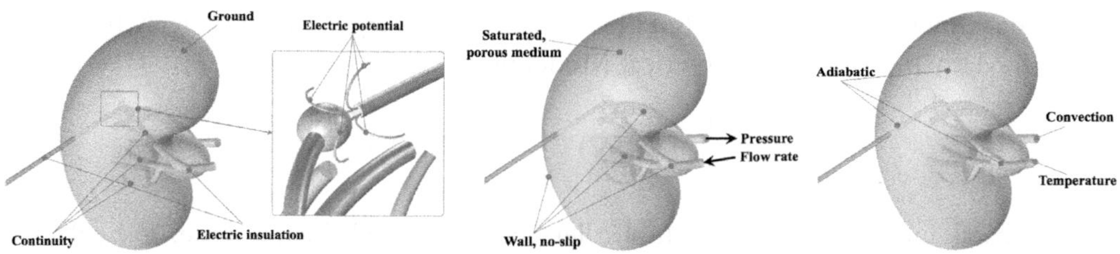

Figure 8.3 Boundary conditions for the RF ablation model. *RF*, Radio frequency (Morega et al., 2020).

The work interaction of the EMF with the renal tissue, or the "resistive" heat source (Joule heating), is readily available once the electric field is solved. Qualitatively RF heating contributes to increasing the internal energy in the ROI, which results in the local temperature increase to the desired level of hyperthermia. The local thermal unbalance triggers the heat transfer, through diffusion and convection, to the neighboring, cooler regions. The blood vessels, from capillaries to arteries and veins, convey a significant part of the heat, and its accurate knowledge is important in adjusting the right level of RF exposure.

The hemodynamic inside the kidney, a saturated porous medium, is slow (Re $<<$ 1); therefore Stokes—Brinkman momentum conservation is recommended. In the larger vessels (arterial and venal trees), the flow is faster Re $\sim$ $O(10^2)$, and Navier—Stokes form of momentum equation may be utilized. Moreover the flow time scale ($\sim$ seconds) is smaller than the heat transfer time scale ($\sim$ minutes), which yields the stationary forms of these momentums equations

$$\rho[(\mathbf{u}\cdot\nabla)\mathbf{u}] = -p\mathbf{I} + \mu\left(\nabla\mathbf{u} + (\nabla\mathbf{u})^T\right), \tag{8.5}$$

$$\nabla\cdot\left[-p\mathbf{I} + \mu\frac{1}{\varepsilon_p}\left(\nabla\mathbf{u} + (\nabla\mathbf{u})^T\right)\right] - \left(\mu\kappa^{-1} + \beta_F|\mathbf{u}| + \frac{Q_{br}}{\varepsilon_p^2}\right) - p\mathbf{I} + \mu\frac{1}{\varepsilon_p}\left(\nabla\mathbf{u} + (\nabla\mathbf{u})^T\right)\mathbf{u} = 0. \tag{8.6}$$

The mass conservation law (incompressible flow)

$$\rho\nabla\cdot\mathbf{u} = Q_{br}, \tag{8.7}$$

adds to the mathematical model. In the above equations $\mathbf{u}$ is the velocity, p the pressure, ρ the mass density, μ the dynamic viscosity, $(\cdot)^T$ the transposition operator, $\mathbf{I}$ the unity matrix, κ the porosity, ε_p the permeability, Q_{br} is a mass source, and β_F is a Forchheimer term (drag coefficient). The larger vessels and the porous medium flows are connected through boundary conditions that refer to pressure and velocity. Moreover at the vascular level of the kidney, the rheology of the blood is presented through a power law, where the dynamic viscosity is $\eta = m\cdot\dot{\gamma}^{n-1}$, Table 8.3 (Morega et al., 2010, 2020; Shibeshi and Collins, 2005). The boundary conditions of the hemodynamic model are presented in Fig. 8.3 (central image).

In the general heat transfer (GHT) model, the energy equation is (Chapter 1: Physical, Mathematical, and Numerical Modeling)

$$\rho C\left[\frac{\partial T}{\partial t} + (\mathbf{u}\cdot\nabla)T\right] = \nabla(k\nabla T) + \dot{q}. \tag{8.8}$$

The basal metabolic rate, much smaller than the RFA heat source, is neglected here. The upstream temperature of the arterial blood is 37°C. Convection heat

Table 8.3 Quantities used in the numerical modeling of the RF kidney ablation.

Symbol	Property	Value
m	Fluid consistency coefficient	$0.017\ \mathrm{Pa} \cdot n$
n	Flow behavior index	0.708
$\dot{\gamma}$	Lower shear rate limit	$0.01\mathrm{s}^{-1}$
κ	Porosity	$10^{-9}\ \mathrm{m}^2$
ε_p	Permeability	0.03
β_F	Forchheimer drag coefficient	0
Q_{br}	Mass source	0
k_b	Thermal conductivity, blood	$0.543\ \mathrm{W/(m \cdot K)}$
k	Thermal conductivity, renal tissue	$0.4812\ \mathrm{W/(m \cdot K)}$
ρ_b	Mass density of blood	$1000\ \mathrm{kg/m^3}$
ρ	Mass density of renal tissue	$1000\ \mathrm{kg/m^3}$
$C_{p,b}$	Specific heat capacity of blood	$4180\ \mathrm{J/(kg \cdot K)}$
C_p	Specific heat capacity at constant pressure renal tissue	$3771\ \mathrm{J/(kg\ K)}$
Ta	Basal temperature	37°C
$\dot{q}$	Specific power (Joule heat source)	$\mathbf{E} \cdot \mathbf{J}\ (\mathrm{W/m^3})$
ω	Blood perfusion rate	$3.4 \times 10^{-3}\mathrm{s}^{-1}$

RF, Radiofrequency.

transfer (no conduction) is set for the vein outlet and continuity conditions for the interfaces. The surface of the kidney is in thermal equilibrium with its surroundings (adiabatic) throughout the RF process (Fig. 8.3 right). The validity of this hypothesis is checked throughout the numerical simulation.

In the bioheat transfer (BHT) model, there is no directional hemodynamic flow, and the kidney is a homogeneous medium with a distributed heat source/sink that accounts for its contribution (convection) to the heat transfer

$$\rho C_p \frac{\partial T}{\partial t} + \rho_b C_{p,b} \omega (T_b - T_a) = \nabla(-k \nabla T) + \dot{q}, \tag{8.9}$$

a form of the energy equation. The hemodynamic and thermal properties are presented in Table 8.3.

For consistency, the input flow rate in the GHT model, related to the renal arterial inlet velocity (0.1 m/s), is sized to match the BHT perfusion rate, ω. Adiabatic boundary conditions close the BHT model too.

Numerical modeling

The mathematical model (4)–(9) was integrated using the finite element method (FEM) (Peralta et al., 2006; Comsol, 2010–2019). The computational domains are meshed using tetrahedral, quadratic, Lagrange elements, Fig. 8.4.

In the GHT model, because there are no couplings, the properties are temperature-independent, the electrokinetic problem (8.4) and the hemodynamic (8.5)–(8.7) problems

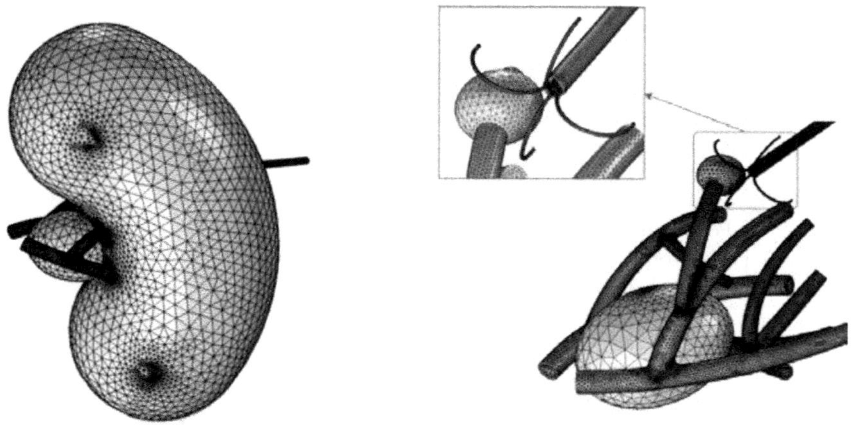

Figure 8.4 The FEM mesh made of ~450,000 tetrahedral, quadratic, and Lagrange elements. *FEM*, Finite element method.

are solved first, only once, independently, and sequentially. For the BHT analysis, the properties are independent of temperature hence the electrokinetic is solved once, in the beginning.

The electric current density provided by the trocar electrodes is seen in Fig. 8.5A. The blood flow presented in Fig. 8.5B and C (Morega et al., 2020) is driven by 49 kPa pressure drop (arterial entrance to the venous exit), which is consistent with available experimental data (Layton, 2013; Peralta et al., 2006).

The RF procedure may be successful provided that the temperature of the targeted ROI is increased to atleast 45°C—a critical threshold. The total (Joule) power received by the tumor is 0.141 W, in the BHT model, and 0.149 W in the GHT one. This slight discrepancy may be caused by the electric current density distribution in the two models. They may be due to the electrical conductivities of blood and renal tissue, which are different.

Figure 8.6 shows the critical isotherm of 45°C, which contains part of the tumor volume (the tumor surface is rendered in green), and the inner isotherms of 60°C and 80°C after approx. 10 min. The progression of the ablated volume may be evaluated by using the Arrhenius model discussed earlier, in this chapter. The two models cast different predictions. Whereas the BHT may suggest that the tumor is completely ablated, the GHT model shows off that only part of the tumor is well addressed. In both situations, the critical isotherm (45°C) contains the volume prone to ablation, consisting of either part (GHT) or the entire (BHT) tumor and also some neighboring, healthy tissue. Depending on the specific conditions, a second electrode array or a better positioning of the electrode array may be advised to reduce the side effects. However, due attention should be devoted to avoiding excessive, unneeded healthy tissue damage.

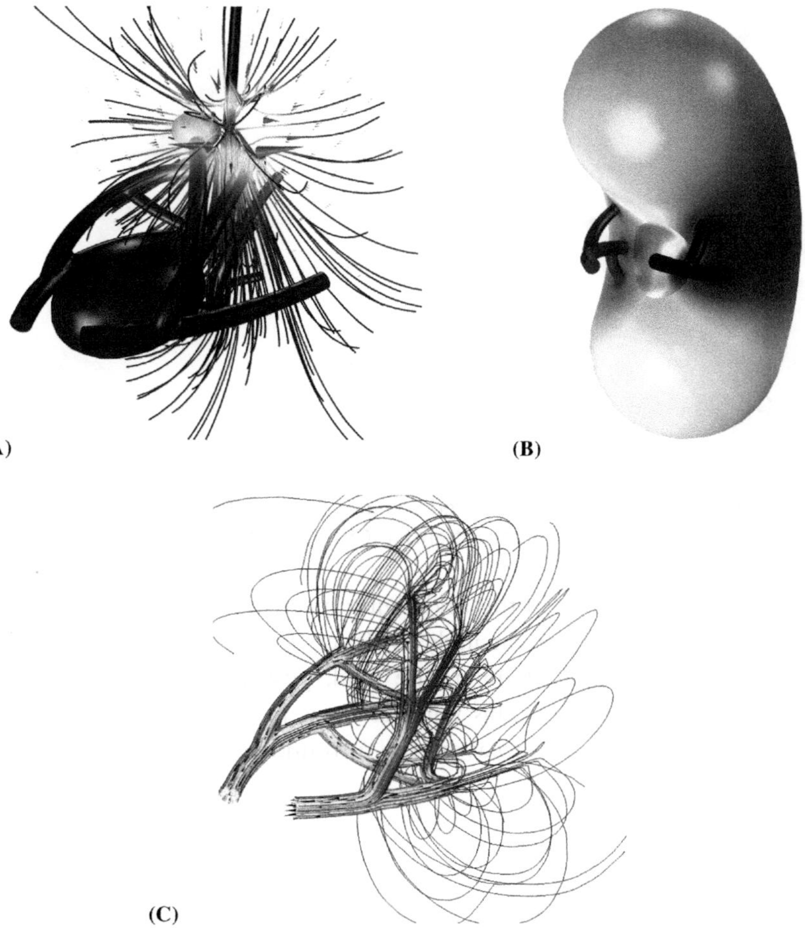

(A) (B) (C)

Figure 8.5 The electric field and the hemodynamic flow in the general heat transfer analysis: (A) electric current density and electric potential (max. 22 V); (B) the pressure field (); (C) the velocity field (inlet velocity, 0.19 m/s).

The volume-averaged temperature of the tumor rise to the critical plateau is predicted by both models, Fig. 8.7, but the BHT model tends to overestimate the RF heating effect (Morega et al., 2020). This finding is important because if the numerical simulation is used to aid the planning of this RF procedure then the bioheat model may be overestimating the RFA success. Joule heating (duration, power level, not to enter into the details of the electrodes and their positioning) that seems to be required for the success of the planned protocol would be, in fact, highly underestimated. This would mean undersizing the input power level. Heating has to be maintained long enough, and the power level should be properly adjusted (increased) because larger vessels are identified as significant enthalpy paths that drain part of the power delivered by the electrode.

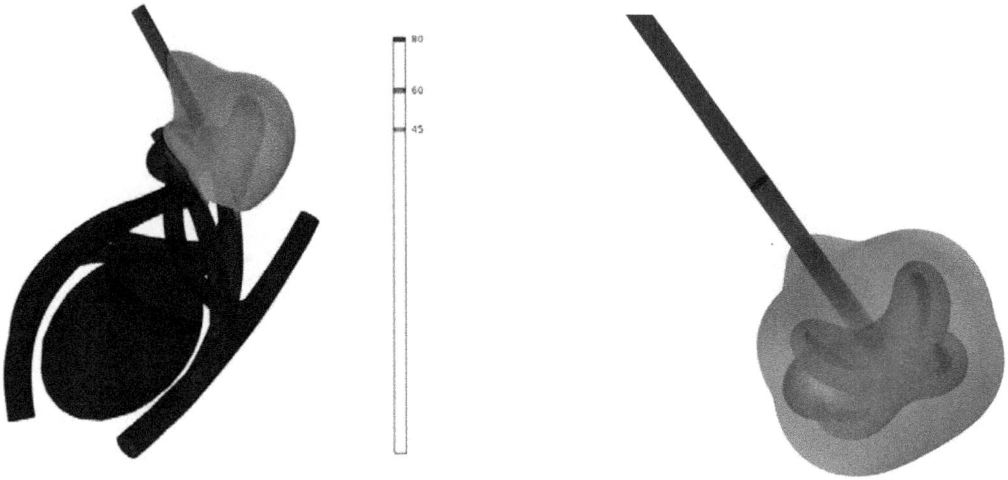

Figure 8.6 The isotherms of 45°C (the ablated volume), 60°C, and 80°C. GHT (left) and BHT models. Values are in degrees Celsius. *BHT*, Bioheat transfer; *GHT*, general heat transfer.

Numerical simulation unveils the underlying mechanisms of heat generation and the transfer of this procedure. It convincingly evidences the discrepancies between the thermal loads in the two models (temperature field, power levels, etc.) and the pending hypothermic or ablative effects. This information is particularly important to thermally approach the tumor volume and to avoid the surrounding healthy tissue perturbation.

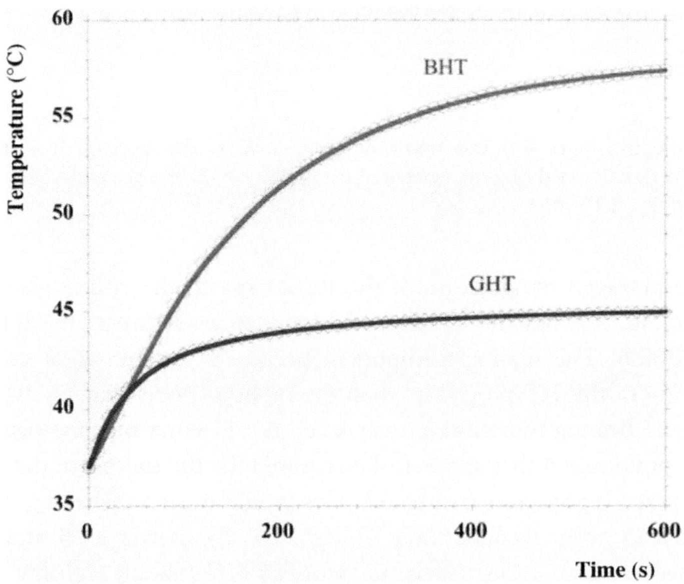

Figure 8.7 The average temperature of the ROI volume. *ROI*, Region of interest.

A key observation is that image-based reconstruction techniques based on the patient's relevant medical scans (DICOM sets), for example (Morega et al., 2010), for the computational domains, are required to implement the GHT as a convenient, cost-efficient, noninvasive tool that may contribute to better understand the RFA procedure, and to its optimization.

Some thermographic considerations

Pennes model belongs to the "two-temperatures" thermal model, with the observation that one of the temperatures (blood) is a reference and not the result of a convective-diffusion process. The benefits and limitations of this bioheat transfer model were discussed in Chapter 1: Physical, Mathematical, and Numerical Modeling. To evidence that its usage comes with some difficulties also in thermographic imagery, we consider the simple model presented in Fig. 8.8, which is a "cut-out" from a tissue vascularized by capillaries and larger blood vessels, and a heat source of spherical shape (e.g., tumor), of 9 mm radius, located in the ROI, spherical and concentric with the tumor, of 8 cm radius.

The solution to heat transfer in large blood vessels requires the solution of a hemodynamic problem and the associated heat transfer solution. We consider the blood flow (here Newtonian fluid) stationary, laminar, and incompressible, with $\rho = 1000\ \mathrm{kg/m^3}$ and $\eta = 0.003\ \mathrm{Pa \cdot s}$. There is no flow between the arterial and venous trees at this ROI level. The boundary conditions for the hemodynamic problem are uniform inlet speed profile (20–50 cm/s), no-slip at the vessel walls, and uniform pressure at the outlet.

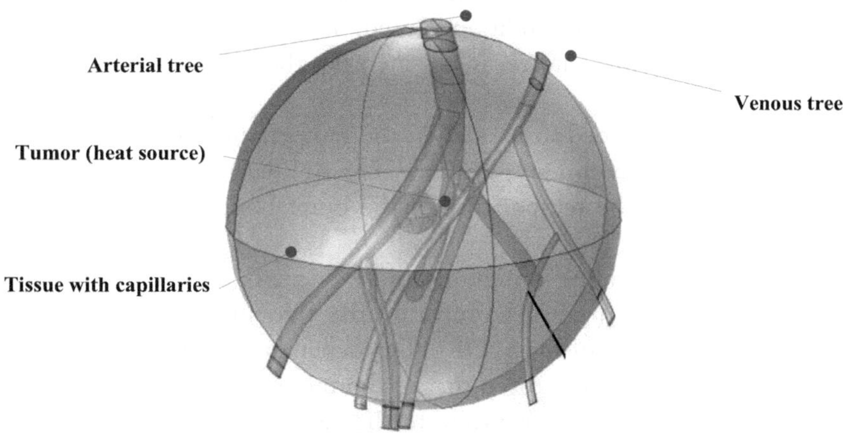

Figure 8.8 The ROI: a spherical volume singled out of a tissue vascularized with capillaries and larger arterial and venous trees that do not communicate. *ROI*, Region of interest.

The boundary conditions for the thermal problem are: metabolic temperature (37°C) for arterial and venous artery inlets and convection heat transfer ($\partial T/\partial n = 0$, $n = |\mathbf{n}|\mathbf{n}$ is the normal unit vector) for the arterial and venous outlets. On the outer boundary, convective heat transfer with a specified heat transfer coefficient, h, is set. The tumor is removed from the computational domain, its thermal contribution is reflected by its surface (inner border), which is considered isothermal at 45°C. This value is higher than the usual values encountered in situations alike (41°C–42°C), but we chose to use this value to highlight the presence of the tumor from a thermal point of view. The metabolic power rate is 1317 W/m^3 (Morega et al., 2015b).

"Model A" refers to the homogenized model and "Model B" to the combined bioheat (for capillaries) and general heat transfer model (for larger vessels). The two flows are not connected here as they were connected previously when a porous medium, Brinkman flow was used for the capillaries. The reason is to show off the discrepancies between the two modes of heat transfer: the homogenization technique (with perfusion, non-directional flow) and the convection heat flow (directional flow).

The boundary condition on the outer surface is convection, and it introduces an average heat transfer coefficient, h. It is not known *a priori*, and some "natural" thermal criterion has to be used to find a realistic estimation for it. The boundary here is imaginary, as the computational domain is "cut" from an organ (e.g., liver). Convection and not conduction is required here because the surface of the ROI is, at any point, in contact with some biological fluid. The reference temperature, under these circumstances, may not be that of the ambient, extracorporeal environment, for example, $T_{amb} = 1\ 8$°C, which is common for thermographic investigations. Using this condition, inadequate for the body inside, may produce vasoconstriction in regions that are not affected by some pathologic inflammatory process (De May et al., 2016).

The tumor is excluded from the computational domain, and its surface is an interior boundary. Here a Dirichlet condition is used, $T_{tumor} = 45$°C, as previously mentioned. The reason for this choice is to increase the thermal contrast with the background.

As the ROI is homogenized in Model A, the temperature on the outer surface of the ROI is, for reasons of symmetry, uniform, Fig. 8.9 (left). The tumor is evidenced by the overall increase in the temperature of the outer boundary that remains, however, isothermal.

In Model B the temperature on the boundary departs from uniform due to the blood vessel trees, Fig. 8.9 (right). Comparing the two models, the tumor thermal "mapping" on the boundary of the ROI differs depending on the modeling assumptions. Moreover, it may be conjectured that in thermography too, the mapping of the tumor as perceived by the camera is prone to the vascularization in between the heat source and the body surface. Here, the temperature outside the ROI, the position, size, and morphology of the tumor, and the structure of the blood vessel network are

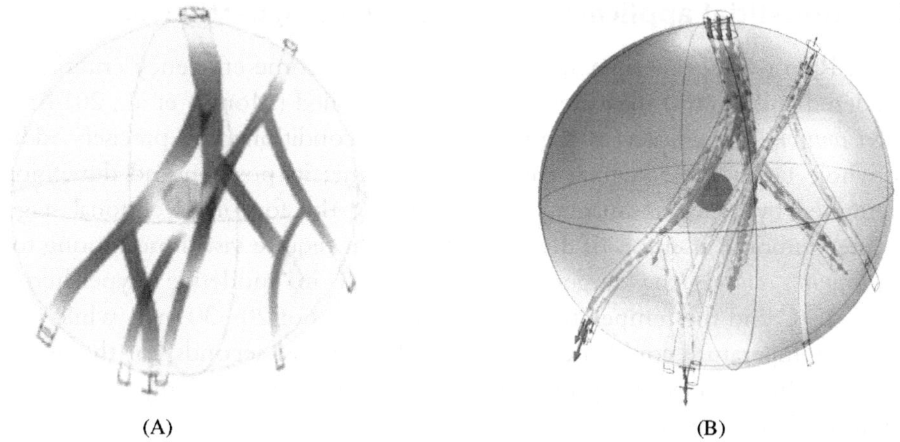

Figure 8.9 The thermographic image of heat source inside the ROI. (A) Model A—The ROI is a homogenized medium. (B) Model B—The ROI is a homogenized medium crossed by arterial and venous trees that do not communicate. *ROI*, Region of interest.

presumed known. The heat transfer to the surrounding tissue coefficient, h, may not be known, and some rationale should be considered to find reasonable values for it.

Figure 8.10 (Morega et al., 2015b) shows the temperature extreme values on the surface of the ROI for the two models, A and B. The shaded area [for $h < 5\ \mathrm{W/(m^2 \cdot K)}$] marks an h interval unlikely to occur due to unphysical temperatures. For $h > 5\ \mathrm{W/(m^2\ K)}$ the difference $\Delta T = T_{max} - T_{min}$ for both models, A and B, becomes almost constant. It follows that the interval $h = 5-10\ \mathrm{W/(m^2 \cdot K)}$ is recommendable. Similar results were also obtained for blood velocity values at the entry of different arterial and venous arteries ($\pm 20\%$), suggesting that this degree of freedom does not significantly influence the thermal effects.

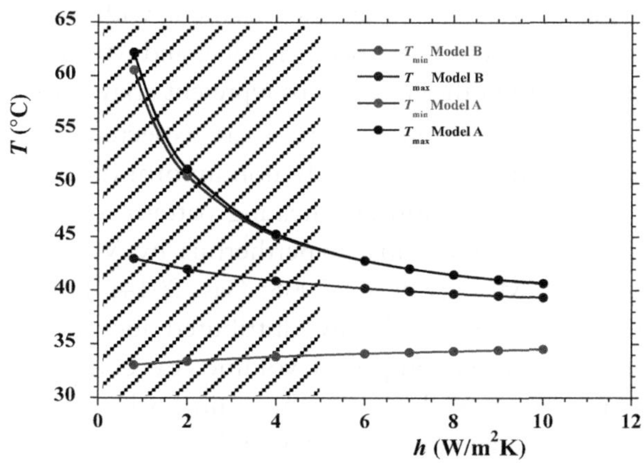

Figure 8.10 The selection of the convection heat transfer coefficient.

8.3 Pin interstitial applicators for microwave hyperthermia

When the RF or MWs antenna applicator is designed, some efficiency criteria need to be considered and control measures consequently applied (Morega et al., 2014):

Target control—the efficacy of the intervention is conditioned by precisely addressing the whole target tissue, which has a finite volume; its position and dimensions are determined by previous imaging analysis, during the fore-interventional stage, but the interventional insertion of the applicator might require visual monitoring too.

Temperature control—the target tissue is heated in moderate hyperthermia to 42°C−45°C and the temperature is held for sessions of 20−30 min, while for ablation the temperature could rise to 80°C−100°C in just seconds, for the burning of the tissue. In all cases, the power of the radiation source and the duration of the action need very careful monitoring.

Current technical solutions for designing interstitial applicators used to efficiently heat small volumes of tissue are achieved through pin-applicators array antennae that are inserted within the target volume or distributed around it.

Thin needle-shaped antennae (pin applicators) were introduced and tested for soft tissue hyperthermia treatment in the early 2000s (Ito et al., 2001, 2002; Saito et al., 2004). Their antenna is derived from a coaxial cable with short-circuited tip and a ring-shaped radiating slot, placed at a short distance from the tip. The original design was intensively tested and adapted for different application conditions, and the dimensions were adapted for different operation frequencies. For example, Morega et al. (2008) and Gas (2014) present optimized applicators models for the 2.45 GHz operating frequency; Gas and Czosnowski (2014) proposes a slightly modified design, namely, the coaxial antenna with two or three radiating slots, while Bertram et al. (2006) complements the design of the pin antenna with an electroconductive choke in an attempt to finely tune radiation best features.

Figure 8.11 (far left) shows the pin antenna element with a particular design, optimized for efficient operation at the Industrial, Scientific, and Medical (ISM) dedicated frequency of 2.45 GHz, while the other pictograms represent arrays of similar pin-antennae.

When a single-pin is used, heated tissue is, as expected, concentrated near the antenna-radiating slot, which leads to a highly nonuniform distribution of the temperature. The assessment of the radiation and heating pattern for a single pin antenna is the first step in the analysis of interstitial hyperthermia by MWs, like the studies presented in (Ito et al., 2002; Morega et al., 2008; Gas, 2014; Trujillo et al., 2018). An array applicator made of several optimally distributed identical pins could provide much uniform heating in the encompassed volume, by cumulative contributions of the individual pin components of the array; the main goal of its design targets to find the balance between the therapeutic efficiency and minimum invasiveness (Ito et al., 2002). Array applicators with several pins (as shown in Fig. 8.11) and their

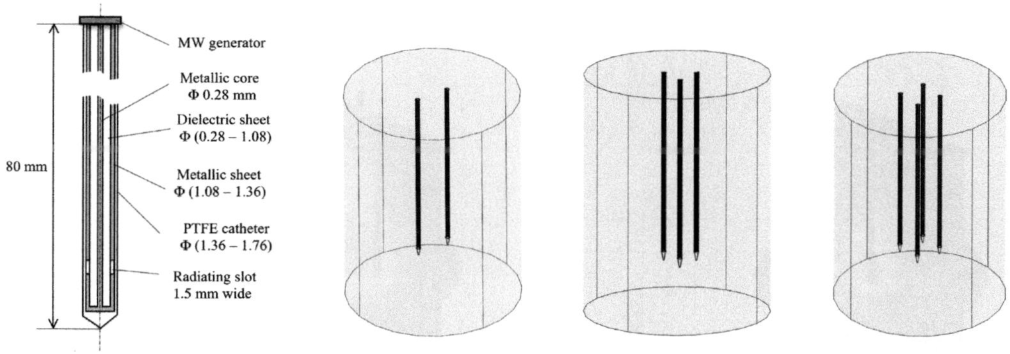

Figure 8.11 Basic models of array MW pin-applicators inserted in the target tissue (here, in the shape of a cylinder); the models are built for FEM numerical analysis (Morega et al., 2014, 2015a). *FEM*, Finite element method.

characteristics were examined in the study of Morega et al., (2008, 2014). Although they provide a good technical solution to the temperature control problem (i.e., the possibility of uniformly heating a predetermined volume) (Ito et al., 2001; Morega et al., 2006; Saito et al., 2004), the array applicators come with the disadvantage of increased discomfort or pain for the patient and contingent tissue damage through bleeding and swelling (NIH); however, these inconveniencies are temporary whereas the proceedings might be decisive.

The optimal design of the applicator translates here in choosing the best number of components and the spacing between adjacent pins, conditions correlated with the particular electrical and thermal behavior of the tissue, expressed by its physical properties for the specific operating frequency and temperature range (Andreuccetti et al., 1997; Rosmann; Karampatzakis et al., 2013). The efficiency of the procedure is furthermore conditioned by other characteristics of the electromagnetic radiation source: waveform, power, phase-shift between individual pin emissions. Various combinations could be tested and assessed with the help of an experimental model and even better by numerical simulation. A numerical analysis was performed that way and the optimal configuration for a three-pin applicator is shown in (Morega et al., 2006). Figure 8.12 illustrates, for comparison, the temperature distributions for a single-pin versus a three-pins array applicator (same pin design), aimed for mild interstitial hyperthermia, but scalable for ablation too.

For the three-pins array, the distance between adjacent pins was determined after an optimization study, and it leads to an almost uniform temperature distribution in the tissue encompassed by the pins. The three-pins are positioned at the vertices of an equilateral triangle at 13 mm distance; the tissue volume of approx. 1,150 mm^3 is quasi-uniformly heated (42°C−43°C) within the therapeutic temperature range for mild hyperthermia (Morega et al., 2006).

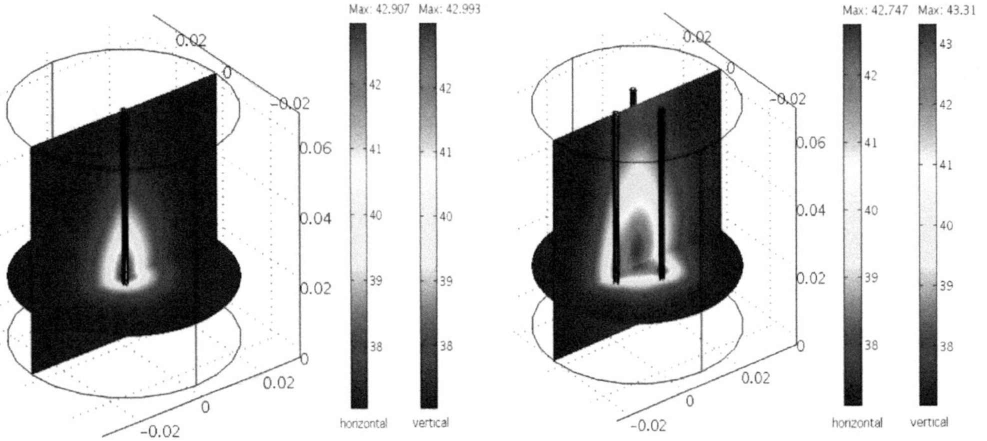

Figure 8.12 Microwave applicators for interstitial hyperthermia of soft tissue—1 W per pin, at 2.45 GHz; dimensions are in meters and temperature is in degrees Celsius. *Adapted from Morega, M., Mogos, L., Neagu, M., Morega, A.M., 2006. Optimal design for microwave hyperthermia applicator. In: Proceedings of the 11th International Conference on Optimization of Electrical and Electronic Equipment—OPTIM 2006, Brasov, Romania, pp. 219–224.*

Numerical analysis of heating when blood flow is taken into account

Temperature and target tissue control are best addressed in preinterventional stages by numerical simulation and analysis. An example of a complex mathematical model is given further, for coupled electromagnetic and thermal problems, with different conditions for blood flow heat transfer: the cooling effect of blood perfusion in soft tissue is compared versus more efficient heat extraction due to the presence of large blood vessels in the interventional tissue volume. Three physical problems are described and transposed in mathematical formulations (Morega et al., 2014):

1. *The EMF problem*—electromagnetic waves is emitted by the pin-antennae of the array; the mathematical expression is given by the waves equation (in terms of either electric field strength **E**, or magnetic field strength **H**). For time-harmonic working conditions [continuous waves (CW)], the equations are presented in their complex form.

$$\nabla\times\left(\frac{1}{\mu_0}\nabla\times\underline{\mathbf{E}}\right)-\omega^2\underline{\varepsilon}\,\underline{\mathbf{E}}=0 \text{ for } \mathbf{E},\ \text{or } \nabla\times\left(\frac{1}{\underline{\varepsilon}}\nabla\times\underline{\mathbf{H}}\right)-\omega^2\mu_0\underline{\mathbf{H}}=0 \text{ for } \mathbf{H}. \quad (8.10)$$

The quantities in Eq. (8.1) are explained in Tables 8.4 and 8.5.

For the anatomical domain, a volume of homogeneous material is considered, with a simple geometrical shape like the representations in Fig. 8.11 (right), or with an anatomical realistic shape, for example an organ (liver, lung, etc.) reconstructed from CT or RMN scans. For interstitial microwaves hyperthermia, the EMF problem is commonly confined to the volume of the organ; the external surface of the

Table 8.4 Physical quantities used in the mathematical models.

Symbol (unit)	Quantity
$\underline{\mathbf{E}}$ (V/m)	Electric field strength (in complex)
$\underline{\mathbf{H}}$ (A/m)	Magnetic field strength (in complex)
μ_0 (H/m)	Free space magnetic permeability
$\underline{\varepsilon}$ (F/m)	Complex permittivity $\underline{\varepsilon} = \varepsilon - j\sigma/\omega$, $\underline{\varepsilon} = \underline{\sigma}/j\omega$
ε (F/m)	Dielectric permittivity $\varepsilon = \varepsilon_r\varepsilon_0$
ε_0 (F/m)	Free space dielectric permittivity
ε_r	Dielectric constant (relative permittivity)
$\underline{\sigma}$ (S/m)	Complex conductivity $\underline{\sigma} = \sigma + j\omega\varepsilon$
σ (S/m)	Electrical conductivity
ω (rad/s)	Angular frequency
j	Complex number $j = \sqrt{-1}$
ρ (kg/m^3)	Mass density
ρ_b (kg/m^3)	Mass density of blood
ω_b (s^{-1})	Blood perfusion rate
C (J/kg · K)	Specific heat coefficient
C_b (J/kg · K)	Specific heat coefficient of blood
T (K)	Tissue temperature
T_b (K)	Blood temperature
k (W/m · K)	Thermal conductivity
Q_{emf} (W/m^3)	Resistive heat (density of the absorbed power)
$\mathbf{u}$ (m/s)	Velocity field
η (Pa · s)	Dynamic viscosity
p (Pa)	Pressure
$\mathbf{I}$	Unity matrix
m (kg)	Mass
W (J)	Electric energy

Table 8.5 Physical properties used in the model (Andreuccetti et al., 1997; Morega et al., 2008).

Hepatic tissue			
	$\underline{\varepsilon} = \varepsilon_0\varepsilon_r - \frac{j\sigma}{\omega}$, where $\varepsilon_0 = \frac{10^{-9}}{36\pi}\frac{\text{F}}{\text{m}}$ and $\omega = 2\pi f$		
	$\sigma = 1.686$ S/m and $\varepsilon_r = 43.035$ for the frequency $f = 2.45$ GHz		
	$\rho = 1060$ kg/m^3	$C = 3600$ J/kg · K	$k = 0.502$ W/m·K
Blood			
	$\sigma_b = 2.545$ S/m and $\varepsilon_{r\ b} = 61$ for the frequency $f = 2.45$ GHz		
	$\rho_b = 1000$ kg/m^3 $\omega_b = 6.4 \times 10^{-3}$ s^{-1}	$C_b = 4180$ J/kg · K $\eta_b = 0.05$ Pa·s	$k_b = 0.543$ W/m·K
Dielectric of coaxial cable $\sigma_{dielec} = 0$, $\varepsilon_{r\ dielec} = 2.03$		PTFE catheter $\sigma_{PTFE} = 0$, $\varepsilon_{r\ PTFE} = 2.6$	

computational domain might be described by scattering, or by low reflecting boundary conditions. The area of penetration of each antenna into the domain could be described by a port boundary condition (coaxial in this case), introducing in that way the source of electromagnetic waves.

2. *The heat transfer problem* is physically described by the balance of energy in the interventional region. When it applies to homogeneous tissue irrigated by a fine capillary blood network, the bioheat equation (Pennes, 1948) is the common mathematical representation, with its term that depends on the local vascularization (Chapter 1: Physical, Mathematical, and Numerical Modeling)

$$\rho C\frac{\partial T}{\partial t} = k\nabla^2 T - \rho_b C_b \omega_b (T - T_b) + Q_{met} + Q_{electric}, \tag{8.11a}$$

while forced convection heat transfer in large blood vessels is described by

$$\rho C\left[\frac{\partial T}{\partial t} + (\mathbf{u}\cdot\nabla)T\right] = k\nabla^2 T + Q_{met} + Q_{electric}. \tag{8.11b}$$

In both heat transfer equations the metabolic energy intake and consumption, Q_{met}, are commonly neglected, while the external heat source, $Q_{electric}$ (W/m^3), is the so-called *resistive heat* (i.e., the absorbed power density); it is generated by the EMF in the exposed tissue and results by processing the solution of the EMF problem

$$Q_{emf} = \frac{1}{2}\mathrm{Re}\left[(\sigma + j\omega\varepsilon)\underline{\mathbf{E}}\cdot\underline{\mathbf{E}}^*\right]. \tag{8.12}$$

It is important to notice that Eqs. (8.11) refer here to the harmonic regime hence the electric and magnetic field strengths are introduced in their complex representation; commercial electromagnetic analysis software commonly operates with the peak values of harmonic quantities.

The interventional region, extended up to adiabatic boundaries is the computational domain for the thermal problem.

3. *The hemodynamic problem* is defined inside the blood vessels; it is representative of the dynamics of blood flow (pulsating, incompressible, laminar) through a large vessel, sufficiently close to the interventional region, to influence the heat transfer. The mathematical form is described by the Navier–Stokes equation, which states for the momentum balance, associated with the mass conservation law; blood is assumed to be a Newtonian fluid for the particular conditions of this study.

$$\rho\left[\frac{\partial \mathbf{u}}{\partial t} + (\mathbf{u}\cdot\nabla)\mathbf{u}\right] = \nabla\left[-p\mathbf{I} + \eta(\nabla\cdot\mathbf{u} + (\nabla\cdot\mathbf{u})^T)\right] \text{ and } \nabla\cdot\mathbf{u} = 0. \tag{8.13}$$

The velocity is set to zero at the walls of the blood vessels.

FEM modeling media (Comsol, 2010–2019) may provide the solution of coupled problems of different physical content with controllable numerical accuracy. The electromagnetic time constants are much smaller than the thermal and flow ones; thus, Q_{emf}, which is a result of the radiation problem, represents a stationary (r.m.s.) heat source. The heat transfer and the hemodynamic flow in large arteries are one-way coupled (Morega et al., 2014).

Thermal analysis in mild hyperthermia of soft tissue

An example of efficiently heating the target volume of tissue for moderate hyperthermia presents the case of a two–pins applicator inserted in soft tissue, with the physical properties of the liver. The physical and mathematical models were presented above, and a comparison between the two cooling mechanisms through blood flow (by the capillary network versus large vessels) is performed. Each antenna follows the model shown in Fig. 8.11 (left), and the physical properties of the model with liver designated as target tissue are given in Table 8.5.

Homogeneous tissue model, with the vascularization provided by a network of capillary blood vessels, is first analyzed, based on the Eqs. (8.10) and (8.11a). The optimal distance of 12 mm between the two pins was determined after a parametric study, following an efficiency criterium: maximum volume of tissue heated as uniform as possible within the temperature therapeutic range. The numerical results presented in Fig. 8.13 show the color spectra of temperature in a longitudinal section plane through the heated volume (left image), and the isothermal surface for 41.4°C (right image), which is the boundary of

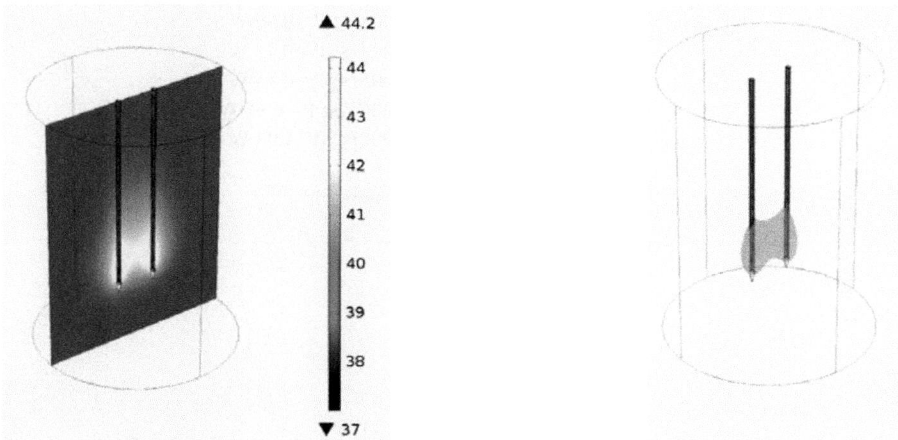

Figure 8.13 Temperature distribution (in Celsius) for the array applicator (left) and the heated volume of tissue, limited by the isothermal surface of 41.4°C (right) (two-pin array at 12 mm distance between axes, 2.45 GHz and 1 W power per antenna). *From Morega, M., Morega, A.M., Diaz, M.I., Sandoiu, A.M., 2014. Percutaneous microwaves hyperthermia study by numerical simulation. In: Proceedings of the Internatinal Conference and Exposition on Electrical and Power Engineering—EPE 2014, Iasi, Romania, pp. 498–503.*

910 mm^3 volume of tissue heated within the range of 41.4°C–44.2°C. The unsteady heating process was also analyzed for further comparison with other heat transfer conditions (see curves in Fig. 8.14).

The physical properties of tissues represent a category of crucial data for numerical simulation and analysis. Not only that all kinds of complications arise when *in vivo* assessments are needed, but the measured values present relatively large dispersion ranges due to inherent factors, like water and other chemical content, homeostasis, or other activity of the living organism, etc. The models show different degrees of sensitivity to various data, and for the hyperthermia problem, the vascularization efficiency quantified by *the blood perfusion rate* is one of the most influential factors of the heating process, while dielectric or thermal properties (see Table 8.4 for a complete list) do not affect in a comparable measure the behavior of the model and the results. A detailed study on this topic is performed in Morega et al. (2015a); the results of a parametric study are further shown for the variation of the perfusion rate in reasonable limits around the reference value previously used $\omega_b = 6.4 \times 10^{-3}$ /s.

The blood perfusion rate depends on the blood flow rate (size of blood vessels and velocity) in the region. Fig. 8.14 compares several temperature distribution profiles computed for some successive the blood perfusion rates, within a common range of $3–8 \times 10^{-3} s^{-1}$; the graphs present the temperature along observation lines drawn horizontally, between the antennae, through the hottest region evidenced also by

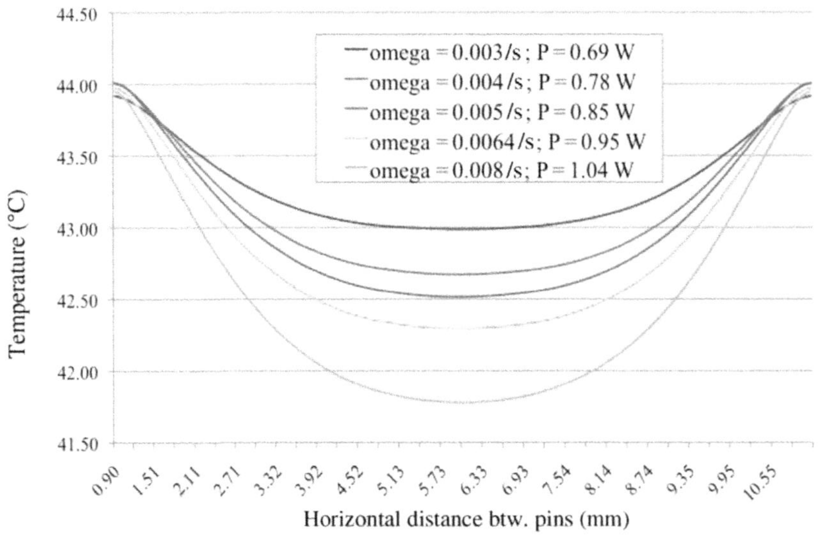

Figure 8.14 The temperature along horizontal observation lines through the heated tissue, for different values of the blood perfusion rate (the feeding power of each pin P is adjusted for heating within the therapeutic range) (Morega et al., 2015a).

Fig. 8.13. It is important to underline that a supplementary restriction is introduced in the study, and it refers to keeping the temperature within the therapeutic range. The maximum temperature is easy to be monitored because it occurs near the radiating slot of each antenna; it is set to 44°C, and the temperature control by the adjustment of the feeding power of the antennae is thus applied if necessary.

For the next step of the analysis, the presence of a large blood vessel in the interventional zone is considered, as the computational domain in Fig. 8.15 is showing. The numerical model is complemented with phenomena associated with the blood flow through a large artery, that is, heat transfer by forced convection introduced in Eq. (8.11b) and flow dynamics described in Eq. (8.13).

Some significant functional points are marked in Fig. 8.15: the Inlet is important because an adequate boundary condition should be specified for the fluid flow at the entrance of the bloodstream into the computational domain; Station Q represents the exit point for the bloodstream; Station P marks the position of a temperature sensor for the monitoring of the heating in the interventional region.

Three timescales govern the dynamics highlighted by the set of equations adopted here: the problem of the electromagnetic wave evolves at high speed, or with the fastest time scale, followed by the flow problem at a slower time scale and finally the heat transfer problem, which has the slowest evolution, that is, slowest time scale. The solving algorithm applied here for moderate hyperthermia does not take into consideration the variation of physical properties with the temperature; in such circumstances, the EMF problem, Eq. (8.10), is solved first and its solution is used for the estimate of the resistive heat Q_{emf}.

Heat transfer and flow, Eqs. (8.11b) and (13), should be solved in one step, due to the convective term coupling. Aiming to get a proper balance between the actual time step size, and the total time required to get the steady-state of the heating

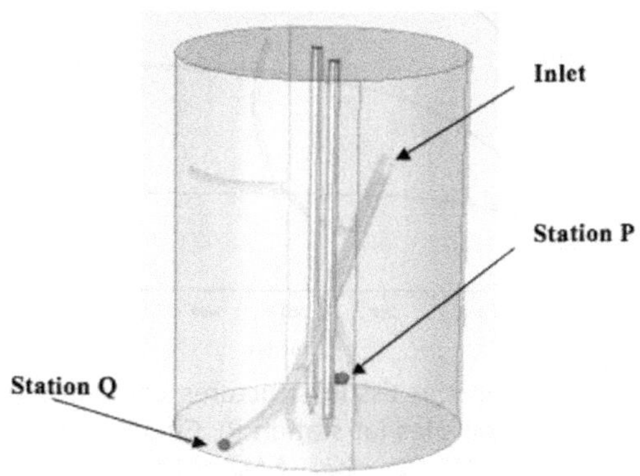

Figure 8.15 Computational domain with an arterial tree segment. Temperature is recorded at Station P.

process, an *equivalent steady flow* was defined, based on the averaging of the inlet velocity over each period τ of the pulsating flow (Morega et al., 2014)

$$U_{av} = \frac{1}{\tau}\int_0^{\tau} U_{inlet}(t)\mathrm{d}t, \text{ where } U_{inlet}(t) = U_0\left[\sin(2\pi ft) + \sqrt{\sin^2(2\pi ft)}\right]. \tag{8.14}$$

In that way, the second step of the numerical procedure is performed. For particular data used in this study, $U_{inlet} = 40$ cm/s, $f = 60$ bpm and $U_{av} = 25.5$ cm/s. The unsteady heat transfer, (8.11b) and (8.13), is finally solved.

A suggestive comparison between the two different heat transfer conditions is made by the temperature dynamics rendered in Fig. 8.16. Curve A shows the temperature rise for the hottest point inside the interventional region; this point is located at the intersection of the Oz axis of the cylindrical volume and the xOy plane, at the mid-height level of the radiating slots. Quasi-steady-state heating is attained after approximately 500 s. The other two graphs show the temperature rise at the same location (Station P marked in Fig. 8.15) for the two heat transfer problems considered in this study: cooling of the tissue by a capillary network (curve B) versus the case of a large artery included in the interventional region (curve C), which leads, of course, to the lowest heating of the spot under observation. In all compared cases, the MWs source provides the same emissions.

Curve C of Fig. 8.16 is the averaged evolution of the temperature; Fig. 8.17A, presents the accurate pulsating temperature rise at Station P positioned inside the interventional region close to the artery (see Fig. 8.15), and influenced by the blood pulsating flow, while Fig. 8.17B shows the cross-sectional average temperature of the blood at the outlet cross-section (Station Q marked in Fig. 8.15).

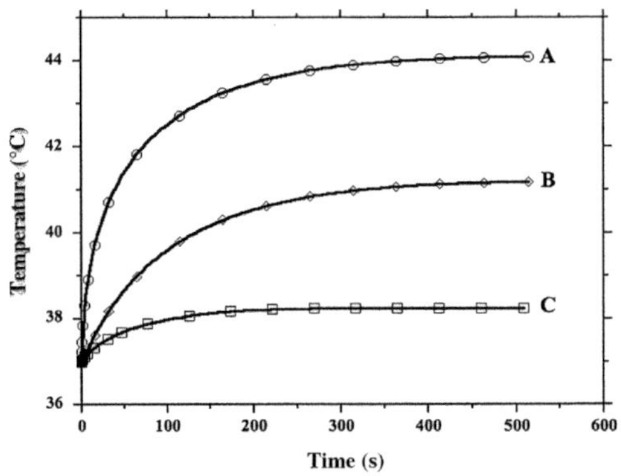

Figure 8.16 Temperature dynamics in the tissue structures compared in the study. A-cooling by capillaries (hot spot); B-cooling by capillaries (at station P); C-cooling by a large artery (at station P). *From Morega, M., Morega, A.M., Diaz, M.I., Sandoiu, A.M., 2014. Percutaneous microwaves hyperthermia study by numerical simulation. In: Proceedings of the Internatinal Conference. and Exposition on Electrical and Power Engineering—EPE 2014, Iasi, Romania, pp. 498–503.*

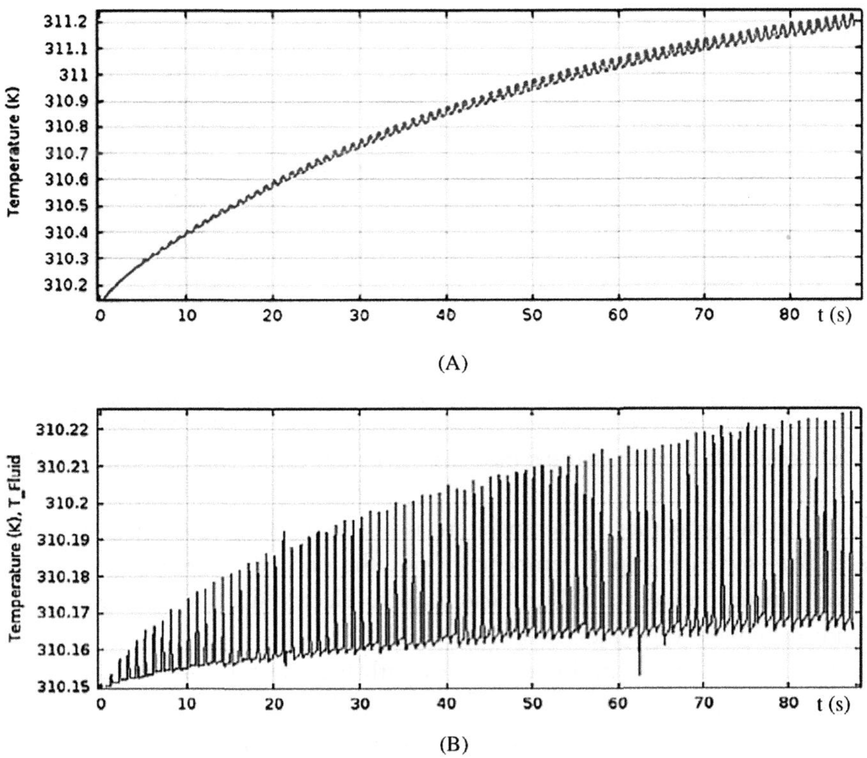

Figure 8.17 Unsteady heating; temperature fluctuations superpose over the general heating trend. (A) Temperature at Station P and (B) Average temperature at Station Q (Morega et al., 2014).

Some more results obtained from the numerical simulation show by blue streamlines in Fig. 8.18 the distributions of the electric (left) and magnetic (right) components, influenced by the dielectric properties of the blood inside the vessels. The influence of the large artery is visible in Fig. 8.19 for the cooling effect and temperature distribution; the image is captured during the heating process, after 480 s.

Temperature-dependent dielectric properties

Heating is monitored by a transient mode. The literature does not provide much information on the temperature dependency for tissues' physical properties in humans. In mild hyperthermia, where the heating does not go beyond 10 degrees, it was shown that variation of dielectric properties of the tissue with the temperature is insignificant to the accuracy of results (Morega et al., 2008); the consequence is that the electromagnetic and heat transfer problems are *one-way coupled*, that is, the first problem is solved for constant dielectric properties and its solution is transferred to the thermal problem by a constant heating source, which steadily contributes to the transient heating process. In ablation, however, the

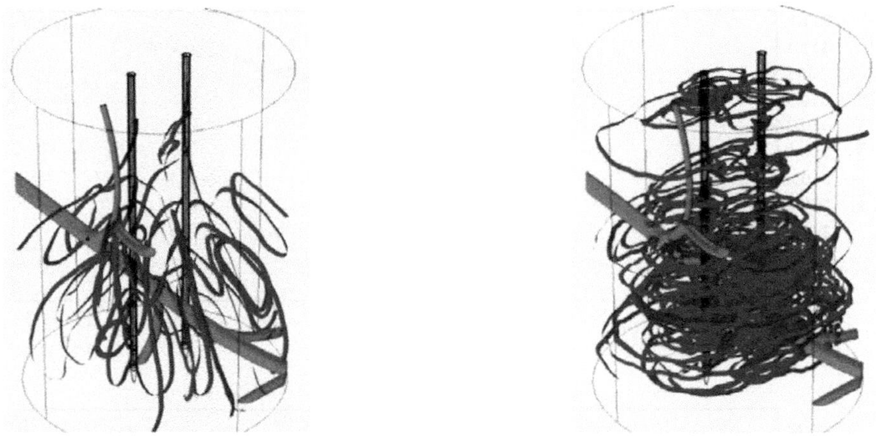

Figure 8.18 Electric field strength (left) and magnetic flux density (right) fields (Morega et al., 2014).

dependency of the properties on temperature is not negligible, and the two problems are *two-way coupled*—for each step of temperature rise, the dielectric properties are adjusted and the electromagnetic problem is solved again, so the electric field and the resistive heat are temperature-dependent quantities.

One could consider, for biological tissues with high water content, the temperature dependence suggested in (Morega et al., 2008; Nikawa, 1995) for dielectric properties—electric conductivity, σ, and relative permittivity, ε_r (T is in kelvins)

$$\sigma(T) = \sigma^*[1 + 0.005(T - 310)]\,\text{S/m and } \varepsilon_r(T) = \varepsilon_r^*[1 - 0.027(T - 310)], \quad (8.15)$$

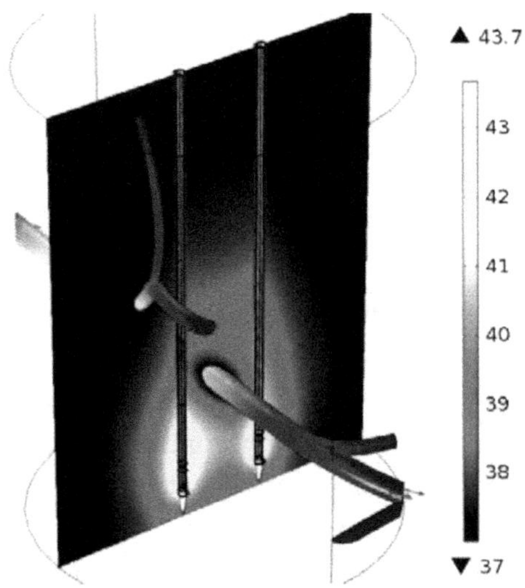

Figure 8.19 Temperature (slice color map in Celsius) and flow (pressure color map) fields ($t = 480$ s) (Morega et al., 2014).

with the reference values σ^* and ε_r^* taken from the well known and cited database (Andreuccetti et al., 1997) at the corresponding working frequency and standard body temperature.

The results of a numerical experiment documented in Morega et al. (2008) allow for a comparison between the evolution of highest temperatures in human liver heated by interstitial mild hyperthermia, performed with one pin-antenna (as shown earlier in Fig. 8.11 left) fed at 2.45 GHz · CW. The numerical analysis was executed as previously described and the reference dielectric properties for liver-like tissue are: $\sigma^* = 1.69$ S/m and $\varepsilon_r^* = 43$. The absorbed power inside the target tissue and the temperature distribution were determined for both coupling cases: unidirectional coupling of the electromagnetic and thermal problems (i.e., constant dielectric properties) and bidirectional coupling of the two problems (i.e., temperature-dependent dielectric properties). Figure 8.20 shows the color temperature spectra in the symmetry section of the coaxial pin-antenna in the two described cases (with constant (A), and temperature-dependent (B) dielectric properties), while Fig. 8.21 compares temperature variations with increasing antenna power in the same two cases.

8.4 Magnetic hyperthermia

More recently increased interest is visible in using either gold (Moran et al., 2009) or magnetic nanoparticles (Giustini et al., 2010) to enhance and improve the selective heating of diseased tissue, thereby obtaining improved hyperthermia treatment. Because tumors have higher heat-sensitivity over normal cells, heating them to 41°C–47°C results in their selective damage. Moreover, an equivalent therapeutic efficacy was observed when combining a

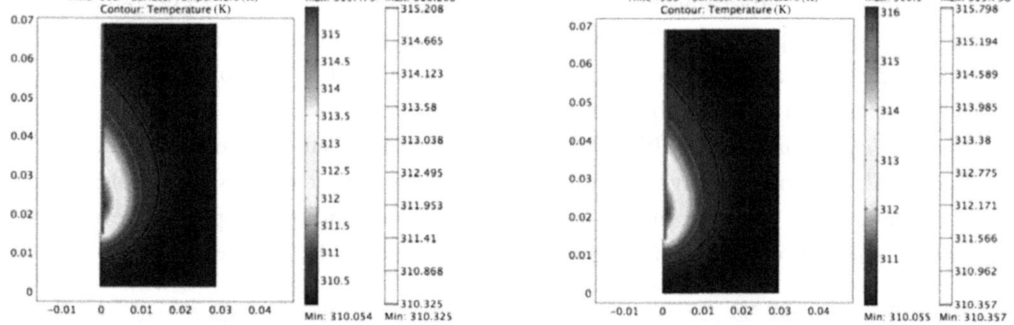

Figure 8.20 Temperature distribution in the plane of axial-symmetry, for constant (A) versus temperature-dependent (B) dielectric properties of heated tissue (2.45 GHz, 0.9 W). *Adapted from Morega, M., Neagu, M., Morega, A.M., 2008. Bidirectional coupling of electromagnetic and thermal processes in radiofrequency hyperthermia. In: Proceedings of the 12th International Conference on Optimization of Electrical and Electronic Equipment—OPTIM 2008, Brasov, Romania, pp. 257–262.*

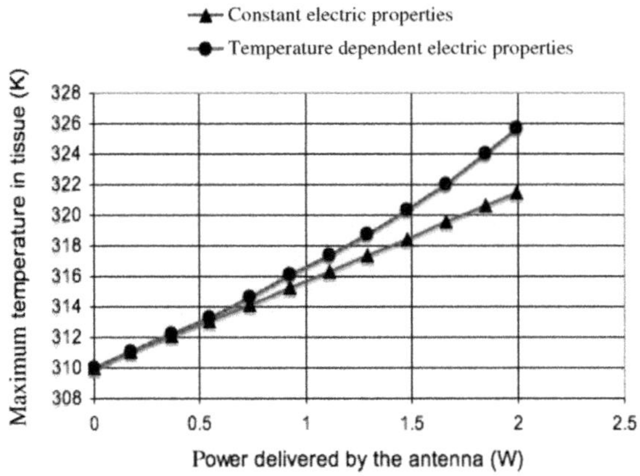

Figure 8.21 The maximum temperature inside the heated tissue as a function of the antenna power for constant versus temperature-dependent dielectric tissue properties. *Adapted from Morega, M., Neagu, M., Morega, A.M., 2008. Bidirectional coupling of electromagnetic and thermal processes in radiofrequency hyperthermia. In: Proceedings of the 12th International Conference on Optimization of Electrical and Electronic Equipment—OPTIM 2008, Brasov, Romania, pp. 257–262.*

smaller dose of radiation with magnetic hyperthermia (then radiation alone) (Johannsen et al., 2010). Several magnetic nanomaterials were considered for their potential for hyperthermia, including Iron oxide nanoparticles (Fe_3O_4 and $\gamma-Fe_2O_3$) stabilized by ligands such as dextran, PEG, and polyvinyl alcohol, to increase their circulation time in the body and to prevent clearance by the *Mononuclear phagocyte system* (MPS) (Ito et al., 2005; Kumar and Faruq, 2011). The Iron oxide MNPs are nontoxic, highly biocompatible, and they are metabolized to form blood hemoglobin, thus maintaining homeostasis of iron inside cells. Other promising applications of MNPs are in the controlled drug release for the treatment of diseases including tumors. MNPs along with therapeutic molecules are encapsulated in a pH or heat-sensitive polymers. Drug release can be triggered by external stimuli. For sensitive polymers, the heat generated within the MNPs may lead to the formation of pores in polymers, resulting in releasing therapeutic molecules.

Magnetic hyperthermia uses MNPs delivered at the tumor site before the application of the EMF. MNPs can be passive, by convection–diffusion mechanisms in vascularized tissues, or active, using targeting surface-attached ligands for binding to appropriate receptors expressed at the tumor site and not expressed by normal cells (Danhier et al., 2010).

The magnetic field work interactions

The external *magnetic field* work interaction with the MNPs depends on the frequency and on the size and other characteristics of the MNPs, and, in ferrofluids, it occurs through two mechanisms: hysteresis (Warburg theorem, Chapter 1: Physical, Mathematical, and Numerical Modeling), because small particles exhibit a large hysteresis loop, and relaxation—Néel relaxation (rotation of the magnetic moments) and Brown(ian) relaxation (rotation of the particles themselves within the fluid)—which occur simultaneously in ferrofluids. In superparamagnetic MNPs only Néel relaxation occurs.

The time constants of these relaxation processes are (Urdaneta, 2015)

$$\tau_{Brownian} = \frac{4\pi\eta r_h^3}{kT}, \quad \tau_{Neel} = \tau_0 e^{KV/\kappa T}, \quad \tau^{-1} = \tau_{Brownian}^{-1} + \tau_{Neel}^{-1}, \tag{8.16}$$

where r_h is the hydrodynamic radius of a particle (its coating is larger than the radius of the MNP), η is the dynamic viscosity of the carrying fluid, T is the temperature, K is an anisotropy constant, V is the magnetic volume, k_B is Boltzmann constant, and $\tau_0 = 10^{-9}$ s. For hyperthermia at near 42°C, the temperature dependence of the relaxation times may be neglected.

Recall (Chapter 1: Physical, Mathematical, and Numerical Modeling) that the temporary magnetization, in complex representation (harmonic working conditions) is $\underline{\mathbf{M}} = \chi\underline{\mathbf{H}}$, where $\underline{\mathbf{H}}$ is the magnetic field strength, and the frequency-dependent magnetic susceptibility is (Rosensweig, 2002) $\chi(f) = \frac{\chi_0}{1 - j\omega\tau_{Browninan}}$, with the equilibrium magnetic susceptibility χ_0, and the angular velocity $\omega = 2\pi f$. Using this constitutive model for the combined Néel and Brown relaxations yields the real and imaginary parts of the magnetic susceptibility

$$\chi' = \text{Re}\{\chi\} = \frac{\chi_0}{1 + (\omega r)^2}, \chi'' = \text{Im}\{\chi\} = \chi_0 \frac{\omega r}{1 + (\omega r)^2}. \tag{8.17}$$

The equilibrium susceptibility is given by

$$\chi_0 = \chi_i \frac{3}{\xi}\left(\coth\xi - \frac{1}{\xi}\right), \quad \chi_i = \frac{\mu_0 \phi M_d^2 V_M}{3k_B T}, \quad \xi = \frac{\mu_0 M_d^2 H_0 V_M}{k_B T}, \tag{8.18}$$

where H_0 is the incident magnetic field, ϕ is the volume fraction of MNPs, and M_d is the domain magnetization.

The elementary change in the internal energy produced by the magnetic work interaction in magnetic media without permanent magnetization is (Chapter 1: Physical, Mathematical, and Numerical Modeling) $dU = \mathbf{H}\cdot d\mathbf{B} = \mu_0\mathbf{H}\cdot d(\mathbf{H} + \mathbf{M}(\mathbf{H}))$ that, for the MNPs of concern, integrated over a cycle yields (Rosensweig, 2002)

$$\Delta U = -\mu_0 \oint \mathbf{M}\cdot d\mathbf{H}, \tag{8.19}$$

which, in the time-domain, $M(t) = \text{Re}\{\chi H_0 e^{j\omega t}\} = H_0(\chi'\cos(\omega t) + \chi'' \sin(\omega t))$, yields

$$\Delta U|_{cycle} = \mu_0 H_0^2 \chi'' \int_0^{2\pi/\omega} \sin^2(\omega t)dt, \tag{8.20}$$

and the dissipated magnetic power density (using the effective value of H_0)

$$\dot{q}_{mg} = f\Delta U|_{cycle} = \mu_0 \omega \chi'' H_0^2. \tag{8.21}$$

This specific heat source contributes to the local heating that adds to the Joule or SAR heating produced by the electric field. For the combined, electric and magnetic heating, the bioheat Eq. (8.11a) then becomes

$$\rho C \frac{\partial T}{\partial t} = k\nabla^2 T - \rho_b C_b \omega_b (T - T_b) + Q_{met} + \underbrace{\sigma_{tissue} E^2}_{Q_{electric}} + \underbrace{\mu_0 \omega \chi'' H^2}_{Q_{magnetic}}. \tag{8.22}$$

For a collection of MNPs, of concentration n and radius r, the magnetic heat source terms become $\left(\frac{4}{3}n\pi r^3\right)\mu_0\omega\chi'' H^2$. When MNPs are injected in the tumor both magnetic and electric fields produce heat that is a superposition of two quadratic forms, the squares of the electric and magnetic fields' strengths.

Microwave magnetic thermal thermotherapy of a hepatic tumor

To exemplify the additional RF heating that happens when MNPs are attached to a tumor, we return to the model discussed in Section 8.3.2 (Fig. 8.13). As it would be practical to safely "contain" the tumor (of complex morphology) into a control volume to be targeted by the procedure, we assume that a spherical volume of a 3 mm radius stands for the tumor. Its electric properties are 10% larger than those of the healthy hepatic tissue, Fig. 8.22.

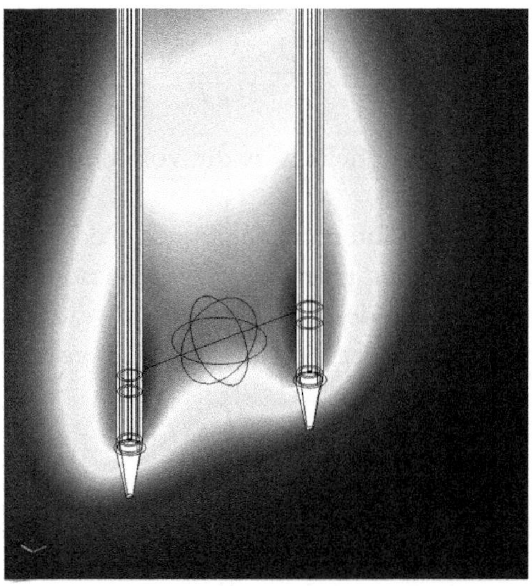

Figure 8.22 The temperature distribution in magnetic MW hyperthermia. Maximum temperature, 46.54°C.

The computational domain was built large enough for its boundary to remain in thermal equilibrium (at 37°C, or adiabatic) with the outside, that is, the cooling effect provided by the perfusion (convection) term in Eq. (8.12) is large enough.

Microwave interstitial hyperthermia is applied to heat the tissue in two situations: in the absence of MNPs (regular hypethermia) and the presence of MNPs (magnetic hyperthermia). It is further assumed that the MNPs are attached to the tumor only. Consequently magnetic heating occurs at the tumor level only, whereas electric heating happens inside and outside the tumor. The concentration of the MNPs is $n = 10^{21}/\mathrm{m}^3$, their susceptibility $\chi'' = 65$, and the hydraulic radius $r = 18$ nm (Urdaneta, 2015). As stationary solutions (EMF and heat transfer) are aimed, an iterative, segregated solver may be used to solve the problems (8.10) (electric field) and (8.22) please call it (heat transfer, stationary form). The temperature distribution is within a relatively tight interval, therefore the electric, magnetic, and thermal properties may be assumed constant.

Figure 8.23 shows the stationary temperature profiles along a horizontal line drawn between antennas and passing through the tumor (see Fig. 8.22), along its horizontal diameter, for three power levels (1.2, 1.4, and 1.6 W) per antenna port. In all cases, the presence of the 6 mm wide MNP-magnetized inclusion is identifiable using the lateral maxima of the temperature curves.

Moreover, the presence of MNPs is seen to require lower levels of exposure (the almost same effect is noticed for 1.4 W, with magnetic work, as of 1.6 W, without magnetic work), and better, deep heating of the tumoral volume.

Patient-specific information may be needed when adequately planning the procedure. Among other input data, the actual organ of concern has to be considered. Its "geometry" may be extracted out of medical scans, and reconstruction techniques

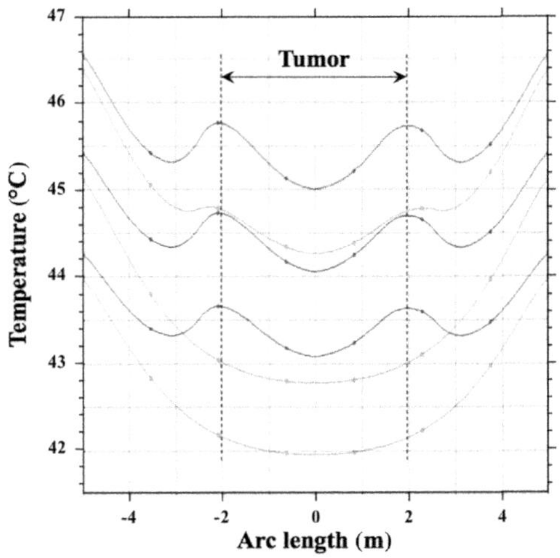

Figure 8.23 The temperature along a horizontal line, between the antennae (Fig. 8.22), with (blue) and without (red) magnetic nanoparticles.

(3DSlicer) and CAD constructs are used to build the computational domain, using fusion techniques (Chapter 3: Computational Domains). The pair of MW antennae are inserted to heat a spherical tumor volume (CAD constructs) of a 3 mm radius inserted within a liver volume (reconstruction technique), Fig. 8.24. The same levels of power are used, that is, 1.2, 1.4, and 1.6 W, for both regular and magnetic hyperthermia. The same MNPs properties are used. For simplification, the bioheat model is used in the entire domain.

To cope with the complex geometry of the domain, the computational domain is FEM meshed using tetrahedral elements (unstructured mesh). First-order Lagrange elements are used to solve the electric field model and second-order elements for the heat transfer problem.

Scattering (EMF) and adiabatic (heat transfer) boundary conditions on the surface of the computational domain (the liver), and metabolic temperature as initial condition close these coupled problems.

The two physics are solved sequentially, first the electric field, Eq. (8.10), and then the thermal field, Eq. (8.22), in its stationary form. Because here the material properties are assumed to be temperature-independent, the two problems are one only way coupled. The validity of the solution, of course, holds as long as this assumption stands, that is, for power levels that induce small temperature gradients in the ROI subjected to MW heating.

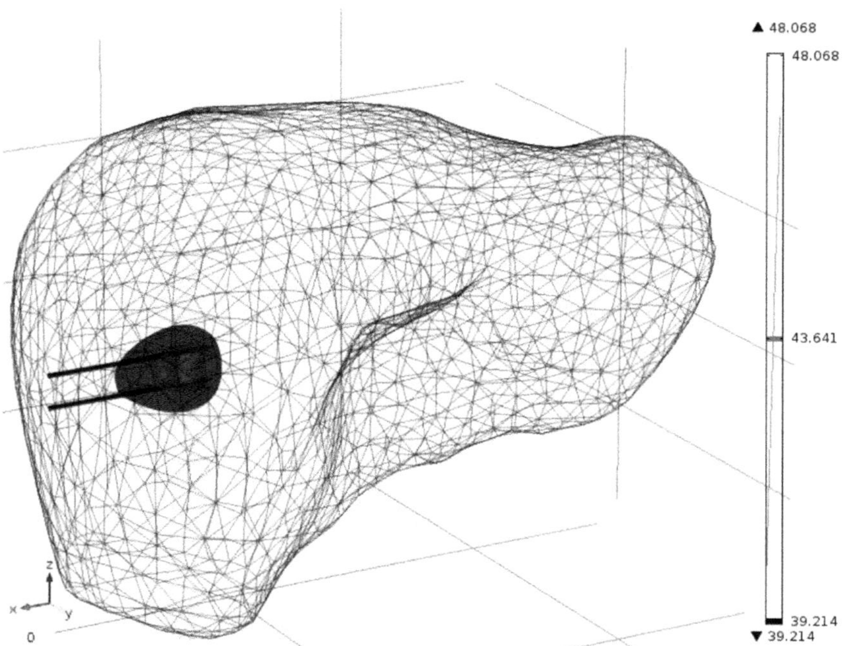

Figure 8.24 The computational domain constructed using image reconstruction (the liver) and CAD blocks (the antennae and the tumor). The isotherms of 39°C, 43.641°C, and 48.068°C.

It is confirmed here that, when compared with the regular hyperthermia, magnetic hyperthermia helps to deliver minimal yet enough power into a close focus that can lead to the required increase in temperature while reducing the unwanted thermal side effects. Figure 8.25 gives a detailed view of the three isotherms that are rendered in Fig. 8.24. For 1.2 W, in magnetic hyperthermia, the temperature inside the tumor is above 46°C whereas in regular hyperthermia it reaches 44°C. Again, the location of the tumor can be identified between the temperature maxima, Fig. 8.26.

Magnetic hyperthermia may be optimized concerning the MNPs size, concentration, properties, etc. and numerical simulation may play a central role. Furthermore, RF hyperthermia may be envisaged with the expected similar beneficial advantages, and numerical modeling may help to cross the procedural and safety development stages that have to be passed.

8.5 Ultrasound thermotherapy

Although thought-about since 1956 as an external method to treat cancer (Hand et al., 1990), the High-Intensity Focused Ultrasound (HIFU) and Magnetic Resonance guided High Intensity Focused Ultrasound (MRgHIFU) (Jolesz, 2009) ablation has received recently increasing attention for the treatment of solid malignant tumors. Without affecting the neighboring healthy tissue, US beams may propagate to produce in focal volume high-intensity mechanical work that results in significant local heating, which may lead to a local increase in temperature able to cause tissue necrosis. US propagation properties and modes of destruction in tissues were investigated in the

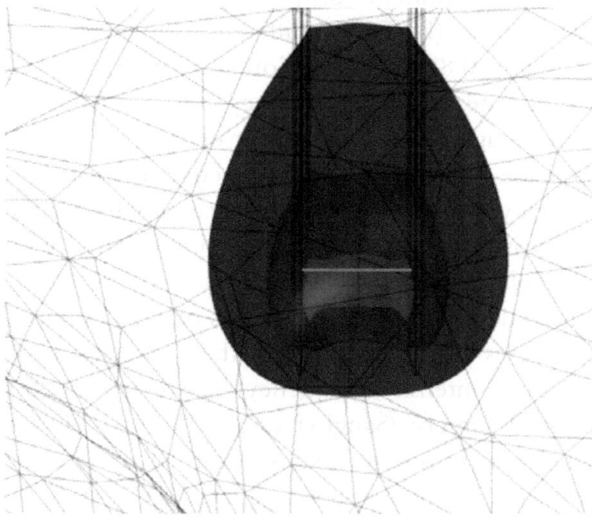

Figure 8.25 The temperature distribution in magnetic MW hyperthermia, detailed view, values are in degree Calsius. *MW*, Microwave.

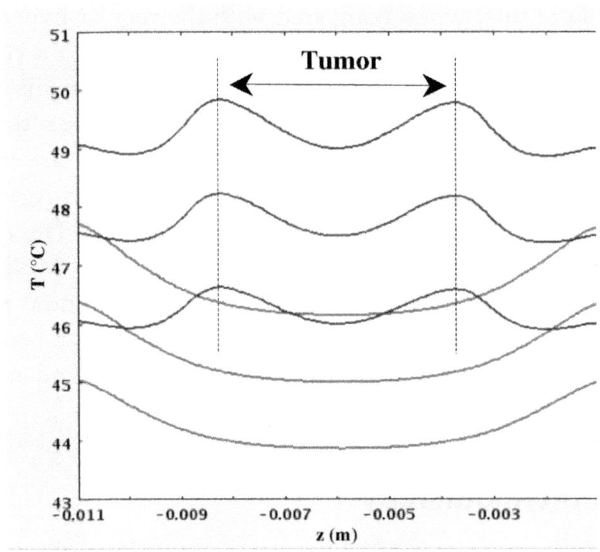

Figure 8.26 The temperature along the horizontal line plotted in Fig. 8.25, between the antennae (Fig. 8.25), with (blue) and without (red) magnetic nanoparticles.

seventies and eighties, and recent years have seen many research venues of potential applications of HIFU in the clinical practice (British Institute of Radiology, 2014). Preclinical, and clinical studies (Furusawa et al., 2007; Maloney and Hwang, 2015; Napoli, 2013), and tests on US devices suggest that HIFU is a safe, effective, and feasible therapy for treating localized breast cancer, which makes HIFU an attractive, non-invasive, potential surgical instrument. US prototype devices were built using ceramic zirconate titanate transducers (PZT), of different diameters (8–10 cm) and shapes (focal lengths), operating at 0.8–1.6 MHz. Some are integrated with diagnostic scanners 3.5 MHz, which operate at higher frequency (e.g., 3.5 MHz).

HIFU, with an encouraging recovery rate, is used to ablate benign and malignant mammary tumors because it is an especially attractive alternative for patients seeking breast preservation therapy. It is completely noninvasive while reducing the number of adverse events and improving the cosmetic outcome compared with surgery (Maloney and Hwang, 2015; Napoli, 2013).

The heating of tissue depends on the properties of the US transducer, its frequency the heat transfer properties of the tissues, and their vascularization. The US dosage parameters (exposure time, intensity, frequency) between the US therapy hyperthermia and the US focused surgery (Sanghvi et al., 1996). In US hyperthermia, with reversible biological effects, tissues are exposed for 10 to 60 min at lower intensity levels to raise and maintain the temperature at 41°C–45°C. In focused surgery intense, short, US bursts (0.1–30 s) are used to raise the temperature at 70°C–90°C within a few seconds, to the US ablate the focused tissue volume.

The ultrasound work interactions

US heating aims to reach a temperature of 42°C−45°C in the tumor tissue, which is large enough to kill tumor cells while minimizing the damage produced to the surrounding tissues (Ter Harr, 2007; Solovchuk et al., 2012). For harmonic working conditions, the US propagation mode may be represented in the complex simplified form of a scalar Helmholtz PDE in the frequency domain, in linear elastic with attenuation media

$$\nabla\left(\frac{1}{\rho_c}(\nabla p_t - \mathbf{q})\right) - \frac{k_{eq}^2 p_t}{\rho_c} = Q, \tag{8.23}$$

where $\rho_c = \frac{\rho c^2}{c_c^2}$ is the complex mass density, ρ is the mass density, c [m/s] is the sound speed, $p_t = p + p_b$, is the pressure, p_b is the background pressure, $k_{eq}^2 = (\omega/c_c)^2$, $c_c = \omega/k$ is the complex speed of sound, $\omega = 2\pi f$ is the angular velocity, f is the frequency, $k = \omega/c - j\alpha$ is the wavenumber [1/m], α is the attenuation coefficient [Np/m], $\mathbf{q}$ [N/m^3] is a dipole source, and Q [s^{-2}] is a monopole source. There is no domain (monopole or dipole) sources in this study.

For the part of the boundary which model the transducer a normal acceleration condition is set, $-\mathbf{n}\cdot(-\frac{1}{\rho_c}(\nabla p_t)) = a_n$, $a_n = -d_0\omega^2$, where d_0 is the maximum displacement, of the order O(nm)—here 4.56 nm. Perfectly matched layers (PML)—cylindrical for the water and torso, and spherical for the cup—with hard sound boundary conditions (no displacement) close the US problem.

Ultrasound ablation of a breast tumor

As in thermal therapies, in general, numerical modeling may contribute to the success of the US therapy, providing useful insights in the preinterventional stage. To this aim, a patient-specific computational domain is required, and the first step is the imaging reconstruction of the breast along with the malignant tumor (3DSlicer). A set of MRI images (e.g., from TCIA) is used as "raw" data. Construction stages, similar to those detailed in Chapter 3: CAD/Medical Image−Based Constructed Computational Domains (Baerov, 2019), are successively leading the volume shown in Fig. 8.27.

The final computational domain, Fig. 8.28, provides the perfectly matched layers (PMLs) that close the model within a conveniently sized space for FEM analysis. The computational domain for the US problem is seen in Fig. 8.28. It comprises the anatomic region (the breast), the surrounding propagation medium (water), the transducer (its trace on the boundary), the back PML (cylindrical) that models the torso, the lateral (cylindrical) and front (spherical) PMLs that model the water reservoir.

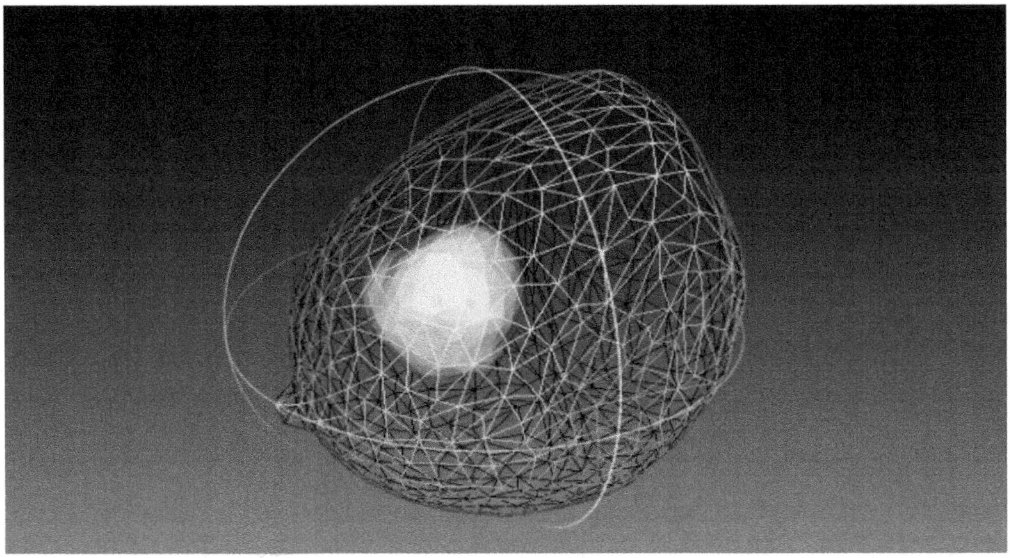

Figure 8.27 The medical image reconstruction part of the computational domain—the breast and the tumoral formation.

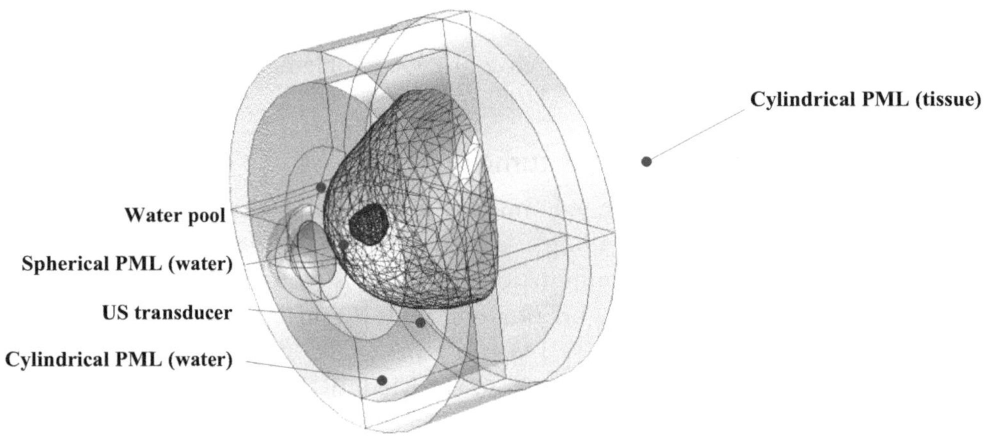

Figure 8.28 The medical image reconstruction part of the computational domain—the breast and the tumoral formation.

The acoustic heating rate, $Q_{US} = 2\alpha I = 2\alpha\,\mathrm{Re}\{\frac{1}{2}pv\}$, where I [W/m^2] is the acoustic intensity, is the heat source term in the bioheat transfer problem

$$\rho C\frac{\partial T}{\partial t} = k\nabla^2 T - \rho_b C_b\omega_b(T - T_b) + Q_{met} + Q_{US}. \tag{8.24}$$

Table 8.6 Material properties used in the US analysis.

Medium	Density, ρ (kg/m^3)	Sound speed, c (m/s)	Attenuation, α (Np/m MHz)	Frequency dependence of α	Specific heat (J/kg K)	Thermal conductivity, k (W/m K)
Water	1000	1522	0.025	f^2	4180	0.59
Tissue	1044	1568	8.550	$f^{1.5}$	3710	0.56
Tumor	1044	1568	8.550	$f^{1.5}$	4070	0.84

US, Ultrasound sources.

The working frequency is 1 MHz. The water pool is a thermostat, maintaining the breast surface temperature at 37°C. Other properties used in the US and heat transfer analyses are listed in Table 8.6 (D'Astous et al., 1986; Ter Harr, 2007; Duck, 1990; wiki; Jin et al., 2014; Preda, 2019). The perfusion rate is taken 6.4×10^{-3} 1/s.

The US problem is solved first, and the heat transfer is integrated next. Some not trivial elements of numerical modeling, for example, the resolution that the FEM meshes for US and heat transfer models have to have, the type and order of the interpolating polynomials that are selected, and the accuracy test that is required for grid-independent numerical solutions.

The acoustic pressure field inside the breast, as seen through orthogonal slices, Fig. 8.29, indicates the US propagation pattern. The US field is focused, with a maximum concentration that is noticeable, perhaps, too deep inside the breast. This region of highest US focalization is subject to work interaction that may result in heating the frontal part. The tumor and some surrounding tissue here are prone to hyperthermia and eventually, ablation provided the therapeutic protocol is set adequately.

Turning to the heat transfer part, once the acoustic intensity is known, the US heating is solved, and Figs. 8.30 and 8.31 unveil the temperature inside the breast after 1800 s of exposure to the US.

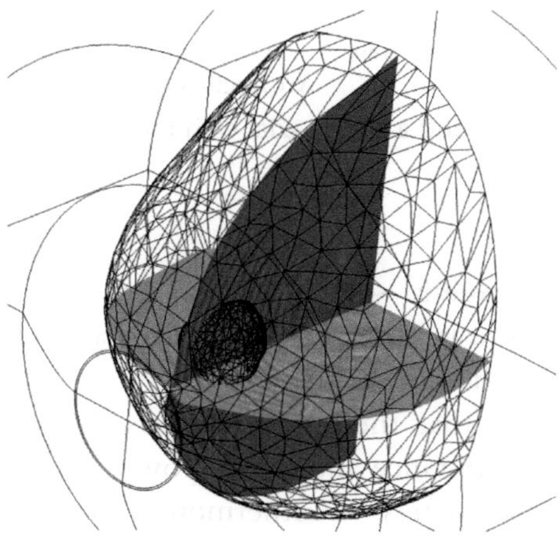

Figure 8.29 The medical image reconstruction part of the computational domain—the breast and the tumoral formation.

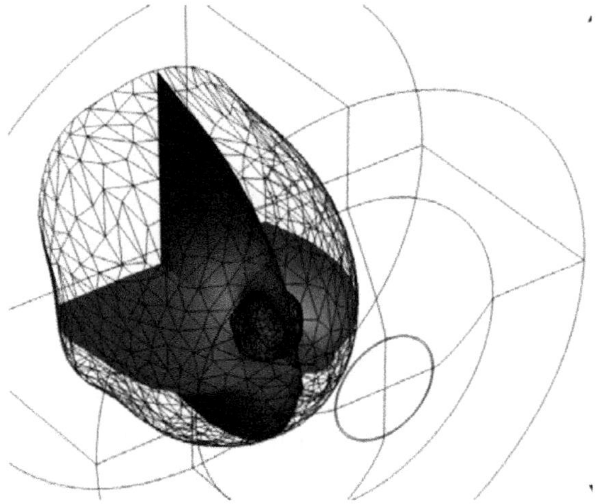

Figure 8.30 The medical image reconstruction part of the computational domain—the breast and the tumoral formation.

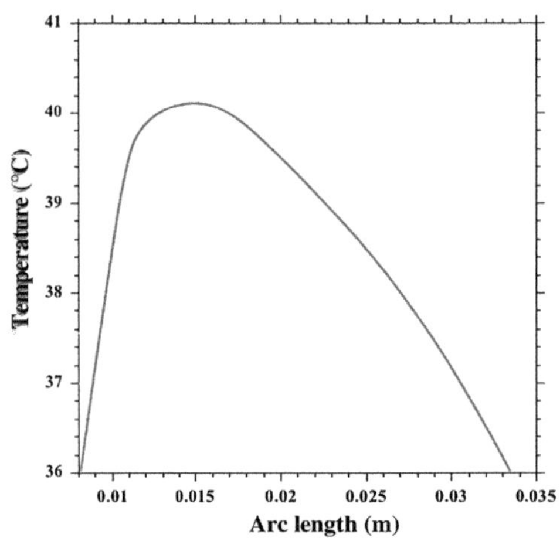

Figure 8.31 Temperature profile along a line coaxial with the cylindrical enclosure, passing through the breast.

The highest computed value is ~41.36°C, and it is found in the tissue, behind the tumor. This thermal exposure is less than required for ablation, but a longer exposure may push this limit beyond the critical value.

The results here suggest that the procedure has to be adjusted—duration, frequency, transducer-to-breast distance and shape, power, etc.—to provide for a satisfactory set of protocol parameters. Furthermore, material properties and local vascularization have to be carefully considered.

References

Andreuccetti, D., Fossi, R., Petrucci, C., 1997. An Internet Resource for the Calculation of the Dielectric Properties of Body Tissues in the Frequency Range 10 Hz–100 GHz. IFAC–CNR, Florence (Italy), Based on data published by Gabriel C. et al. in 1996. [Online]. <http://niremf.ifac.cnr.it/tissprop/>.

Bachmann, D., Haasner, E., 1965. cited by *Illustrated Encyclopedia of Human Anatomic Variation: Opus II: Cardiovascular System: Arteries: Thorax Intercostal Artery*, Ronald A. Bergman, PhD, Adel K. Afifi, MD, MS, Ryosuke Miyauchi, MD Truncus Intercosto–Renalis, 102. Fortschr. Geb. Rontgenstr. Nuklearmed, pp. 712–713.

Baerov, R., 2019. *Magnetic Stimulation of the Lumbar Spine – Planning Assisted by Numerical Modeling* (in Romanian), Diploma Project. The Faculty of Medical Engineering, "Politehnica" University of Bucharest, Romania.

Baronzio, G., Gramaglia, A., Fiorentini, G., 2006. Hyperthermia and immunity. A brief overview. Vivo 20, 689–696.

Bhowmick, S., Coad, J.E., Swanlund, D.J., Bischof, J.C., 2004. In vitro thermal therapy of AT–1 Dunning prostate tumors. Int. J. Hyperth. 20 (1), 73–92.

Bejan, A., 2000. Shape and Structure, from Engineering to Nature. Cambridge University Press.

Bejan, A., Zane, J.P., 2012. Design in Nature. How Constructal Law Governs Evolution in Biology, Physics, Technology, and Social Organization. Anchor Books, Random House Inc, New York.

Bertram, J.M., Yang, D., Converse, M.C., Webster, J.G., Mahvi, D.M., 2006. Antenna design for microwave hepatic ablation using an axisymmetric electromagnetic model. Biomed. Eng. OnLine 5, 15.

Boston Scientific Corporation, 2013. Boston Scientific Advancing Since for life, Natick, MA, USA, <https://www.bostonscientific.com/content/dam/bostonscientific/pi/portfolio-group/rfa/RFA-Family-Brochure-ENDO-149636-AA.pdf>

Brace, C., 2011. Thermal tumor ablation in clinical use. IEEE Pulse 2 (5), 28–38.

British Institute of Radiology, 2014. <https://www.birpublications.org/doi/abs/10.1259/bjr/17150274>

Brown, S.L., Hunt, J.W., Hill, R.P., 1992. Differential thermal sensitivity of tumor and normal tissue microvascular response during hyperthermia. Int. J. Hyperth. 8 (4), 501–514.

Christensen, D.A., Durney, C.H., 1981. Hyperthermia production for cancer therapy: a review of fundamentals and methods. J. Microw. Power 16, 89–105.

Comsol Multiphysics, 2010–2019. AB, v. 3.5a. . .5.3a.

Danhier, F., Feron, O., Préat, V., 2010. To exploit the tumor microenvironment: passive and active tumor targeting of nanocarriers for anti-cancer drug delivery. J. Controlled Rel. 148, 135–146.

Dassault Systemes, 2016. Solidworks Corporation, France.

Dahl, O., Overgaard, J., 1995. A century with hyperthermic oncology in Scandinavia. Acta Oncologica 34 (8), 1075–1083.

De Mey, G., Strakowska, M., Wiecek, B., 2016. Macroscale heat transfer in human tissues. In: 23rd International Conference Mixed Design of Integrated Circuits and Systems, 23–25 June 2016, èodz, Poland.

Dewhirst, M.W., Lora-Michiels, M., Viglianti, B.L., Dewey, W.C., Repacholi, M., 2003a. Carcinogenic effects of hyperthermia. Int. J. Hyperth. 19, 236–251.

Dewhirst, M.W., Viglianti, B.L., Lora-Michiels, M., Hanson, M., Hoopes, P.J., 2003b. Basic principles of thermal dosimetry and thermal thresholds for tissue damage from hyperthermia. Int. J. Hyperth. 19 (3), 267–294. May–June.

Dickson, J.A., Calderwood, S.K., 1980. Temperature range and selective sensitivity of tumors to hyperthermia: a critical review. Ann. N.Y. Acad. Sci. 335, 180–205.

Dobre A.A., Morega A.M., Virlan L.D., Sandoiu A.M., Morega M., 2017. Planning of renal tumor ablation aided by numerical modeling. In: Proceedings of the 10th International Symposium on Advanced Topics in Electrical Engineering, *ATEE 2017*, pp. 274–278, Bucharest, Romania.

Duck, F.A., 1990. Physical Properties of Tissue. Academic Press.

Durlofsky L., Brady J.F., 1987. Analysis of the Brinkman equation as a model for flow in porous media. Phys. Fluids, **30** (11), 3329–3341, ISBN 0031–9171 /87/113329–13.

Dutreix, J., Le Bourgeois, J.P., Salama, M., 1978. The treatment of the tumors by hyperthermia. J. Radiol. Electrol. Med. Nucl. 59, 323–334.

D'Astous, F.T., Foster, F.S., 1986. Frequency dependence of ultrasound attenuation and backscatter in breast tissue. Ultrasound Med. Biol. 12, 795–808.

Feijóo, 2000. Computational methods in biology. In: Proceedings of the 2nd Summer School LNCC/MCT, Petrópolis, Brazil.

Furusawa, H., Namba, K., Nakahara, H., Tanaka, C., Yasuda, Y., Hirabara, E., et al., 2007. The evolving non-surgical ablation of breast cancer: MR guided focused ultrasound (MRgFUR). Breast Cancer 14, 55–58.

Gas, P., 2014. Study on interstitial microwave hyperthermia with multi-slot coaxial antenna. Revue Roumaine Sci. Tech. Electrotechn. et. Energ. 59 (2), 215–224. 2014.

Gas, P., Czosnowski, J., 2014. Calculation of the coaxial–slot antenna characteristics used for the interstitial microwave hyperthermia treatment. Przegląd Elektrotech. 90 (3), 176–178.

Germer, C.T., Roggan, A., Ritz, J.P., Isbert, C., Albrecht, D., Muller, G., et al., 1998. Optical properties of native and coagulated human liver tissue and liver metastases in the near infrared range. Lasers. Surg. Med. 23, 194–203.

Giustini, A.J., Petryk, A.A., Cassim, M., Tate, J.A., Baker, I., Hoopes, P.J., 2010. Magnetic nanoparticle hyperthermia in cancer treatment. Nano Life 1, 1–2. March–June.

Goldberg, S.N., Charboneau, J.W., Dodd III, G.D., Dupuy, D.E., Gervais, D.A., Gillams, A.R., et al., 2003. Image-guided tumor ablation: proposal for standardization of terms and reporting criteria. Radiology 228, 335–345.

Habash, R.W.Y., Bansal, R., Krewski, D., Alhafid, H.T., 2006. Thermal therapy, Part 1: an introduction to thermal therapy, Critical Reviews TM in Biomedical Engineering, Begell House Inc, 36, 6, pp. 459–489 <http://begellhouse.com>.

Haemmerich, D., Laeseke, P.F., 2005. Thermal tumor ablation: devices, clinical applications and future directions. Int. J. Hyperth. 21, 755–760.

Haemmerich, D., Lee Jr, F.T., Schutt, D.J., Sampson, L.A., Webster, J.G., Fine, J.P., et al., 2005. Large-volume radiofrequency ablation of ex vivo bovine liver with multiple cooled cluster electrodes. Radiology 234, 563–568.

Hand, J.W., Hynynen, K., Saylor, T.K., Shrivastava, T.K., 1990. In: Gauthier, M. (Ed.), Methods of External Hyperthermic Heating. Clinical Terminology. Subseries Terminology. Springer Verlag, Heidelberg.

Hasgall P.A., Di Gennaro F., Baumgartner C., Neufeld E., Gosselin M.C., Payne D., et al., 2015. IT'IS Database for Thermal and Electromagnetic Parameters of Biological Tissues, V. 3.0. <http://www.itis.ethz.ch/database>.

Hildebrandt, B., Wust, P., Ahlers, O., Dieing, A., Sreenivasa, G., Kerner, T., et al., 2002. The cellular and molecular basis of hyperthermia. Crit. Rev. Oncol. Hematol. 43 (1), 33–56.

Horsman, M.R., Overgaard, J., 2007. Hyperthermia: a potent enhancer of radiotherapy. Clin. Oncol. 19 (6), 418–426.

ITIS Foundation, 2019. <https://itis.swiss/virtual-population/tissue-properties/database/thermal-conductivity/> (accessed July 2019).

Ito, K., Saito, K., Taniguchi, T., Okabe, S., Yoshimura, H., 2001. Minimally invasive thermal therapy for cancer treatment by using thin coaxial antennas. In: Proceedings of the 23rd Annual International Conference of the IEEE EMBS, October 25–28, 2001, Istanbul, Turkey.

Ito, K., Saito, K., Taniguchi, T., Yoshimura, H., 2002. Temperature distribution in and around array applicator for interstitial microwave hyperthermia combined with interstitial radiation therapy. In: Proceedings of the 27th International URSI General Assembly, Maastricht.

Ito, A., Shinkai, M., Honda, H.,, Kobayashi, T., 2005. Medical application of functionalized magnetic nanoparticles. J. Biosci. Bioeng. 100 (1), 1–11.

Jacques, S., Rastegar, S., Thomsen, S., Motamedi, M., 1996. Nonlinear finite-element analysis. The role of dynamic changes in blood perfusion and optical properties in laser coagulation of tissue. IEEE J. Sel. Top. Quantum Electron. 2 (4), 922–933.

Jain, T.K., Morales, M.A., Sahoo, S.K., Leslie-Pelecky, D.L., Labhasetwar, V., 2005. Iron oxide nanoparticles for sustained delivery of anticancer agents. Mol. Pharm. 2 (3), 194–205.

Jin, S., Wang, X., Yuan, M., Zheng, A., 2014. Hyperthermia ablation of breast tumors using ultrasoundMay 9 BEE 4530 – Computer-Aided Engineering: Applications to Biomedical Processes. Cornell University.

Johannsen, M., Thiesen, B., Wust, P., Jordan, A., 2010. Magnetic nanoparticle hyperthermia for prostate cancer. Int. J. Hyperth. 26 (8), 790–795.

Jolesz, F.A., 2009. MRI-guided focused ultrasound surgery. Annu. Rev. Med. 60, 417–430.

Karampatzakis, A., Kühn, S., Tsanidis, G., Neufeld, E., Samaras, T., Kuster, N., 2013. Heating characteristics of antenna arrays used in microwave ablation: A theoretical parametric study. Comput. Biol. Med. 43 (10), 1321–1327.

Knavel, E.M., Brace, C.L., 2013. Tumor ablation: common modalities and general practices. Tech. Vasc. Interv. Radiol 16 (4), 192–200.

Koda, M., Tokunaga, S., Matono, T., Sugihara, T., Nagahara, T., Murawaki, Y., 2011. Comparison between different thickness umbrella–shaped expandable radiofrequency electrodes (SuperSlim and CoAccess): experimental and clinical study. Exp. Therapeutic Med. 2, 1215–1220.

Kumar, C.S.S.R., Faruq, M., 2011. Magnetic nanomaterials for hyperthermia-based therapy and controlled drug delivery. Adv. Drug. Delivery Rev. 63 (9), 789–808.

Layton, A.T., September 2013. Mathematical modeling of kidney transport. Rev. Syst. Biol. Med. 5 (5), 557–573. Wiley Interdiscip.

Lepock, J.R., 2003. Cellular effects of hyperthermia: relevance to the minimum dose for thermal damage. Int. J. Hyperth. 19, 252–266.

Lepock, J.R., 2005. How do cells respond to their thermal environment? Int. J. Hyperth. 21, 681–687.

Maeshima, Y., Makino, H., 2010. Angiogenesis and chronic kidney disease. Fibrogenesis & Tissue Repair. 3 (13), 1–17.

Maloney, E., Hwang, J.H., 2015. Emerging HIFU applications in cancer therapy. Int. J. Hyperth. 31 (3), 302–309.

Mayo Clinic, Cryoablation of Cancer. <https://www.mayoclinic.org/tests-procedures/cryoablation-for-cancer/about/pac-20385216>.

Mayo Clinic, <https://www.mayoclinic.org/tests-procedures/radiofrequency-ablation/about/pac-20385270>. Accessed in August 2020.

Michaelson, S., Lin, J.C., 1987. Biological Effects and Health Implications of Radio Frequency Radiation. Plenum, New York.

Miles, C.A., 2006. Relating cell killing to inactivation of critical components. Appl. Environ. Microbiol. 72, 914–917.

Moran, C.H., Wainerdi, S.M., Cherukuri, T.K., Kittrell, C., Wiley, B.J., 2009. Size-dependent Joule heating of gold nanoparticles using capacitively coupled radiofrequency fields. Nano Res. 2, 400–405.

Morega, M., Mogos, L., Neagu, M., Morega, A.M., 2006. Optimal design for microwave hyperthermia applicator. In: Proceedings of the 11th International Conference on Optimization of Electrical and Electronic Equipment—OPTIM 2006, Brasov, Romania, pp. 219–224.

Morega, M., Neagu, M., Morega, A.M., 2008. Bidirectional coupling of electromagnetic and thermal processes in radiofrequency hyperthermia. In: Proceedings of the 12th International Conference on Optimization of Electrical and Electronic Equipment—OPTIM 2008, Brasov, Romania, pp. 257–262.

Morega, A.M., Dobre, A.A., Morega, M., 2010. Numerical simulation of magnetic drug targeting with flow structural interaction in an arterial branching region of interest. In: *Comsol Conferene*, Versailles, France, 17–19 Nov., 2010.

Morega, M., Morega, A.M., Diaz, M.I., Sandoiu, A.M., 2014. Percutaneous microwaves hyperthermia study by numerical simulation. In: Proceedings of the Internatinal Conference. and Exposition on Electrical and Power Engineering—EPE 2014, Iasi, Romania, pp. 498–503.

Morega, M., Morega, A.M., Sandoiu, A.M., 2015a. Sensitivity to parameters variation in numerical simulation of microwave thermotherapy. In: The 9th International Symposium on Advanced Topics in Electrical Engineering, *IEEE–ATEE*, May 7–9, Bucharest, Romania.

Morega, A.M., Sandoiu, A.M., Morega, M., 2015b. On the thermography mapping of anatomic media with capillaries and larger vessels blood flow. In: IEEE International Conference on e-Health and Bioengineering, EHB 2015, November 19–21, Iasi, Romania.

Morega, A.M., Dobre, A.A., Sandoiu, A.M., Morega, M., 2019. Numerical modeling of hemodynamic heat transfer in hyperthermia tumor ablation. In: *Proceedings of the Romanian Academy, Series A* **17**, *3*, 2016, 259–266.

Mulier, S., Miao, Y., Mulier, P., Dupas, B., Pereira, P., de Baere, T., et al., 2005. Electrodes and multiple electrode systems for radiofrequency ablation: a proposal for updated terminology. Eur. Radiol. 15, 798–808.

Murray, C.D., 1926. The physiological principle of minimum work: I. The vascular system and the cost of blood volume. Proc. Nat. Acad. Sci. USA 12 (3), 207–214.

Napoli, A., 2013. MR-guided ultrasound offers noninvasive treatment for breast cancer. In: *Radiological Society of North America*, News, December 4, 2013. <https://www.sciencedaily.com/releases/2013/12/131204091409.htm>.

NIH, National Cancer Institute. <https://www.cancer.gov/about-cancer/treatment/types/surgery/hyperthermia-fact-sheet>. Accessed in August 2020.

NIH, 2019. Your Kidneys and How They Work, National Institute of Health, National Institute of Diabetes and Digestive and Kidney Diseases, U.S. Department of Health and Human Services. <https://www.niddk.nih.gov/health-information/kidney-disease/kidneys-how-they-work> (accessed January 2019).

NIH, National Cancer Institute, Hyperthermia in Cancer Treatment, Fact Sheet. <http://www.cancer.gov/cancertopics/factsheet/Therapy/hyperthermia> (reviewed 31.08.2011)

Nikawa, Y., 1995. Temperature depending SAR distribution in human body during hyperthermia treatment. IEICE Trans. Electron. E78-C (8), August.

Nikfarjam, M., Muralidharan, V., Christophi, C., 2005a. Focal hyperthermia produces progressive tumor necrosis independent of the initial thermal effects. J. Gastrointest. Surg. 9, 410–417.

Nikfarjam, M., Muralidharan, V., Christophi, C., 2005b. Mechanisms of focal heat destruction of liver tumors. J. Surg. Res. 127, 208–223.

Osepchuk, J.M., Petersen, R.C., 2001. Safety and environmental issues. In: Golio, M. (Ed.), Modern Microwave and RF Handbook. CRC Press, Boca Raton, FL, pp. 3.28–3.43.

Osteaux, M., Jeanmart, L. Kidney tumor vascularization: morphology and angiogenesis, a microangiographic experimental study, In: Lohr, E. (Ed.), Renal and Adrenal Tumors, 1979, Springer-Verlag, Berlin, Heidelberg, New York , pp. 69–77.

Overgaard, K., Overgaard, J., 1972. Investigations on the possibility of a thermic tumor therapy. I. Short-wave treatment of a transplanted isologous mouse mammary carcinoma. Eur. J. Cancer 8, 65–78.

Pankhurst, Q.A., Connolly, J., Jones, S.K., Dobson, J., 2003. Applications of magnetic nanoparticles in biomedicine. J. Appl. Phys. D: Appl. Phys. 36, 13.

Pennes, H.H., 1948. Analysis of tissue and arterial blood temperature in the resting human forearm. J. Appl. Phys. 1, 93–122.

Petryk, A.A., Giustini, A.J., Ryan, P., Strawbridge, R.A., Hoopes, P.J., 2009. Energy–based treatment of tissue and assessment, V 25–26, 7181:71810N. In: Proceedings of SPIE, Progress in Biomedical Optics and Imaging, San Jose, California, vol. **10**, p. 21.

Peralta, C.A., Whooley, M.A., Ix, J.H., Shlipak, M.G., 2006. Kidney function and systolic blood pressure new insights from Cystatin C: Data from the heart and soul study. Am. J. Hypertens. 19 (9), 939–946.

Pop, M., Molckovsky, A., Chin, L., Kolios, M.C., Jewett, M.A., Sherar, M.D., 2003. Changes in dielectric properties at 460 kHz of kidney and fat during heating: importance for radio–frequency thermal therapy. Phys. Med. Biol. 48, 15.

Poulou, L.S., Botsa, E., Thanou, I., Ziakas, P.D., Thanos, L., 2015. Percutaneous microwave ablation vs radiofrequency ablation in the treatment of hepatocellular carcinoma. World J. Hepatol. 7, 18 May.

Preda, S.A., 2019. *The Therapy with High-Intensity Ultrasounds for the Ablation of Tumoral Formations*, (in Romanian) Diploma Project. The Faculty of Medical Engineering, University, "Politehnica" of Bucharest, Romania.

Putnam D.F., 1971. Composition and concentrative properties of human urine, MCDonnell Douglas Astronautics Company – Western Division Huntington Beach, California 92647 for Langley Research Center, National Aeronautics and Space Administration Washington, D.C. <https://ntrs.nasa.gov/archive/nasa/casi.ntrs.nasa.gov/19710023044.pdf>.

RadiologyInfo.org, <https://www.radiologyinfo.org/en/info.cfm?pg = rfakidney#overview>.

Ramanathan, R., Leveillee, R.J., 2010. Ablative therapies for renal tumors. Ther. Adv. Urol. 2 (2), 51–68.

Rosen, A., Rosen, H.D., 1995. New Frontiers in Medical Device Technology. Wiley, New York.

Rosen, A., Stuchly, M.A., Vander Vorst, A., 2002. Applications of RF/microwaves in medicine. IEEE Trans. Microw. Theory Tech. 50, 963–974.

Rosensweig, R.E., 2002. Heating magnetic fluid with alternating magnetic field. J. Magn. Magn. Mater. 252, 370–374.

Rossmann, C., Haemmerich, D., 2016. Review of temperature dependence of thermal properties, dielectric properties, and perfusion of biological tissues at hyperthermic and ablation temperatures. Crit. Rev. Biomed. Eng. 42 (6), 467–492.

Ryu, K.S., Kim, J.H., Ko, H.S., Kim, J.W., Ahn, W.S., Park, Y.G., et al., 2004. Effects of intraperitoneal hyperthermic chemotherapy in ovarian cancer. Gynecol. Oncol. 94 (2), 325–332.

Saito, K., Yoshimura, H., Ito, K., Aoyagi, Y., Horita, H., 2004. Clinical trials of interstitial microwave hyperthermia by use of coaxial–slot antenna with two slots. IEEE Trans. Microw. Theory Tech. 52 (8), 1987–1991.

Sands, J.M., Layton, H.E., 2000. Urine concentrating mechanism and its regulation. In: Seldin, D.W., Giebisch, G. (Eds.), The Kidney: Physiology and Pathophysiology, 3. Lippincott Williams and Wilkins, Philadelphia.

Sanghvi, N.T., Hynynen, K., Lizzi, F.L., 1996. New developments in therapeutic ultrasound. IEEE Eng. Med. Biol. Mag. 15 (6), 83–92.

Seegenschmiedt, M.H., Vernon, C.C., 1995. A historical perspective on hyperthermia in oncology. In: Seegenschmiedt, M.H., Fessenden, P., Vernon, C.C. (Eds.), Thermoradiotherapy and Thermochemotherapy, 1. Springer Verlag, Berlin.

Sekhar, K.R., Sonar, V.N., Muthusamy, V., Sasi, S., Laszlo, A., Sawani, J., et al., 2007. Novel chemical enhancers of heat shock increase thermal radiosensitization through a mitotic catastrophe pathway. Cancer Res. 67 (2), 695–701.

Shibeshi, S.S., Collins, W.E., 2005. The rheology of blood flow in a branched arterial system. Appl. Rheol. 15 (6), 398–405.

Solovchuk, M.A., Sheu, T.W.H., Thiriet, M., Lin, W.L., 2012. Effect of acoustic streaming on tissue heating due to high-intensity focused ultrasound Int. Congress on UltrasonicsAIP, Conf. Proc. 1433, 589-592.

3DSlicer *v. 4.10*, <https://www.slicer.org>.

Stauffer, P.R., Goldberg, S.N., 2004. Introduction: thermal ablation therapy. Int. J. Hyperth. 7, 671–677.

Stauffer, P.R., 2005. Evolving technology for thermal therapy of cancer. Int. J. Hyperth. 21, 731–744.

Strezer, F., 2002. Microwave medical devices. IEEE Microw. Mag. 3, 65–70.

TBQ Editors, 2013. Anatomy and Physiology, vol. **3**, Text Book Equity, Inc., OpenStax College, Rice Univeristy, HU, Texas, USA, pp. 22–28.

TCIA, The Cancer Imaging Archive <https://www.cancerimagingarchive.net/>. Accessed in August 2020.

Ter Harr, G., Coussios, C., 2007. High intensity focused ultrasound: Physical principles and devices. Int. J. Hyperth. 23 (2), 89–104.

Trujillo, C.J., Rico, G., Leija, L., Vera, A., Gutiérrez, J., 2018. Micro–coaxial slot antenna to treat bone tumors by thermal ablation: theoretical and experimental evaluation. IEEE Lat. Am. Trans. 16 (11), 2731–2737.

Truskey, G.A., Yuan, F., Katz, D.F., 2004. Transport Phenomena in Biological Systems. Pearson Prentice Hall, ISBN 0–13–042204–5.

Tsutsui, H., Nishiguchi, S., 2014. Importance of Kupffer cells in the development of acute liver injuries in mice. Int. J. Mol. Sci. 15, 7711–7730.
Urdaneta, M., 2015. Enhanced Microwave Hyperthermia using Nanoparticles. PhD Thesis). Department of Electrical Engineering and Computer Science, University of Central Florida. Spring Term, Orlando, Florida.
Van der Zee, J., 2002. Heating the patient: a promising approach? Ann. Oncol. 13 (8), 1173–1184.
Vander Vorst, A., Rosen, A., Kotsuka, Y., 2006. RF/Microwave Interaction with Biological Tissues. John Wiley & Sons.
Wacker, F.K., Lippert, H., Pabst, R., 2018. Atlas der arteriellen Variationen, Bauch und Becken, **11**, Nierenarterie (A. Renalis).
Westermark, F., 1898. Über die behandlung des ulcerirenden cervixcarcinoms mittels konstanter warme. Zentralblatt für Gyniikologie 22 (I), 335–339.
Welc, S.S., Phillips, N.A., Oca-Cossio, Wallet, S.M., Chen, D.L., Clanton, T.L., 2012. Hyperthermia increases interleukin-6 in mouse skeletal muscle. Am. J. Physiol. Cell Physiol. 303 (4), C455–C466. Aug. 15.
Wust, P., Hildebrandt, B., Sreenivasa, G., Rau, B., Gellermann, J., Riess, H., et al., 2003. Hyperthermia in combined treatment of cancer. Lancet Oncol. 3, 487–497.
Xu, F., Seffen, K.A., Lu, T.J., 2008. Temperature-dependent mechanical behaviors of skin tissue. IAENG Int. J. Computer Sci. 35, 1.
Zagoria, R.J., 2004. Imaging–guided radiofrequency ablation of renal masses. RadioGraphics 24, 5971.
Zhang, J., Misra, R.D., 2007. Magnetic drug–targeting carrier encapsulated with thermosensitive smart polymer: core-shell nanoparticle carrier and drug release response. Acta Biomater. 3 (6), 838–850.

Index

Note: Page numbers followed by "*f*" refer to figures.

K

L

M

9780128178973